STEM Education 2.0

STEM Education 2.0

Myths and Truths – What Has K-12 STEM Education Research Taught Us?

Edited by

Alpaslan Sahin and Margaret J. Mohr-Schroeder

BRILL
SENSE

LEIDEN | BOSTON

All chapters in this book have undergone peer review.

The Library of Congress Cataloging-in-Publication Data is available online at http://catalog.loc.gov

Typeface for the Latin, Greek, and Cyrillic scripts: "Brill". See and download: brill.com/brill-typeface.

ISBN 978-90-04-40538-7 (paperback)
ISBN 978-90-04-40539-4 (hardback)
ISBN 978-90-04-40540-0 (e-book)

Copyright 2019 by Koninklijke Brill NV, Leiden, The Netherlands.
Koninklijke Brill NV incorporates the imprints Brill, Brill Hes & De Graaf, Brill Nijhoff, Brill Rodopi, Brill Sense, Hotei Publishing, mentis Verlag, Verlag Ferdinand Schöningh and Wilhelm Fink Verlag.
All rights reserved. No part of this publication may be reproduced, translated, stored in a retrieval system, or transmitted in any form or by any means, electronic, mechanical, photocopying, recording or otherwise, without prior written permission from the publisher.
Authorization to photocopy items for internal or personal use is granted by Koninklijke Brill NV provided that the appropriate fees are paid directly to The Copyright Clearance Center, 222 Rosewood Drive, Suite 910, Danvers, MA 01923, USA. Fees are subject to change.

This book is printed on acid-free paper and produced in a sustainable manner.

Printed by Printforce, the Netherlands

Contents

Foreword IX
   Joseph Krajcik
Acknowledgements XII
List of Figures and Tables XIII
Notes on Contributors XVI

PART 1
Setting Stage for STEM Education: Updates on STEM Education

1. STEM Literacy: Where Are We Now? 3
   Maureen Cavalcanti and Margaret J. Mohr-Schroeder

2. Getting to the Bottom of the Truth: STEM Shortage *or* STEM Surplus? 22
   Cathrine Maiorca, Micah Stohlmann and Emily Driessen

3. Underrepresentation of Women and Students of Color in STEM 36
   Dionne Cross Francis, Kerrie G. Wilkins-Yel, Kelli M. Paul and Adam V. Maltese

4. Teaching and Learning Integrated STEM: Using Andragogy to Foster an Entrepreneurial Mindset in the Age of Synthesis 53
   Louis S. Nadelson and Anne L. Seifert

5. National Reports on STEM Education: What Are the Implications for K-12? 72
   Sarah Bush

PART 2
How to Provide Rigorous STEM Education

6. The Role of Interdisciplinary Project-Based Learning in Integrated STEM Education 93
   Alpaslan Sahin

7 Inclusive STEM Schools: Origins, Exemplars, and Promise   104
   Sharon J. Lynch

8 Inclusive STEM School Models: A Review of Characteristics & Impact   132
   Justin M. Bathon

9 Informal Learning in STEM Education   143
   Soledad Yao and Margaret J. Mohr-Schroeder

10 Bringing out the "T" in STEM Education   153
   Bulent Dogan and Susie Gronseth

PART 3
Bringing out the "E" in STEM in K-12 Settings

11 Engineering Education in K-12: A Look Back and Forth   177
   Christine Guy Schnittka

12 Classroom Assessment in the Service of Integrated STEM Education Reform   189
   Carol L. Stuessy and Luke C. Lyons

13 A Shared Language: Two Worlds Speaking to One Another through Making and Tinkering Activities   228
   Amber Simpson, Jackie Barnes and Adam V. Maltese

14 Educational Robotics as a Tool for Youth Leadership Development and STEM Engagement   248
   Kathleen Morgan, Bradley Barker, Gwen Nugent and Neal Grandgenett

PART 4
Factors Affecting Students' Choice of STEM Majors

15 Factors Affecting Students' STEM Choice and Persistence: A Synthesis of Research and Findings from the Second Year of a Longitudinal High School STEM Tracking Study   279
   Adem Ekmekci, Alpaslan Sahin and Hersh Waxman

CONTENTS VII

16 Reaching Youth with Science: A Look at Some Data on When Science Interest Develops and How it Might Be Sustained   305
    Robert H. Tai

PART 5

Community Partnerships and Innovation to Improve K-12 Students' 21st Century Skills

17 Crossing Borders and Stretching Boundaries: A Look at Community-Education Partnerships and Their Impact on K-12 STEM Education   319
    Brett Criswell, Theodore Hodgson, Carol Hanley and Kimberly Yates

18 What Skills Do 21st Century High School Graduates Need to Have to Be Successful in College and Life?   337
    Kristina Kaufman

19 An International View of STEM Education   350
    Brigid Freeman, Simon Marginson and Russell Tytler

20 Conclusions and Next Directions   364
    Margaret J. Mohr-Schroeder and Alpaslan Sahin

Foreword

We live in a science and technology driven world where useable knowledge of science, engineering, technology and mathematics (STEM) is critical for a person's well-being and for the economy and prosperity of all nations. On a personal level, an individual needs useable-knowledge of STEM to make daily decisions to participate in daily life from maintaining and improving their personnel health to maintaining and supporting the health of the environment. STEM also provides an individual with a wealth of fulfilling career opportunities and hobbies. Although society needs to have individuals in the arts and humanities, STEM careers provide a range of fulfilling opportunities essential to the growth and prosperity of the nation. At the societal level, STEM drives the economic well-being and security of a nation. Without a skilled, well-prepared and knowledgeable workforce that can accomplish technical jobs, innovate, think creatively, and develop new solutions, a society cannot flourish. STEM is also vital to a sustainable world. Each nation needs to work together to find solutions to global problems including various health issues, globe warming, and maintaining clean water.

The series of chapters presented in *STEM Education 2.0 – Myths and Truths: What has K-12 STEM Education Research Taught Us?* provides a range of research and examples on the very topics our society, especially K-12 education, should be addressed in STEM Education. These topics span from exploring the impact of policy to how to design learning environments that foster development of STEM knowledge; from creating high quality STEM assessments to utilizing a variety of instructional techniques in STEM; from encouraging and supporting women and minorities in STEM to what does high quality STEM learning look like. STEM Education 2.0 takes stock of where the field has come and where we are headed.

More importantly, STEM Education 2.0 explores STEM topics from an integrated perspective. Many problems that face society require a more interdisciplinary approach to STEM teaching and learning where individuals need to integrate knowledge of science disciplines, engineering, mathematics and technology in order to solve them. This richer, more productive vision of STEM education (Krajcik & Delen, 2017) provides students with the knowledge-in-use (Pellegrino & Hilton, 2012) needed to solve pressing individual and societal problems. This vision of integrated STEM education places STEM on a continuum from students learning each of the fields separately to learning about the disciplines in an integrated manner. Integration provides a more productive and powerful vision but can be harder to accomplish.

To develop usable knowledge of STEM is essential for K-12 education, as it lays the foundation for individuals to solve personal problems and make decisions, go deeper into the study STEM, and provides the foundation for solving complex problems and thinking critically and creatively. STEM Education 2.0's sections two and three, addressing rigorous STEM school models and innovative curriculum, center on the need for K-12 students to experience STEM learning around big ideas of the various fields. These big ideas, as presented in the book, are integrated with the practice of the field to allow learners to develop useable knowledge of STEM. Students need to apply their STEM knowledge in new contexts to solve problems, think critical and creatively, and to collaborate and communicate ideas and solutions.

STEM Education 2.0 provides research and examples on how to support K-12 students in building deep, integrated and useable knowledge of STEM so that learners have the knowledge and problem solving skills necessary to live in and improve the world. For example, project-based learning, Maker Spaces, and engineering design solutions provide three viable contexts for such integrated STEM learning. Each of these learning environments place the learner in meaningful contexts where they solve problems and construct products or artifacts to represent what they have accomplished. Project-based learning (PBL) and engineering design are an approach to teaching STEM that focus learners on investigating questions or problems that they find meaningful and engaging and that spark wonderment and curiosity. Maker Spaces take an interdisciplinary approach to promote creativity and collaborative engagement for learners in STEM. Makers spaces help to integrate both informal (e.g., science and art museums) and formal (e.g., school-based Maker Spaces) contexts. Maker Spaces have the potential to appeal to a great diversity of learners – a theme expressed throughout the book. However, as designers and teachers of Maker Spaces, project-based environments and design spaces, it is essential that we ensure important learning goals are at the forefront in the design of the environment and not left aside. Engineers, computers scientists and scientists have deep knowledge of science, technology, engineering and mathematics and we need to prepare all of our students to have useable knowledge. Students need to leave our schools armed with a tool set of knowledge and science and engineering practices that will serve them as they continue in their schooling or in their personal lives.

STEM Education 2.0 stresses that all students should have access to high-quality learning opportunities in STEM. Project-based learning environments, design environments and Maker Space that focus students on critical learning goals are essential to fostering and developing learners from diverse background. Such environments will also allow learners to experience the joy of discovery and innovation that will provide the intrinsic motivation to continue to learn STEM when it becomes challenging.

STEM EDUCATION 2.0 provides a broad perspective on STEM and STEM education and shows how to promote all students in developing useable knowledge of STEM. Engaging learners in STEM education, away from rote memorization to doing STEM, will require shifts in teaching practices and new ways of thinking about how to support students in learning, but the benefits of meeting these challenges will allow us as a nation to develop a population of learners who have knowledge-in-use, curiosity, creativity and critical thinking abilities necessary to invent the solutions of tomorrow, to live a fruitful life and to help develop a sustainable world. STEM education can promote women, people of color and English Language Learners remaining markedly underrepresented in many areas of STEM.

References

Fortus, D., Dershimer, C. R., Krajcik, J. S., & Marx, R. W. (2004). Design-based science and student learning. *Journal of Research in Science Teaching, 41*(10), 1081–1110.

Krajcik, J., & Delen, I. (2017). How to support learners in developing usable and lasting knowledge of STEM. *International Journal of Education in Mathematics, Science and Technology, 5*(1), 21–28. doi:10.18404/ijemst

Pellegrino, J. W., & Hilton, M. L. (Eds.). (2012). *Education for life and work: Developing transferable knowledge and skills in the 21st century*. Washington, DC: The National Academies Press.

Joseph Krajcik
CREATE for STEM, Michigan State University

Acknowledgements

Writing a book is always a complex task with many people behind the scenes. We are thankful for all the people who directly or indirectly contributed to this project through their encouragement and ideas.

We would first like to thank the authors of the chapters contained in this book for their valuable contributions: Cathrine Maiorca – California State University, Long Beach; Micah Stohlmann – University of Nevada, Las Vegas; Emily Driessen – University of Kentucky; Maureen Cavalcanti – Ohio State University; Sarah Bush – University of Central Florida; Adam Maltese – Indiana University; Dionne Cross Francis – Indiana University; Kerrie G. Wilkins-Yel – Indiana University; Kelli M. Paul – Indiana University; Louis Nadelson – University of Central Arkansas; Anne L. Seifert – GATE Idaho; Sharon Lynch – George Washington; Justin Bathon – University of Kentucky; Bulent Dogan – University of Houston; Susie Gronseth – University of Houston; Christine Guy Schnittka – University of Auburn; Carol Stuessy – Texas A&M University; Luke C. Lyons – Texas A&M University; Amber M. Simpson – Binghamton University; Jackie Barnes – Children's Museum of Pittsburg; Kathleen Morgan – FIRST LEGO League; Bradley Barker – University of Nebraska-Lincoln; Gwen Nugent – University of Nebraska-Lincoln; Neal Grandgenett – University of Nebraska at Omaha; Adem Ekmekci – Rice University; Hersh Waxman – Texas A&M University; Robert Tai – University of Virginia; Brett Criswell – University of Kentucky; Theodore Hodgson – Northern Kentucky University; Carol Hanley – University of Kentucky; Kristina J. Kaufman – North Park University; Brigid Freeman – University of Melbourne; Russell Tytler – Deakin University; and Simon Mariginson – University College London.

A special thanks go to Professor Sharon Lynch and Professor Carol Stuessy, for taking time from their busy schedule and reading the proposal and chapter ideas for the book. Their promise of writing a chapter for the book was also an unbelievable push to carry out the project.

We would like to thank Joseph Krajcik, Lappan-Phillips Professor in the department of teacher education at Michigan State University, who agreed to write the foreword for this book although, we're sure, he has been swamped with his own projects and responsibilities.

Finally, we would like to thank our families for their support; without the extra help at home, deadlines would not have been met.

Figures and Tables

Figures

4.1   The STEM integration spectrum.   57
4.2   Tool for determining knowledge and process of STEM disciplines needed or used in integrated STEM projects.   58
4.3   STEM processes and knowledge identification tool completed using the parabolic hotdog cooker as the project.   59
4.4   The zone of optimal learning where assignment complexity and students' capacity align.   60
9.1   Example of bridging between formal school learning and informal learning via an after school program (adapted from Noam et al., 2002).   145
10.1   Students working on PBL projects in S.O.S model.   160
10.2   A digital story created for a PBL project.   161
10.3   A website created for a PBL project.   162
10.4   A brochure design for a PBL project in the S.O.S model.   163
10.5   A circuit bug created in STEM Art class as part of the iTECH-STEM program.   164
10.6   Students working on stop-motion Digital Storytelling projects during iTECH-STEM.   165
10.7   Students working on programming projects in iTECH-STEM.   166
10.8   3D Modeling in iTECH-STEM.   167
10.9   Students learn how to use 3D-printers in iTECH-STEM.   167
12.1   An excerpt from the *Next Generation Science Standards* for two performance expectations and their elaborations from the Life Science Topic 1-LS 1, entitled From Molecules to Organisms: Structure and Processes.   194
12.2   An adaptation of the *Next Generation Science Standards* to specify dinosaurs as organisms in a bundle of performance expectations for K-2 learners.   195
12.3   The three elements involved in conceptualizing assessment as a process of reasoning from evidence (from Pellegrino et al., 2014, p. 49, reproduced with permission).   200
12.4   A scenario written in students' language situating the SCLE within a "story" in which students think and act like paleontologists. The guiding question for the SCLE refers to the scientists' work that students will simulate during four sequential lessons. The performance task is indicated (in italics) as part of the story and naturally includes evidence that students have met the student expectation for the SCLE.   201
12.5   Developmental rubric for assessing the iSTEM performance task.   202

12.6 Leading questions (LQ) and lesson-level performance expectations (LPE) for the four lessons in the SCLE, *How We Know What We Know about Dinosaurs.* 203

12.7 General SCLE Lesson Template used to construct the sequence of lessons comprising an iSTEM Performance TASK. Note the dual focus on what both teacher and students do. 204

12.8 Storyline for Lesson 1 in *How Do We Know What We Know about Dinosaurs?* 205

12.9 Schematic of the "revise and reflect clock diagram," indicating the process occurring at all stages of the 12-step process of designing an iSTEM learning performance. In the initial conception of the SCLE (dark arrows), completion of Step 2 after Step 1 requires the designer to revisit Step 1 and to check for coherence between the two steps (indicated by the dual arrow). The process occurs throughout the more or less linear process of moving through the 12 steps. Once all 12 steps are completed, the designer reviews the steps once again and adjusts decision outcomes at each step when necessary to maintain coherence in the system. 206

12.10 A linear depiction of the 12-step process of iSTEM performance task design for classroom assessments. 208

13.1 Image of Bailey's construction of a car using LEGOs and litteBits. 232

13.2 Bailey continuing construction of his car from Day 1. 235

14.1 Predicted rescaled leadership importance ratings for gender and experience groups. 262

14.2 Predicted current leader capacity for significant effect variables. 264

16.1 Project crossover questions 18 and 19. 307

16.2 Question 18, When did you first become interested in science, in general? 307

16.3 Question 19, When did you first become interested in chemistry/physics [your career disipline]? 308

16.4 FOCIS learning activities. 309

16.5 FOCIS instrument items. 310

16.6 Future career aspiration question. 311

Tables

1.1 Standards for Mathematical Practice (NGA Center & CCSSO, 2010) and Related NGSS Science and Engineering Practice (NGSS Lead States, 2013). 10

2.1 Percentages of US schools that don't offer core mathematics and science classes. 29

2.2 Percentage of 8th grade students proficient in math and science according to the NAEP. 29

5.1	Recommendations of national reports on STEM education.	75
7.1	Critical components for Inclusive Stem High Schools ISHSs in Opportunity Structures for Preparation and Inspiration in STEM (OSPrI Study) developed deductively and inductively.	115
12.1	Adapted 12-step approach to designing classroom task assessments aligned with the Next Generation Science Standards.	198
12.2	Levels of student-centered instructional strategies in integrated STEM learning environments.	210
14.1	Ingredients for leading innovation.	253
14.2	Leadership importance scale question means.	261
14.3	Multilevel model for leadership importance.	262
14.4	Question means for current leadership capacity and change in leadership capacity scales.	263
14.5	Multilevel regression for current leadership capacity – Significant variables.	264
15.1	Impacts of school and out of school-related activities on STEM-major intentions.	287
15.2	Impacts of Pygmalion effect variables on STEM-major intentions.	289
15.3	Impacts of self-expectation and math and science efficacy on STEM major intentions.	290
15.4	Independent samples t-test comparing positive and negative changers' average changes in school, Pygmalion, and motivational factors.	291
16.1	Comparison of descriptive statistics among youth participants in Grades 3–8.	312
16.2	Learning activity preference comparisons between Science & Engineering Aspiring Youth (SEA) versus non-Science & Engineering Aspiring (non-SEA) Youth.	313
18.1	Principals' perceptions on partnership outcomes.	343
19.1	Percentage of students in tertiary education enrolled in STEM tertiary education programs (both sexes) (2011–2015).	354
19.2	Percentage of students in STEM tertiary education programs (both sexes), by discipline (2015).	355

Notes on Contributors

Bradley Barker
is a Professor and Youth Development Specialist with 4-H Extension. Dr. Barker is spent eight years with Nebraska Educational Telecommunications where he was an Interactive Media Producer. Dr. Barker has directed media productions for the CLASS project, the Nebraska Law Enforcement Training Center, and the Nebraska National Guard. Dr. Barker has been the Principal Investigator on several National Science Foundation including the Nebraska 4-H Robotics and GPS/GIS project, the Nebraska 4-H Wearable's Technologies (WearTec) Wearable Technologies Grant, and most recently the Nebraska Innovative Maker Co-Laboratory (NiMC). Dr. Barker received his Ph.D. in Administration, Curriculum and Instruction in the area of Instructional Technology in 2002 from the University of Nebraska-Lincoln.
bbarker1@unl.edu

Jackie Barnes
has studied educational games and other new technologies in the context of classrooms for a decade. She has taught game design and educational courses for audiences from undergraduate to elementary students. She is currently working within an independent school to support innovative curricular experiences, both with and without the use of technology.
jbarnes@comday.org

Justin M. Bathon
(J.D./Ph.D.) is an Associate Professor and Director of Innovative School Models at the University of Kentucky, College of Education. He is the co-developer of the UK Center for Next Generation Leadership and the STEAM Academy High School in Lexington, KY. Justin works directly with schools and school leaders to reform the systems that support the learner experience in school. In so doing, his research focuses on the underlying code of education at the intersections of educational law, technological architectures, systemic norms of deeper learning, and the promotion of equity for all students.
justin.bathon@uky.edu

Sarah Bush
is an Associate Professor of K-12 STEM Education and Program Coordinator of the Mathematics Education PhD at the University of Central Florida in Orlando, FL. Dr. Bush's scholarship and research focuses on deepening student and teacher understanding of mathematics through transdisciplinary STE(A)M

problem-based inquiry and mathematics and STE(A)M education professional development effectiveness. Dr. Bush is a member of the National Council of Teachers of Mathematics (NCTM) Board of Directors (2019–2022).
Sarah.Bush@ucf.edu

Maureen Cavalcanti
earned a BS in Mathematics and a MEd from the University of California, Los Angeles; and PhD in Education Sciences with a specialization in STEM Education from the University of Kentucky. Dr. Cavalcanti began her career as a high school Mathematics teacher in Lawndale, CA and Boston, MA, before transitioning to higher education. She worked as a Master Teacher for Mathematics for CS*Uteach*, a STEM teacher preparation program at Cleveland State University. There she taught and co-taught numerous teacher education courses and supervised prospective secondary mathematics teachers. Her experiences at Cleveland State University shaped her pursuit for a doctoral degree in STEM Education, and continued commitment to academics. While at the University of Kentucky, Dr. Cavalcanti helped develop an instrument to measures student STEM literacy, and engaged in research on the impact of informal learning on student knowledge, skills, and attitudes toward STEM, particularly for groups historically underrepresented in STEM. She worked as an Educational Consultant for Battelle, facilitating professional development on problem based learning for high school teachers. Currently, she is the Director of the Office of Curriculum and Scholarship at The Ohio State University College of Medicine. She and her team engage health sciences faculty in educational scholarship; utilize education analytics to evaluate program effectiveness and identify actionable insights for medical education programs; facilitate faculty development on curriculum and instruction; and collaborate with faculty on designing their instructional content.
maureen.cavalcanti@gmail.com

Brett Criswell
taught high school chemistry for 15 years before completing his PhD in Curriculum & Instruction with a Science Education emphasis. During his 10 years in higher education, he has focused on science teacher preparation, including running undergraduate and graduate programs in this area. His research has focused on the use of video in science teacher preparation and on the development of [STEM] teacher leaders. Currently, he is working on a national project with NBPTS around the use of the ATLAS video library, as well as on a Noyce research project focused on studying teacher leadership development.
brett.criswell@uky.edu

Dionne Cross Francis
is an associate professor of mathematics education in the Department of Curriculum and Instruction at Indiana University and the Director of the Center for P-16 Research and Collaboration. She has a BA in Mathematics from the University of the West Indies, Jamaica and a PhD in Educational Psychology from the University of Georgia. Her research interests include investigating the relationships among psychological constructs such as beliefs, identity and emotions and how the interplay between these constructs influence teachers' instructional decision-making prior to and during the act of teaching mathematics independently, and within the context of STEM.
dicross@indiana.edu

Bulent Dogan
is a Clinical Assistant Professor of Instructional Technology at College of Education, University of Houston (UH), Houston, TX. Dr. Dogan has 6 years of public-school teaching and 5 years of school administration experience as a principal, head principal, and superintendent in STEM focused and high-performing Title-I public schools. Dr. Dogan's research interest include STEM instructional projects, social media in education, educational uses of digital storytelling, teaching emerging technologies including 3D Printing, Programing/ Coding, Mobile App Development, and Robotics to youth. Through his DISTCO (Digital Storytelling Contests) project since 2008, a series of annual STEM contests for K-12 students and teachers, Dr. Dogan has been the pioneer in his field by having students to create STEM projects with digital storytelling. Dr. Dogan is also the director of iTECH-STEM (Innovative Technology Challenges for STEM) program which aims to develop an innovative STEM program, by promoting STEM through a series of activities for 1st–5th grade students, with priority given to underrepresented minority, economically disadvantaged, and female students. iTECH-STEM includes teaching emerging technologies such as 3D Printing, Programing/Coding, Mobile App Development to youth. Dr. Dogan has also have special interest in teaching interactive online classes in higher education and has been awarded Outstanding Teaching Awards in 2014, 2016 and 2018.
bdogan@uh.edu

Emily Driessen
is a Minneapolis, MN native who earned a B.S. in microbiology with a minor in chemistry at North Dakota State University. She then continued on to earn a Master's in STEM Education at the University of Kentucky where she focused on middle-level student understanding of engineering. She is now pursuing a PhD in Biology at Auburn University where she will focus on improving undergraduate biology education. When she is not at school, she can be found

running outside training for marathons, playing with her German Shorthair, Lou, or reading a good book.
Epdoo16@tigermail.auburn.edu

Adem Ekmekci

(Ph.D.) is RUSMP Director of Research and Evaluation, and Clinical Assistant Professor of Mathematics, Wiess School of Natural Sciences. Dr. Ekmekci received his doctorate in STEM Education (Mathematics Education focus) from the University of Texas at Austin in 2013 and completed his post-doctoral training at Rice University in 2016. Dr. Ekmekci has developed research and evaluation designs for and conducted research and evaluation of several programs/projects funded by private, non-profit, and federal institutions including National Science Foundation, U.S. Department of Education, and Spencer Foundation. Dr. Ekmekci's research focus on student's motivational and achievement outcomes in STEM education, STEM teacher quality, and professional development of STEM teachers.
ekmekci@rice.edu

Dionne Cross Francis

is an associate professor of mathematics education in the Department of Curriculum and Instruction at Indiana University and the Director of the Center for P-16 Research and Collaboration. She has a BA in Mathematics from the University of the West Indies, Jamaica and a PhD in Educational Psychology from the University of Georgia. Her research interests include investigating the relationships among psychological constructs such as beliefs, identity and emotions and how the interplay between these constructs influence teachers' instructional decision-making prior to and during the act of teaching mathematics independently, and within the context of STEM.
dicross@indiana.edu

Brigid Freeman

is an education researcher with the Australia India Institute, University of Melbourne in Australia. Her research is comparative and primarily focuses on internationalization and higher education policy and governance. Brigid worked with Professors Marginson and Tytler on the STEM: Country Comparisons project and co-edited *The Age of STEM: Educational Policy and Practice in Science, Technology, Engineering and Mathematics Across the World*. Brigid also worked on the Humanities in the Asia Region Project with the Australian Academy of the Humanities, completed an international admissions policy consultancy for UNESCO, and been a visiting scholar with the University of California, Berkeley.
brigid.freeman@unimelb.edu.au

Neal Grandgenett
is the Dr. George and Sally Haddix Community Chair of STEM Education in the College of Education at UNO, where he teaches undergraduate and graduate courses in STEM education, interdisciplinary learning, and research methods. Dr. Grandgenett's interests include the development and evaluation of technology-based learning environments in STEM Education, and he has authored over 150 articles and research papers related to these interests, as well as six book chapters and one book. Dr. Grandgenett has received more than $18,000,000 in federal grants while at UNO.
ngrandgenett@unomaha.edu

Susie Gronseth
is a Clinical Associate Professor in the Learning, Design, and Technology program area in the College of Education at the University of Houston. She specializes in learning technologies, educational multimedia, teaching strategies, instructional design, and applications of Universal Design for Learning (UDL) to address diverse learner needs in online, face-to-face, and blended contexts. She has a Ph.D. in instructional systems technology from Indiana University. Her research interests include use of learning technologies in ways that engage learners, represent content in a variety of ways, and provide opportunities for learners to demonstrate their knowledge and skills.
slgronseth@uh.edu

Carol Hanley
worked as a science educator for over 30 years. She taught high school science in Fayette County for 13 years and worked at the Kentucky Department of Education to develop Kentucky's science content standards. While at the University of Kentucky (UK), she has been an extension specialist in 4-H youth development, Director of Education and Communications at the Tracy Farmer Institute for Sustainability and Environment, and Assistant Director of International Programs in the College of Agriculture, Food and Environment. Currently, she is a PhD candidate in Quantitative and Psychometric Methods in the College of Education at the University of Kentucky (UK).
chanley@uky.edu

Theodore Hodgson
is a professor of mathematics education in the Department of Mathematics & Statistics at Northern Kentucky University and Faculty Associate at the Kentucky Center for Mathematics. At NKU, he teaches courses for pre-service teachers and has directed numerous professional learning programs

for in-service teachers throughout Kentucky. One recent project, NKY-FAME, embedded teachers in advanced manufacturing firms for short-term externships, with follow-up support to develop and implement lessons in the K-12 classroom. Dr. Hodgson has a BS in mathematics from Indiana University, MS in statistics from the University of California, and Ph.D. in mathematics education from Indiana University.
hodgsont1@nku.edu

Kristina Kaufman
is a professor with an academic and occupational blend in the fields of education and business. With a doctorate degree in education combined with an M.B.A., she teaches, consults, and researches at the intersection of marketing, advertising, and education. She has conducted research on corporate influence and business partnerships with K-12 schools as well as her innovative and unconventional practices teaching in the post-secondary classroom. Dr. Kaufman has taught several business and marketing courses at the university level, a literacy course for pre-service teachers, and also has experience teaching students at the elementary and middle levels.
krisjkaufman@gmail.com

Sharon J. Lynch
(PhD) is a science educator and researcher at the George Washington University Institute for Public Policy, and Professor Emerita in the Department of Curriculum and Pedagogy in the Graduate School of Education at George Washington University. She has a BS in education, an MS in biology, and a PhD in education from Wayne State University in Detroit, MI. She has been a secondary school teacher of science in grades 7–12. Lynch was Director for Science and Mathematics Education for Johns Hopkins University Center for Talented Youth, but after serving as a Fulbright Fellow in Poland, turned her scholarly attention to issues of equity and excellence in the U.S. She is Professor Emerita of curriculum and instruction at the George Washington University Graduate School of Education and Human Development, and now is a research professor at the GW Institute for Public Policy. Lynch's book, *Equity and Science Education Reform* (2000) is still widely cited and she is the author of many research articles and book chapters. She served as President of the National Association for Research on Science Teaching, and is on the Editorial Board of *Science Education*. After serving as Program Director at the National Science Foundation in 2010, she turned her attention to research on inclusive STEM high schools. She was Principal Investigator of the *Opportunity Structures for Preparation and Inspiration* (OSPrI) research study that focused on identifying the critical components of exemplar STEM high schools. That research has

been discussed in publications including *Nature, Scientific American and US News & World Report*, as well as scholarly publications. Lynch was also co-PI on the *iSTEM* companion study that demonstrated the efficacy of inclusive STEM high schools in three states. She currently works for a portion of her time as an Intermittent Expert/Program Director for the National Science Foundation. She also maintains an active role in writing and lecturing about inclusive STEM schools nationally and internationally.
slynch@gwu.edu

Luke C. Lyons
is a former high school science teacher and completed his PhD in Curriculum & Instruction, Science Education, in May of 2018 at Texas A&M University. Lyons currently is teaching physiology and co-directing a program for undergraduate research at Texas A&M. His research emphases include science teacher preparation and enhancement through the use of innovative curriculum designs such as learning progressions, high-interest topics (e.g., dinosaurs) in science and the impacts of undergraduate research. A common thread is his use of authentic assessment through real science practices in student-centered learning environments in classrooms from Kindergarten to collegiate science courses.
lukelyons@tamu.edu

Cathrine Maiorca
is an assistant professor of mathematics education at California State University, Long Beach. Her research interests include how preservice/inservice teachers incorporate mathematical modeling and the engineering design process into their mathematics classrooms, informal STEM learning environments, preservice/inservice teachers and students perceptions of STEM, and issues of equity in STEM education.
cathrine.maiorca@csulb.edu

Adam V. Maltese
teaches courses in secondary science methods and graduate seminars at the School of Education at Indiana University around making and the development of interest in STEM education. In addition, he leads seminars for doctoral students in STEM fields who plan to pursue academic careers and are interested in improving their teaching practices based on research. His current research involves collection and analysis of both quantitative and qualitative data regarding student experiences, performance and engagement in science education from elementary school through graduate school.
amaltese@indiana.edu

Simon Marginson

is Professor of Higher Education at the University of Oxford, Director of the ESRC/OFSRE Centre for Global Higher Education (CGHE), and Editor-in-Chief of the journal *Higher Education*. Simon's research is focused primarily on global and international higher education, and higher education and social inequality. He is currently preparing an integrated theorization of higher education. His most recent books are *Higher Education in Federal Countries*, edited with Martin Carnoy, Isak Froumin and Oleg Leshukov (Sage, 2018) and 'High Participation Systems of Higher Education', edited with Brendan Cantwell and Anna Smolentseva (Oxford University Press, 2018).
simon.marginson@education.ox.ac.uk

Margaret Mohr-Schroeder

is a Professor of STEM Education and Associate Dean in the College of Education at the University of Kentucky. Her research interests include the transdisciplinary nature of STEM education and how they can be applied to innovative preservice teacher education and K12 school models. Further, she is interested in ways to broaden participation in STEM, especially of underrepresented populations and the effects these mechanisms have on their STEM literacy. Through this work, she has gained perspective on how to create opportunity and access to STEM activities to populations that normally would not have the opportunity and have witnessed and studied the statistically significant effects these mechanisms have.
m.mohr@uky.edu

Kathleen Morgan

has served as a Partner Services Manager for FIRST LEGO League and FIRST LEGO League Jr. since 2013. In this role, she supports program partners in the northeast United States and Canada and manages judging for the global FIRST LEGO League program. Prior to joining the FIRST staff, Kathleen was the Project Manager for 4-H Youth Development GEAR-Tech-21 at the University of Nebraska-Lincoln, where she launched FIRST in Nebraska and earned her Master of Applied Science in Leadership Education. Kathleen also holds a Bachelor of Arts in Math and Physics from Colorado College.
kmorgan@firstinspires.org

Louis S. Nadelson

is an associate professor and chair in Leadership Studies at University of Central Arkansas. He has a BS from Colorado State University, a BA from the Evergreen State College, a MEd from Western Washington University, and a PhD in educational psychology from UNLV. Nadelson uses his over 20 years of high

school and college math, science, computer science, and engineering teaching to frame his research on STEM teaching and learning. His scholarly interests include all areas of STEM teaching and learning, leadership program evaluation, entrepreneurship, transdisciplinary research, and conceptual change.
lnadelson1@uca.edu

Gwen Nugent
is a Research Professor in the Center for Research on Children, Youth, Families, and Schools at the University of Nebraska-Lincoln. She coordinates development and research projects focusing on the impact of technology to improve student learning and teacher competencies and has led projects funded by the National Science Foundation and U. S. Department of Education. She has over 30 years' experience in the design, production, and evaluation of mediated instruction and has designed over 300 multimedia projects in a variety of subject areas and for a variety of audiences.
gnugent@unl.edu

Kelli M. Paul
worked for 10 years as an independent consultant evaluating STEM-focused projects involving middle and high school students, especially women and minority students. She currently is a post-doctoral fellow at Indiana University. Her research interests include students' STEM identity and interest in STEM careers, the influence of role models on STEM identity and interest, and the development of instruments/tools to assess these constructs.
kelpaul@iu.edu

Alpaslan Sahin
is Ph.D. Research Scientist at Harmony Public Schools, Houston, Texas. He was previously employed as a Research Scientist at Aggie STEM Center at Texas A&M University. His work at the Aggie STEM spring boarded him into the Harmony Public Schools where he has carefully studied and helped teachers and administrators implement and embrace the STEM SOSTM (Students on the Stage) model. Over the past several years, his works appeared in a variety of books and peer-reviewed journals. His research interests include teachers' questioning techniques, STEM education, informal STEM learning, 21st century skills, charter schools, and educational technology.
asahin@harmonytx.org

Christine Guy Schnittka
is an associate professor of science education in Auburn University's College of Education with a joint appointment in the College of Engineering. Her

current research involves developing and evaluating engineering design-based curriculum units that target key science concepts through the contextual lens of environmental issues that engage us all. Her curricula have been used by teachers in over 33 states and 13 countries. Prior to receiving her Ph.D. in science education at the University of Virginia, Dr. Schnittka was a middle school teacher and administrator for 10 years, and prior to that, worked as a mechanical engineer and musician.
schnittka@auburn.edu

Anne Seifert
is a Capacity Builder for the Idaho Capacity Builder Project, coaching and assisting low performing schools and a team lead for AdvanceEd STEM School Certification. She holds a BS degree in elementary education, an MA in Education Administration, an EdS in Educational Leadership, and is a 30 year veteran teacher, school administrator and cofounder of the i-STEM network. Seifert's research interests include STEM education, inquiry and project-based instruction with the incorporation of 21st century learning, change practices, and cultural influences on school effectiveness.
anneseif@yahoo.com

Amber Simpson
is an Assistant Professor of Mathematics Education in the Department of Teaching, Learning, and Educational Leadership, Binghamton University, Binghamton, NY. Her research includes understanding the role of making and tinkering in formal and informal education settings with an emphasis on mathematical play. She also conducts research on understanding the interplay of voices shaping and embodying individual's STEM identity.
asimpson@binghamton.edu

Micah Stohlmann
is an associate professor of Mathematics/STEM Education at the University of Nevada, Las Vegas. His research program focuses on mathematical modeling and STEM integration through open-ended problems.
micah.stohlmann@unlv.edu

Carol L. Stuessy
earned a B.A. in Biology from the University of Texas at Austin; and a B.S. and Ph.D. in Science Education from The Ohio State University. Dr. Stuessy began her academic career at New Mexico State University in Las Cruces before coming to Texas A&M. She has served as a member of the Texas A&M University Graduate Faculty for 28 years, chairing 44 doctoral students during

her academic career. She has been active in international and national science education research organizations for 35 years, serving on numerous boards of directors and as President of the School Science and Mathematics Association. She received the meritorious Mallinson award for outstanding service to that organization. In 2016, Dr. Stuessy received an Award of Distinction from the Ohio State University; and in 2017, a Transformational Leader award by the Dean of the College of Education at Texas A&M University. Most recently, Dr. Stuessy served as Director of the Online EdD in Curriculum and Instruction and as co-director of the Center for Science and Mathematics Education at Texas A&M. She has received over two dozen grants for external funding amounting to about $12M to support her research interests in high school science education policy reform, innovative science curriculum design, and the effects of engaging scientists and engineers in the online mentoring of high school students. In addition, Dr. Stuessy has developed a mixed methods classroom observation protocol, which has been widely used for estimating the complexity of teaching and learning in science classrooms.
c-stuessy@tamu.edu

Robert H. Tai
is an associate professor of science education at the Curry School of Education at the University of Virginia since 2001. Prior to this, he taught high school physics, earned an Ed.D. from the Harvard Graduate School of Education, and he taught at the College of Staten Island of the City University of New York. His current research includes longitudinal research on young engagement in STEM learning and understanding and assessing the effectiveness of informal science education programs. In 2018, he was named by the National Afterschool Association to be among the Most Influential in Research and Evaluation.
rht6h@virginia.edu

Russell Tytler
is Alfred Deakin Professor and Chair in Science Education at Deakin University, Melbourne. He has researched and written extensively on student learning and reasoning in science. His interest in the role of representation in reasoning and learning in science extends to pedagogy and teacher and school change. He researches and writes on student engagement with science and mathematics, school-community partnerships, and STEM curriculum policy and practice.
russell.tytler@deakin.edu.au

Hersh Waxman
is a Professor in the Department of Teaching, Learning, and Culture (TLAC), Director of the Texas A&M University Education Research Center, and

Co-Director of the Center of Mathematics and Science Education. He received the Distinguished Alumnus Award from the College of Education at University of Illinois-Chicago, and Outstanding Research Awards from the American Educational Research Association, Southwest Educational Research Association, and the Society for Information Technology and Teacher Education. He has written more than 150 research articles in the areas of teacher and school effects, classroom learning environments, and students at risk of failure.
hwaxman@tamu.edu

Kerrie Wilkins-Yel
is an Assistant Professor in the Department of Counseling and Educational Psychology at Indiana University Bloomington. She received her B.S. in Experimental Psychology from the University of South Carolina Upstate, and both her M.A. and Ph.D. in Counseling Psychology from Arizona State University. Dr. Wilkins-Yel's research broadly focuses on promoting academic persistence and career advancement among women and individuals from marginalized backgrounds. Currently, she takes an intersectional approach to examining the influential factors that enhance career persistence among women of color enrolled in science, technology, engineering, and mathematics (STEM) disciplines. Dr. Wilkins-Yel has received approximately $1M in federal and institutional funding to support her research. She has also collaboratively developed empirically based STEM Initiatives that apply psychological science and culturally responsive approaches to the advancement of STEM persistence among women from diverse backgrounds.
kgwilkin@indiana.edu

Soledad Yao
(PhD) is a research assistant in the STEM Education Master's Program at the University of Kentucky's College of Education. She received her PhD in Chemistry from the University of Kentucky's Department of Chemistry. For over thirty years, she taught Chemistry at two Universities in the Philippines-the University of the Philippines Manila and the University of the East.
slyao11211@gmail.com

Kimberly Yates
is an Assistant Professor in the Department of Teacher Education at Northern Kentucky University. She is also the Director of the NKU Center for Environmental Education. Her teaching and research are focused on the areas of science education and environmental education. In addition to teaching, Dr. Yates provides professional development to preservice teachers, in-service

teachers, and non-formal educators. She is interested in providing educators with the skills needed to create an engaging and challenging learning environment for their students, especially in the areas of STEM education and environmental education, and in the intersection of those two fields. yatesk2@nku.edu

PART 1

*Setting Stage for STEM Education:
Updates on STEM Education*

∴

CHAPTER 1

STEM Literacy: Where Are We Now?

Maureen Cavalcanti and Margaret J. Mohr-Schroeder

Abstract

STEM literacy becomes a synergy of applying the knowledge and skills of STEM to "increase students' understanding of how things work and improve their use of technologies" (Bybee, 2010, p. 1) to "study the grand challenges of our era" (Bybee, 2010, p. 2). Addressing the issues that will present this generation will require the development of the STEM literacy of current students (Zollman, 2012). Developing STEM literacy must be a multi-faceted. Collaborative of industry and professionals, government support, education with highly qualified teachers, and a society that embraces STEM all feed into creating a STEM literate workforce (Augustine, 2005). Increasing STEM literacy of society, especially the youngest generation, is essential for addressing the issues that face humanity.

1 Introduction

The search for understanding literacy in the context of science, technology, engineering, and mathematics (STEM) is not a novel endeavor. Knowledge has been a central theme in defining *disciplinary literacy* (e.g., Shanahan & Shanahan, 2012), where concepts of knowledge have encompassed problem-solving and critical thinking to make informed decisions (Garmire & Pearson, 2006; National Research Council, 1999). In thinking about the application of knowledge, learning across science, technology, engineering, and mathematics (STEM) has resulted in new literacy demands (i.e., daily workplace; Asunda, 2012). Arguably, Bybee (2010) and the National Research Council (NRC, 2011) brought the term *STEM literacy* into the spotlight, but the groundwork to uncover the layers of literacy within and across STEM disciplines has been laid by stakeholder throughout the 20th and 21st centuries. It is the combined effort of stakeholders which has led us to where we are now in our understanding of STEM literacy and its implications for ensuring learners can meet societal demands.

Literacy has been broadly defined as "the ability to negotiate and create texts in appropriate ways or in ways other members of a discipline would recognize as 'correct' or 'viable'" (Draper & Siebert, 2010, p. 30), whereby learners possess skills that "place them in a particular place on the literacy continuum" (Lemke et al., 2004, p. 2) It can be viewed as an outcome or goal for education (AAAS, 1993; Asunda, 2012; Bybee, McCrae, & Laurie, 2009; Gardner, 1983; Tout, 2000) STEM provides a meaningful context for thinking about how educational systems can "transition to 'new literacy skills'" to achieve literacy (Schleicher, 2010, p. 433). Educators, industry, and policymakers generally agree STEM literacy should be an educational priority (e.g., Bybee, 2010; DOE, 2016) to engage learners in STEM experiences, broaden participation in STEM, and prepare learners for college and career pathways. The work in the field of STEM education has addressed issues of defining literacy and describing what it means for an individual to be literate in a specific discipline. The integration of STEM disciplines in curricula could support the development of STEM literacy and pathways to STEM-related careers (Asunda, 2012) and is important to help students make connections between STEM concepts (Honey et al., 2014; NRC, 2011).

Before the concept of STEM literacy gained traction, stakeholders worked to characterize literacy within and across STEM disciplines. An integrative approach has been undertaken in varied ways including grounding notions of literacy across STEM disciplines in a single discipline (AAAS, 1993) and defining literacy more broadly (e.g., DeBoer, 2000). The integrated perspective has remained relevant within the current climate to engage learners in STEM by arming they with the knowledge, skills, an attitude to apply relevant practices and processes in authentic settings. Undoubtedly, there is great overlap between literacy within and across STEM areas and 21st century skills (Mohr-Schroeder, Cavalcanti, & Blyman, 2015). Existing definitions of STEM literacy (e.g., Balka, 2011; Bybee, 2013). have been widely accepted and referenced within educational platforms. The widespread efforts to define and operationalize STEM literacy in educational and workplace settings call for existing definitions to be revisited in the context of historical perspectives on literacy in STEM disciplines, current practices and policies in STEM education, and workforce needs. Doing so allows us to draw upon 20th and 21st century perspectives and outcomes of empirical studies to uncover a comprehensive model for conceptualizing STEM literacy.

A definition of STEM literacy rooted in the relationship between STEM disciplines can inform efforts to address concerns over shortages in the STEM workforce identified in national reports such as the National Science Board's (NSB) report, *Revisiting the STEM Workforce* (2015), and The Committee on STEM

Education National Science and Technology Council's strategic plan, *Federal Science, Technology, Engineering, and Mathematics (STEM) Education: 5-Year Strategic Plan* (2013). Aligning diverse perspectives on literacy can help address many of the current needs in the United States educational system and inform a new direction for STEM literacy. The socio-political climate in the United States has influenced shifts in education throughout history. For instance, in mathematics, Gutiérrez (2013) identified the increasing importance of identity and power in learning mathematics. The launch of Sputnik in 1957 spurred reforms in mathematics, science, and engineering rooted in a desire to grow the capacity for Americans to contribute to STEM fields. A synthesis of diverse disciplinary perspectives can help educators and policymakers design and implement educational experiences that support positive outcomes of STEM literacy for learners. Doing so, has the potential to prepare individuals for future college and career pathways, even those unrelated to STEM. A number of initiatives focus on developing a STEM-literate population including, the Common Core State Standards (CCSS), STEM-focused schools, and STEM learning networks have been identified in the literature (e.g., Mohr-Schroeder et al., 2015). As an example, Means, Wang, Young, Peters, and Lynch (2016) found a positive impact of enrollment in STEM-focused high school on student interest and achievement in STEM. Students at the STEM-focused high school experienced integrated curriculum, informal learning opportunities, and interactions with professionals.

A major consequence of the understanding STEM literacy is the move from theory to practice. A clear conceptual understanding of STEM literacy is needed to design and implement integrated STEM learning experiences. Kelley and Knowles (2016) have taken such an approach in their conceptual framework for integrated STEM teaching and learning. The purpose of this chapter is to draw connections to current efforts to define and enact STEM literacy in educational settings. Our process to build a definition of STEM literacy using history and research is intended to offer an integrated view of STEM literacy inclusive of knowledge, skills, and dispositions relevant within and across STEM disciplines.

2 The Status of STEM Literacy

STEM literacy has been referred to frequently in publications within the last 10 years (e.g., Asunda, 2012; Bybee, 2010, 2013; Gonzalez & Kuenzi, 2012; Hayford, Blomstrom, & DeBoer, 2014; NRC, 2011; Vasquez, Sneider, & Comer, 2013; Zollman, 2012). Much of the research on STEM literacy has focused on the definition or application of STEM literacy to classroom teaching and learning (e.g., Zollman, 2012). The most widely cited come from Balka (2011), Bybee (2010), the National Governor's Association (NGA; 2007), and the National Research Council (NRC, 2011), which each refer to, to some extent, as the knowledge and skills related to STEM for the purpose of application and problem-solving. The concept of integration is implicit or explicit amongst the definitions of STEM literacy. A recent definition cited by the Department of Education's (DOE) national report, *STEM 2026: A Vision for Innovation in STEM Education* (DOE, 2016), of STEM literacy focused on foundational skills in STEM areas for all adults, and advanced knowledge and skills for STEM professionals (Business Roundtable and Change the Equation, 2014)

Current perspectives on STEM literacy can be additionally understood by considering the characteristics of an individual who is STEM-literate (Abts, 2011; Meeder, 2014). Abts (2011) identified four attributes of someone who is STEM-literate: (1) problem-solver, (2) interdisciplinary thinker, (3) self-reliant, and (4) technology capable. Meeder (2014) also noted the problem solving aspect, but additionally identified conceptual knowledge of STEM subjects, connecting STEM content to STEM careers, and psychological aspects of achieving in STEM. Our work is, in part, a response to Balka's (2011) recommendation to look at different components of STEM literacy. The historical perspectives on literacy and efforts to integrate learning in STEM inform the way components of STEM literacy are identified.

3   Emergent Themes in Perspectives of Literacy in STEM Disciplines

The STEM disciplines have similar needs and therefore, literacy should be able to be defined to unify themes that arise across definitions of literacy for the subjects. *Science for All Americans* (AAAS, 1989) identified criteria related to the common core of learning in science, mathematics, and technology. Those criteria include utility, social responsibility, intrinsic value of knowledge, philosophical value, and childhood enrichment. These criteria have arisen in descriptions of literacy in the STEM areas before and after the 1989 publication. In addition to these criteria, scientific and technological change, role of the standards, and decision-making have also been addressed in discussions of literacy.

To meet the needs of society, young people in the United States must become literate in the STEM fields. Politically, there have been incentives to prepare scientists, engineers and teachers (Riechard, 1985) for the growing technical workforce. Those incentives have only grown as conversations have focused on funding, innovation, equity in STEM, and pathways to STEM careers. In the K-12 setting, efforts to promote literacy in STEM areas have been tied to curriculum reform, and now since the beginning of the 21st century, to content standards.

The relationship between literacy, although sometimes described in more general terms, and making decisions is a recurrent theme across the STEM disciplines (AAAS, 1989; Asunda, 2012; Betz, 1948; NGA Center & CCSSO, 2010; Everitt, 1944; Gorham et al., 2003; Miller, 1983; NGSS Lead States, 2013; NRC, 1996; Steen, 1999). The result has been practices in mathematics and science calling for students to use judgment and make arguments as they use the knowledge and processes of the content areas to communicate informed decisions.

4   Integrating STEM Literacy

The standards and practices (e.g., International Society for Technology Education, 1998, 2007, 2016); NGA Center & CCSSO, 2010; NGSS Lead States, 2013) offer an entry point for incorporating literacy and proficiency across subject areas into the outcomes for other subject areas. Additionally, they can be viewed as outcomes for developing the core learning in STEM areas that support pathways to STEM literacy. While the standards are largely geared toward the K12 setting, other stakeholders have expanded the role of literacy such as accrediting agencies for post-secondary work (e.g., Accreditation Board for Engineering and Technology (ABET), 2014). Together, the associations between the standards, outcomes, and practices related to literacy across all STEM domains and contexts lead to a unified definition of STEM literacy. Included, the alignment of standards and practices (e.g., Jackson & Mohr-Schroeder, 2018), literacy, and integrated content can inform ways to think about and subsequently advance STEM literacy in diverse learning environments (e.g., AAAS, 1989). Asunda (2012) claimed clear connection between each of the disciplines is missing from definitions of STEM literacy. Given the attention placed on national standards, practices and processes, literacy, and assessment, Asunda's claim may no longer hold true. Further, it may be argued the connections are present but the appropriate lens must be used to think about those connections. The evolution of content standards to the potential for using the standards and practices to address 21st century skills through integrative STEM education can serve as such a lens.

Standards for individual STEM disciplines as learning outcomes have held an important place within U.S. education for almost 30 years. Lederman (1998) described the national need to assess learning using standards writing, "The nation has the challenge to ensure that all America's children have the opportunity to learn and understand science, mathematics and technology at the higher levels defined by national standards" (p. 2). Collaboration between policymakers, schools, teachers, and professional organizations has been essential to designing and disseminating standards.

The standards-based environment received initial momentum from national efforts in mathematics and science education. Since the National Council of Teacher of Mathematics (NCTM) published *Curriculum and Education Standards* in 1989, the first of three NCTM standards documents, stakeholders have placed targeted attention on establishing standards for learning in education. The interpretations of the standards brought to light possible meanings of standards-based practices (NRC, 1997, p. 11). In an effort to "preserve the main messages of the original Standards, while bringing together the 'classroom' parts of the three Standards documents into a single documents" (NRC, 1997, p. 13), NCTM put forth the *Principles and Standards for School Mathematics* (2000) which marked "a new phase in the standards movement" (p. 14). The NRC with the Center for Science, Mathematics, and Engineering Education spearheaded movement toward standards-based science education (NRC, 1997) using the *National Science Education Standards* (NSES; NRC, 1996). The birth of the Common Core State Standards (CCSS) and Next Generation Science Standards (NGSS) further capitalized on content and concepts alongside the practices intended to support deeper understanding of content and application of knowledge to new settings. Include are explicit and implicit connections to literacy.

Standards for technology have evolved in a similar way since the International Society for Technology Education (ISTE) standards were introduced in 1998, first as outcomes for learning how to use technology and later (2007, 2016) as outcomes aligning to goals for 21st century learning, namely to be active and engaged learners in a global and technological society. There has, however, always been an absence of national standards for engineering. In 2010, the Committee for Standards for K-12 Engineering Education advised against creating such standards because of the "evolving status of K-12 engineering education" (NRC, 2010, p. 37). Instead, they recommended ways to integrate outcomes for engineering education into mathematics, technology, and science using "infusion" and "mapping" approaches. Even so, some states have independently developed engineering standards (Dugger, 2010). Massachusetts Department of Education (MA-DOE) developed the *Massachusetts Science and Technology/Engineering Curriculum Framework* (MA-DOE, 2001, 2006), with the most recent revisions published in 2016 (MA-DOE, 2016).

The Standards for Mathematical Practice (NGA Center & CCSSO, 2010) and the Science and Engineering Practices (NGSS Lead States, 2013) are intended to support a common goal of acquiring knowledge and the ability to use knowledge to solve problems; these goals being consistent with current definitions of STEM literacy. Commonalities across the Standards for Mathematical Practice and Science and Engineering Practices build understanding of shared practices related to literacy. Cheuk (2013) represented convergence in practices across CCSS-ELA, CCSS-M, and NGSS in a Venn diagram while noting that highlighted relationships are not definitive and other relationships could be argued (2012). Cheuk's work coupled with our reevaluation of the language of the practices was the basis for the relationships between the Standards for Mathematical Practice (SMP) and the Science and Engineering Practices resulted in broader linkages between the two (Table 1.1). As an example, the description of SMP4 ("Models with mathematics") includes phrases such as "analyze a problem," "solve a design problem," "map their relationship," and "draw conclusion" (NGA Center & CCSSO, 2010, para. 5). NGSS described relationship between the Science and Engineering practices using similar language: "For example, the practice of "asking questions" may lead to the practice of "modeling" or "planning and carrying out an investigation," which in turn may lead to "analyzing and interpreting data" (NGSS Release, 2013, p. 3).

The *Standards for Technological Literacy* (STL; ITEA, 2007) offer a pathway for integrating technology across STEM disciplines. Gorham et al. (2003) compare the STLs (ITEA, 2007) and ABET Outcomes (2014) by identifying implied and direct relationships between concepts outlined in each. For example, ABET asserted identifying formula and solving engineering problems (ABET Criteria 3 Student Outcome e) relates to 14 of the 20 STL standards including applying the design product (STL Standard 11) and understanding the role of society in advancing technology (STL Standard 6). Neidorf et al.'s (2016) compared NGSS to NAEP science, NAEP technology and engineering literacy (TEL), and NAEP mathematics frameworks. As intended by the development of the NGSS, the NGSS and NAEP frameworks have similar foci. It is essential to expand work to make clear connections between the standards, outcomes, and practices related to literacy across all STEM domains.

More recently has been the expansion of content that falls within the STEM reach and the application of integrated STEM content. The Maker Movement (e.g., Honey & Kanter, 2013), embodied initially by fab labs and makerspaces, exploded into a national movement for all ages, and through the *making process*, makers can foster STEM literacy practices (e.g., Tucker-Raymond, Gravel, Kohberger, & Browne, 2018). Computer and data sciences has seen significant attention within the last five years including calls to expand STEM and

TABLE 1.1   Standards for Mathematical Practice (NGA Center & CCSSO, 2010) and related NGSS Science and Engineering Practice (NGSS Lead States, 2013)

Standards for Mathematical Practice (CCSSO, 2010)	1. Asking questions and defining problems	2. Developing and using models	3. Planning and carrying out investigations	4. Analyzing and interpreting data	5. Using mathematics and computational thinking	6. Constructing explanations and designing solutions	7. Engaging in argument from evidence	8. Obtaining, evaluating, and communicating information
1. Make sense of problems and persevere in solving them.	X	X	X			X		
2. Reason abstractly and quantitatively.				X	X			
3: Construct viable arguments and critique the reasoning of others.							X	X
4: Model with mathematics.	X	X		X	X			X
5: Use appropriate tools strategically.		X	X			X		
6: Attend to precision.	X							X
7: Look for and make use of structure.					X			
8: Look for and make use of regularity in repeated reasoning.				X	X			

Note: SMP1–SMP8 represent the CCSS Standards for Mathematical Practice
SOURCE: CHEUK (2013)

computer science educational offerings (i.e., US DOE 2018 SEED grants) and the increased workforce demand for data scientists (e.g., LinkedIn Workforce Report, August 2018). Davenport and Patil (2012) highlighted the need for data scientists in their *Harvard Business Review* article titled, "Data scientist:

The sexiest job of the 21st century" (October 2012). The Association for Computing Machinery, Code.org, Computer Science Teachers Association, Cyber Innovation Center, and National Math and Science Initiative in partnership with states and districts developed the 2016 *K-12 Computer Science Framework*. Within the framework, the authors define computer literacy as "the general use of computers and programs, such as productivity software" including "performing an Internet search and creating a digital presentation" (p. 13) and identify the goal of literacy within the framework as "foundational literacy in computer science" (p. 16).

5   Toward a Common Definition of STEM Literacy

Understanding STEM literacy involves a comprehensive analysis of many facets of literacy within and across STEM disciplines and contexts. Zollman (2012) recommends we think about STEM literacy by incorporating historical perspectives of literacy in individual STEM disciplines alongside the dynamic nature of STEM literacy. The definition of STEM literacy must remain dynamic to reflect the constantly changing technological demands of society. Meyers, Erickson, and Small (2013) have done a good job of describing this balance that needs to be struck in defining STEM literacy, "All literacies are built on a foundation of the traditional literacy skills of reading, writing, speaking, and listening. However, in today's digital, globalized world, a much broader definition of literacy is required" (p. 361).

It needs to be made clear how to think about literacy within different STEM disciplines to contribute to growing literature on how STEM literacy from an integrated perspective can be defined, and how stakeholders can come to understand how literacy has been defined within and across science, technology, engineering, and mathematics. The question of whether literacy within individual STEM literacy should be defined broadly or within the specific context of the field can be raised in terms of an integrated view of STEM literacy. Bybee (2010) defined STEM literacy as the "conceptual understandings and procedural skills and abilities for individuals to address STEM-related personal, social, and global issues" (p. 31). This definition is more closely aligned to the broadly defined line of thinking, and does generalize some of the common themes that have been discussed. NRC (2011) has also offered a definition for STEM literacy, influenced by NRC's 1996 definition of scientific literacy – "The knowledge and understanding of scientific and mathematical concepts and processes required for personal decision making, participation in civic and cultural affairs, and economic productivity for all students" (p. 5).

Using the substantive literature on literacy within the STEM disciplines, it may behoove stakeholders to place more targeted attention to the themes of utility, social responsibility, response to change, communication, decision-making, and knowledge to define STEM literacy in a way that can be understood and applied across a variety of contexts.

This chapter ultimately aimed to chronicle how existing definitions of STEM literacy and related efforts could give rise to a unified definition of STEM literacy. Included, we aimed to propose a dynamic definition of STEM literacy, one informed by coordinated modern and historical perspectives and research outcomes related to disciplinary literacies within and across STEM disciplines. In doing so, we observed synergy in existing definitions that could be coordinated to highlight the knowledge, skills, and attitudes that are quintessential to the definition of STEM literacy. As such, STEM literacy may be defined as the

> conceptual understandings and procedural skills and abilities for individuals to address STEM-related personal, social, and global issues (Bybee, 2010, p. 31); the ability to engage in STEM specific discourse; a positive disposition toward STEM (e.g., Wilkins, 2000, 2010, 2015), including a willingness to engage and persist in STEM-related areas (e.g., Wilkins, 2000, 2010, 2015); an understanding of the utility of applying STEM concepts to solve real world problems; and, an appreciation of how the processes and practices of STEM areas change as technologies and demands of modern society change. (Cavalcanti, 2017, pp. 65–66)

STEM literacy further subsumes the characteristics of STEM professionals. For the school years (K12), this definition of STEM literacy proposes STEM literacy development exists along a continuum (Lemke et al., 2004). By engaging in STEM literacy practices in authentic environments, individuals become more STEM-literate; their level of literacy increasingly reflects that of a STEM professional.

6 Conclusions and Future Directions

It is important to think about how literacy can be supported to ensure positive outcomes of efforts to produce a STEM-literate society. Our work on the emergence of STEM Literacy through the exposure of high-quality STEM experiences in informal learning environments has confirmed our more holistic definition of STEM literacy: (a) self-efficacy/perception of ability, (b) attitude and interest (willingness to engage, career belief, disposition), (c) role and utility of STEM in society, and (d) sense of community (Cavalcanti, 2017; Delaney, Cavalcanti, Jackson, & Mohr-Schroeder, 2017; Roberts et al., 2018). The role of curriculum has been touched on, but support also involves the individuals facilitating the work to support STEM literacy. Collaboration among teachers (NRC, 1996), including the use of a shared language, are important. That importance rings as true today as it did during the mid 20th century. Betz very wisely asserted "If we had only an ounce of a more genuine type of cooperation from the educators and administrators we should have had truly functional mathematical curricula long ago" (1948, p. 197). When such cooperation exists among stakeholders, young minds are given opportunities to develop the STEM literacy imperative for success in the 21st century.

Acknowledgements

This work was supported by the National Science Foundation under Grant Numbers 1348281. Any opinions, findings, and conclusions or recommendations expressed in this material are those of the author(s) and do not necessarily reflect the views of the National Science Foundation.

References

Ainley, J., Fraillon, J., Schulz, W., & Gebhardt, E. (2016). Conceptualizing and measuring computer and information literacy in cross-national contexts. *Applied Measurement in Education, 29*(4), 291–309.

American Association for the Advancement of Science (AAAS). (1989). *Science for all Americans: A project 2061 report on literacy goals in science, mathematics, and technology* (AAAS Publication 89-01S). Washington, DC: Author.

American Association for the Advancement of Science (AAAS). (1993). *Benchmarks for science literacy*. New York, NY: Oxford University Press.

Asunda, P. A. (2012). Standards for technological literacy and STEM education delivery through career and technical education programs. *International Journal of Science Education, 32*, 143–172.

Becker, K., & Park, K. (2011). Effects of integrative approaches among Science, Technology, Engineering, and Mathematics (STEM) subjects on students' learning: A preliminary meta-analysis. *Journal of STEM Education: Innovations and Research, 12*(5–6), 23–37. Retrieved from http://search.proquest.com/openview/865fa033fdc4c40d761bf9f300ce5469/1?pq-origsite=gscholar

Berman, S. L. (1945). Some thoughts on tomorrow's mathematics. *The Mathematics Teacher, 38*(6), 269–273.

Betz, W. (1948). Functional competence in mathematics – Its meaning and its attainment. *The Mathematics Teacher, 41*(5), 195–206.

Bishop, J. P. (2012). "She's always been the smart one. I've always been the dumb one": Identities in the mathematics classroom. *Journal for Research in Mathematics Education, 43*(1), 34–74.

Broughton, M. A., & Fairbanks, C. M. (2003). In the middle of the middle: Seventh-grade girls' literacy and identity development. *Journal of Adolescent & Adult Literacy, 46*(5), 426–435.

Bybee, R. W. (2000). Achieving technological literacy: A national imperative. *The Technology Teacher, 60*(1), 23–28.

Bybee, R. W. (2010). Advancing STEM education: A 2020 vision. *Technology and Engineering Teacher, 70*(1), 30–35.

Bybee, R. W. (2013). *The case for STEM education: Challenges and opportunities.* Arlington, VA: National Science Teachers Association.

Bybee, R. W., McCrae, B., & Laurie, R. (2009). PISA 2006: An assessment of scientific literacy. *Journal of Research in Science Teaching, 46*(8), 865–883.

Cheuk, T. (2013). *Relationships and convergences among the mathematics, science, and ELA practices. Refined version of diagram created by the Understanding Language Initiative for ELP Standards.* Palo Alto, CA: Stanford University.

Cockcroft, W. H. (1982). *Mathematics counts.* London: HM Stationery Office.

Crowther, G. S. (1959). *The Crowther report.* Retrieved from http://www.educationengland.org.uk/documents/crowther/crowther1959-1.html

DeBoer, G. E. (2000). Scientific literacy: Another look at its historical and contemporary meanings and its relationship to science education reform. *Journal of Research in Science Teaching, 37*(6), 582–601.

Delaney, A., Cavalcanti, M., Jackson, C., & Mohr-Schroeder, M. J. (2017). Opening access to all students: STEMing self-efficacy. In E. Galindo & J. Newton (Eds.), *Proceedings of the 39th annual meeting of the North American Chapter of the International Group for the Psychology of Mathematics Education* (pp. 1099 – 1102). Indianapolis, IN: Hoosier Association of Mathematics Teacher Educators.

De Lange, J. (2003). Mathematics for literacy. In B. L. Madison & L. A. Steen (Eds.), Quantitative literacy: Why numeracy matters for schools and colleges (pp. 75–89). Princeton, NJ: National Council on Education and the Disciplines. Retrieved from http://www.maa.org/ql/pgs75_89.pdf

Dimmel, J. K., & Herbst, P. G. (2015). The semiotic structure of geometry diagrams: How textbook diagrams convey meaning. *Journal for Research in Mathematics Education, 46*(2), 147–195.

Doherty, R. E. (1939). Social responsibility of the engineer. *Electrical Engineering, 58*(9), 367–371.

Draper, R. J., & Broomhead, G. P. (Eds.). (2010). *(Re)imagining content-area literacy instruction.* New York, NY: Teachers College Press.

Engineering Accreditation Commission of The Accreditation Board for Engineering and Technology (ABET). (2014). *Criteria for accrediting engineering programs.* Baltimore, MD. Retrieved from http://www.abet.org/wp-content/uploads/2015/05/E001-15-16-EAC-Criteria-03-10-15.pdf

Everitt, W. L. (1944). The Phoenix-A challenge to engineering education. *Proceedings of the IRE, 32*(9), 509–513.

Fives, H., Huebner, W., Birnbaum, A. S., & Nicolich, M. (2014). Developing a measure of scientific literacy for middle school students. *Science Education, 98*(4), 549–580. doi:10.1002/sce.21115

Gagel, C. W. (1997). Literacy and technology: Reflections and insights of technological literacy. *Journal of Industrial Teacher Education, 34*(3), 6–34. Retrieved from http://scholar.lib.vt.edu/ejournals/JITE/v34n3/Gagel

Gardner, D. P. (1983). *A nation at risk.* Washington, DC: The National Commission on Excellence in Education, US Department of Education.

Garmire, E., & Pearson, G. (Eds.). (2006). *Tech tally: Approaches to assessing technological literacy.* Washington, DC: The National Academies Press.

Gee, J. P. (1999). *An introduction to discourse analysis: Theory and practice.* London & New York, NY: Routledge.

Gee, J. P. (2000). Identity as an analytic lens for research in education. *Review of Research in Education, 25,* 99–125.

Gonzalez, H. B., & Kuenzi, J. J. (2012, August). Science, Technology, Engineering, and Mathematics (STEM) education: A primer. *Congressional Research Service, Library of Congress.* Retrieved from http://digital.library.unt.edu/ark:/67531/metadc122233/m1/1/high_res_d/R42642_2012Aug01.pdf

Gorham, D., Newberry, P. B., & Bickart, T. A. (2003). Engineering accreditation and standards for technological literacy. *Journal of Engineering Education, 92*(1), 95–99.

Hammack, R., Ivey, T. A., Utley, J., & High, K. A. (2015). Effect of an engineering camp on students' perceptions of engineering and technology. *Journal of Pre-College Engineering Education Research (J-PEER), 5*(2), 10–21.

Hannigan, A., Gill, O., & Leavy, A. M. (2013). An investigation of prospective secondary mathematics teachers' conceptual knowledge of and attitudes towards statistics. *Journal of Mathematics Teacher Education, 16*(6), 427–449.

Hayford, B., Blomstrom, S., & DeBoer, B. (2014). STEM and service-learning: Does service-learning increase STEM literacy?. *The International Journal of Research on Service-Learning and Community Engagement, 2*(1), 32–43. ISSN: 2374-9466

Hazari, Z., Sonnert, G., Sadler, P. M., & Shanahan, M. C. (2010). Connecting high school physics experiences, outcome expectations, physics identity, and physics career choice: A gender study. *Journal of Research in Science Teaching, 47*(8), 978–1003.

Hekimoglu, S., & Sloan, M. (2005). A compendium of views on the NCTM standards. *Mathematics Educator, 15*(1), 35–43.

Heywood, J. (1993). *Engineering literacy for non-engineers K-12: A curriculum conundrum for the engineering profession.* Paper presented at the 23rd Annual Frontiers in Education conference on engineering education: Renewing America's technology, Washington, DC.

Hollomon, J. (1965, June 24). *The future of engineering education.* Presented at the ASEE World Congress on Engineering Education, Illinois Institute of Technology, Chicago, IL.

Honey, M., & Kanter, D. E. (Eds.). (2013). *Design, make, play: Growing the next generation of STEM innovators.* New York, NY: Routledge.

Hurd, P. D. (1958). Science literacy: Its meaning for American schools. *Educational Leadership, 16*(1), 13–16.

Hurd, P. D. (1998). Scientific literacy: New minds for a changing world. *Science Education, 82*(3), 407–416.

Israel, M., Maynard, K., & Williamson, P. (2013). Promoting literacy-embedded, authentic STEM instruction for students with disabilities and other struggling learners. *Teaching Exceptional Children, 45*(4), 18–25. doi:10.1177/004005991304500402

Jackson, C. D., & Mohr-Schroeder, M. J. (2018). Increasing STEM literacy via an informal learning environment. *Journal of STEM Teacher Education, 53*(1), 43–52. doi:10.30707/JSTE53.1Jackson

Johanning, D. I. (2008). Learning to use fractions: Examining middle school students' emerging fraction literacy. *Journal for Research in Mathematics Education, 39*(8), 281–310.

Laugksch, R. C. (2000). Scientific literacy: A conceptual overview. *Science Education, 84*(1), 71–94.

K-12 Computer Science Framework. (2016). Retrieved from http://www.k12cs.org

Kamberelis, G., Gillis, V. R., & Leonard, J. (2014). Disciplinary literacy, English learners, and STEM education. *Action in Teacher Education, 36*(3), 187–191.

Kelley, T. R., & Knowles, J. G. (2016). A conceptual framework for integrated STEM education. *International Journal of STEM Education, 3*(1), 1–11. doi:10.1186/s40594-016-0046-z

Kirsch, I. S. (1993). *Adult literacy in America: A first look at the results of the national adult literacy survey.* Washington, DC: US Government Printing Office, Superintendent of Documents. (Stock No. 065-000-00588-3)

Kiuhara, S. A., & Witzel, B. S. (2014). Focus on inclusive education: Math literacy strategies for students with learning difficulties. *Childhood Education, 90*(3), 234–238.

Koomen, M. H., Weaver, S., Blair, R. B., & Oberhauser, K. S. (2016). Disciplinary literacy in the science classroom: Using adaptive primary literature. *Journal of Research in Science Teaching, 53*(6), 847–894.

Kragten, M., Admiraal, W., & Rijlaarsdam, G. (2013). Diagrammatic literacy in secondary science education. *Research in Science Education, 43*(5), 1785–1800.

Krajcik, J. S., & Sutherland, L. M. (2010). Supporting students in developing literacy in science. *Science, 328*(5977), 456–459.

Lemke, M., Sen, A., Pahlke, E., Partelow, L., Miller, D., Williams, T., Kastberg, D., Jocelyn, L. (2004). *International outcomes of learning in mathematics, literacy and problem solving: PISA results from the U.S. Perspective.* Washington, DC: National Center for Education Statistics.

LinkedIn Economic Graph Team. (2018, August). *LinkedIn workforce report | United States | August 2018.* Retrieved December 1, 2018, from https://economicgraph.linkedin.com/resources/linkedin-workforce-report-august-2018

Mallozzi, F., & Heilbronner, N. (2013). The effects of using interactive student notebooks and specific written feedback on seventh grade students' science process skills. *Electronic Journal of Science Education, 17*(3).

Marcarelli, K. (Ed.). (2010). *Teaching science with interactive notebooks.* Thousand Oaks, CA: Corwin Press.

McCright, A. M. (2012). Enhancing students' scientific and quantitative literacies through an inquiry-based learning project on climate change. *Journal of the Scholarship of Teaching and Learning, 12*(4), 86–101.

McCurdy, R. C. (1958). Toward a population literate in science. *The Science Teacher, 25*(7), 366–408.

Meyers, E. M., Erickson, I., & Small, R. V. (2013). Digital literacy and informal learning environments: An introduction. *Learning, Media and Technology, 38*(4), 355–367. doi:10.1080/17439884.2013.783597

Michaels, S. (2013, February 12). *Connections between practices in NGSS, common core math, and common core ELA* [PPT]. Retrieved from https://learningcenter.nsta.org/products/symposia_seminars/NGSS/webseminar17.aspx

Miller, J. D. (1983). Scientific literacy: A conceptual and empirical review. *Daedalus, 112*(2), 29–48.

Moje, E. B. (2011). Developing disciplinary discourses and identities: What's knowledge got to do with it. *Discourses and identities in contexts of educational change.* New York, NY: Peter Lang.

Moje, E. B., & Luke, A. (2009). Literacy and identity: Examining the metaphors in history and contemporary research. *Reading Research Quarterly, 44*(4), 415–437.

Napp, J. B. (2004). Survey of library services at engineering news record's top 500 design firms: Implications for engineering education. *Journal of Engineering Education, 93*(3), 247–252.

Nathan, M. J., Tran, N. A., Atwood, A. K., Prevost, A., & Phelps, L. A. (2010). Beliefs and expectations about engineering preparation exhibited by high school STEM teachers. *Journal of Engineering Education, 99*(4), 409–426.

National Academy of Sciences-National Research Council, Washington, DC. Mathematical Sciences Education Board. (1989). *Everybody counts. A report to the nation on the future of mathematics education.* ERIC Clearinghouse.

National Assessment of Educational Progress (NAEP). (2014). *Technology and engineering literacy.* Retrieved from http://www.nationsreportcard.gov/tel_2014/#results/overall

National Governors Association Center for Best Practices & Council of Chief State School Officers (CCSSO). (2010). *Common core state standards.* Washington, DC: Authors. Retrieved from http://www.corestandards.org

National Research Council (NRC). (1996). *National science education standards.* Washington, DC: The National Academies Press.

National Research Council. (2010). *Standards for K-12 engineering education?* Washington, DC: National Academy of Science.

National Research Council (NRC). Committee on Highly Successful Schools or Programs for K-12 STEM Education. (2011). *Successful K-12 STEM education: Identifying effective approaches in science, technology, engineering, and mathematics.* Washington, DC: National Academies Press. doi:10.17226/13158

National Science Teachers Association (NSTA). (1971). *NSTA position statement on school science education for the 70s.* The Association.

Newsom, J. (1963). *Half our future. A report of the central advisory council for education.* London: HMSO.

NGSS Lead States. (2013). *Next Generation Science Standards: For states, by states.* Washington, DC: The National Academies Press.

O'Neill, K., & Polman, J. (2004). Why educate "little scientists?" Examining the potential of practice-based scientific literacy. *Journal of Research in Science Teaching, 41*(3), 234–266.

Ozgen, K., & Bindaka, R. (2011). Determination of self-efficacy beliefs of high school students towards math literacy. *Educational Sciences: Theory and Practice, 11*(2), 1085–1089. Retrieved from http://files.eric.ed.gov/fulltext/EJ927392.pdf

Pecen, R., Humston, J. L., & Yildiz, F. (2012). Promoting STEM to young students by renewable energy applications. *Journal of STEM Education: Innovations and Research, 13*(3), 62–72.

Penick, J. D. (1993). *Scientific literacy: An annotated bibliography.* Paris: The ICASE/UNESCO Conference Project 2000+: Science Litearcy for All.

Price, C. A., & Lee, H. S. (2013). Changes in participants' scientific attitudes and epistemological beliefs during an astronomical citizen science project. *Journal of Research in Science Teaching, 50*(7), 773–801.

Riechard, D. E. (1985). Politics and scientific literacy. *Education, 106*(1).

Roberts, O. T., Jackson, C., Mohr-Schroeder, M. J., Bush, S. B., Maiorca, C., Cavalcanti, M., Schroeder, D. C., Delaney, A., Putman, L., & Cremeans, C. (2018). Students' perceptions of STEM learning after participating in a summer informal learning experience. *International Journal of STEM Education, 5*(35). doi:10.1186/s40594-018-0133-4

Robinson, M., & Kenny, B. (2003). Engineering literacy in high school students. *Bulletin of Science, Technology & Society, 23*(2), 95–101.

Romine, W. L., Sadler, T. D., & Kinslow, A. T. (2016). Assessment of scientific literacy: Development and validation of the Quantitative Assessment of Socio-Scientific Reasoning (QuASSR). *Journal of Research in Science Teaching, 46*(3), 309–327.

Roper, T., Threlfall, J., & Monaghan, J. (2005, June). In D. Hewitt (Ed.), *Proceedings of the British society for research into learning mathematics, 25*(2), 19–24.

Sanders, M. E. (2008, December/January, 20–26). Stem, stem education, stemmania. *The Technology Teacher.* Retrieved from https://vtechworks.lib.vt.edu/bitstream/handle/10919/51616/STEMmania.pdf?sequence=1

Sanders, M. E. (2009). Integrative STEM education: Primer. *The Technology Teacher, 68*(4), 20–26.

Scharf, D. (2014, March). Instruction and assessment of information literacy among STEM majors. In *Integrated STEM Education Conference (ISEC), 2014 IEEE* (pp. 1–7). IEEE.

Shanahan, T., & Shanahan, C. (2008). Teaching disciplinary literacy to adolescents: Rethinking content-area literacy. *Harvard Educational Review, 78*(1), 40–59.

Shanahan, T., & Shanahan, C. (2012). What is disciplinary literacy and why does it matter?. *Topics in Language Disorders, 32*(1), 7–18.

Shanahan, L. E., McVee, M. B., Slivestri, K. N., & Haq, K. (2016). Disciplinary literacies in an engineering club exploring productive communication and the engineering design process. *Literacy Research: Theory, Method, and Practice, 65*(1), 404–420.

Shen, B. S. (1975, May–June). Science literacy and the public understanding of science. *American Scientist, 63*(3), 265–268.

Steen, L. A. (1997). *Why numbers count: Quantitative literacy for tomorrow's America.* New York, NY: The College Board.

Steen, L. A. (1999). Numeracy: The new literacy for a data-drenched society. *Educational Leadership, 57*, 8–13.

Steen, L. A., Turner, R., & Burkhardt, H. (2007). Developing mathematical literacy. In *Modelling and applications in mathematics education* (pp. 285–294). New York, NY: Springer US.

Stevens, R., O'Connor, K., Garrison, L., Jocuns, A., & Amos, D. M. (2008). Becoming an engineer: Toward a three dimensional view of engineering learning. *Journal of Engineering Education, 97*(3), 355–368.

Sullivan, F. R. (2008). Robotics and science literacy: Thinking skills, science process skills and systems understanding. *Journal of Research in Science Teaching, 45*(3), 373–394. doi:10.1002/tea.20238

Technology for All Americans Project, & International Technology Education Association (ITEA). (2000/2002/2007). *Standards for technological literacy: Content for the study of technology*. International Technology Education Association. Retrieved from https://www.iteea.org/File.aspx?id=67767&v=b26b7852

Tildsley, J. L. (1932). The thirty-first yearbook of the national society for the study of education. Part I. A program for teaching science. *Journal of Chemical Education, 9*(5), 962–966. New York, NY: Teachers College.

Tout, D. (2000). Numeracy up front: Behind the international life skills survey. *ARIS Resources Bulletin, 11*(1), 1–5. Retrieved from http://files.eric.ed.gov/fulltext/ED439290.pdf

Tucker-Raymond, E., Gravel, B. E., Kohberger, K., & Browne, K. (2017). Source code and a screwdriver: STEM literacy practices in fabricating activities among experienced adult makers. *Journal of Adolescent & Adult Literacy, 60*(6), 617–627.

Tytler, R. (2014). Attitudes, identity, and aspirations toward science. In N. G. Lederman & S. K. Abell (Eds.), *Handbook of research on science education* (pp. 82–103). New York, NY: Routledge.

Urban-Lubrain, M., & Weinshank, D. J. (2001). Do non-computer science students need to program?. *Journal of Engineering Education, 90*(4), 535–541.

United States Department of Education. (2018, April). *U.S. Department of Education announces STEM, computer science education grant opportunities*. Retrieved December 1, 2018, from https://www.ed.gov/news/press-releases/us-department-education-announces-stem-computer-science-education-grant-opportunities

Varelas, M., Pieper, L., Arsenault, A., Pappas, C. C., & Keblawe-Shamah, N. (2014). How science texts and hands-on explorations facilitate meaning making: Learning from Latina/o third graders. *Journal of Research in Science Teaching, 51*(10), 1246–1274.

Vasquez, J. A., Sneider, C. I., & Comer, M. W. (2013). *STEM lesson essentials, grades 3–8: Integrating science, technology, engineering, and mathematics*. Portsmouth, NH: Heinemann. ISBN:978-0325043586

Waetjen, W. B. (1987). The autonomy of technology as a challenge to education. *Technology & Society, 7*, 28–35.

Wertz, R. E., Purzer, Ş., Fosmire, M. J., & Cardella, M. E. (2013). Assessing information literacy skills demonstrated in an engineering design task. *Journal of Engineering Education, 102*(4), 577–602.

Wilkins, J. L. (2000). Special issue article: Preparing for the 21st century: The status of quantitative literacy in the United States. *School Science and Mathematics, 100*(8), 405–418.

Wilkins, J. L. (2010). Modeling quantitative literacy. *Educational and Psychological Measurement, 70*(2), 267–290.

Wilkins, J. L. (2015). Standards-based mathematics curricula and the promotion of quantitative literacy in elementary school. *International Journal of STEM Education, 2*(1), 1.

Wilkins, J. L., Zembylas, M., & Travers, K. J. (2002). Investigating correlates of mathematics and science literacy in the final year of secondary school. In *Secondary analysis of the TIMSS data* (pp. 291–316). Dordrecht: Springer Netherlands.

Williams, D. C., Ma, Y., Prejean, L., Ford, M. J., & Lai, G. (2007). Acquisition of physics content knowledge and scientific inquiry skills in a robotics summer camp. *Journal of Research on Technology in Education, 40*(2), 201–216.

Wilson-Lopez, A., & Gregory, S. (2015). Integrating literacy and engineering instruction for young learners. *The Reading Teacher, 69*(1), 25–33.

Wulf, W. (2000). The standards for technological literacy: A National Academies perspective. *The Technology Teacher, 59*(6), 10–12.

Zollman, A. (2012). Learning for STEM literacy: STEM literacy for learning. *School Science and Mathematics, 112*(1), 12–19.

Zucker, A., Staudt, C., & Tinker, R. (2015). Teaching graph literacy across the curriculum. *Science Scope, 38*(6), 19–24.

CHAPTER 2

Getting to the Bottom of the Truth: STEM Shortage *or* STEM Surplus?

Cathrine Maiorca, Micah Stohlmann and Emily Driessen

Abstract

This chapter will evaluate data from multiple reports concerning the reported shortage and reported surpluses in the STEM areas. Is STEM really in dire need of job applicants? What really is the STEM career projection and how are we going to fill that pipeline?

1   Introduction

National policy movements, national and international student achievement data, and the technologically based data driven world in which we live in has made STEM (Science, Technology, Engineering, and Mathematics) education and STEM careers of great importance. There have been an abundance of United States policy documents related to STEM and STEM education published in the last decade (e.g., President's Council of Advisors on Science and Technology [PCAST], 2010; National Academies Press [NAP], 2014; National Research Board [NRB], 2015; National Research Council [NRC], 2007, 2009). A main theme woven throughout these documents is that if we are going to continue as a prosperous nation, it is imperative to improve STEM education in the U.S. and develop future generations of STEM professionals. For example, a capable STEM workforce is necessary to sustain the United States' innovation, global competitiveness, and national security (NRC, 2007; NSB, 2003). In the past decade the National Science Foundation has made significant investments in STEM-related grants intended to improve students' educational experiences and impact their decisions to pursue STEM-related careers. As recently as 2016 the Department of Education has reported that STEM careers are rapidly increasing and that there is a need for STEM workers. Despite the numerous reports claiming a STEM worker shortage, there is some debate on if there is actually a STEM worker shortage or a STEM worker surplus. Depending on who you ask the answer could be different. This chapter describes areas where there is a STEM worker shortage, what future projections are for STEM workers, and the benefits of a quality STEM education for students to be adaptable and successful in the job market.

The answer to the surplus-shortage question of STEM workers is not a straightforward "yes" or "no." It depends on different factors such as the specific STEM field being discussed and the geographic location (e.g. different states, rural areas, and/or high-technology corridors). According to the National Science Board (2015), surplus-shortage debate cannot be simplified to 'enough' or 'not enough STEM workers'. The diversity of the workforce, the quantity of workers actually needed, the educational level of the needed workers or a combination of these, needs be considered when trying to answer the shortage-surplus question. There does appear to be consensus on the value of STEM skills in jobs across the U.S. economy; as well as the need to improve access to and the quality of STEM education at all levels including addressing the lack of racial/ethnic and gender diversity in the STEM workforce (NSB, 2015).

2   STEM U.S. Occupations by Numbers

The occupations qualified as STEM, per Fayer, Lacey, and Watson (2017) for the U.S. Bureau of Labor Statistics, include: mathematical science occupations; architects, surveyors, and cartographers; STEM-related postsecondary teachers; physical scientists; life scientists; life and physical science technicians; STEM-related sales; STEM-related management; drafters, engineering technicians, and mapping technicians; engineers; and computer occupations. As of May 2015, there were close to 8.6 million STEM jobs in the United States, which is 6.2% of all employment in the United States. Of those 8.6 million jobs, 45% of them were computer jobs, 19% of them were engineering positions, and less than 4% were those held by cartographers, architects, and surveyors, combined. With the large percentage of STEM jobs involving computer work, it is not surprising that seven of the ten most common STEM occupations involved information systems and computers (e.g., applications software developers, computer systems analysts, computer programmers, computer user support specialists). The three positions included in the top ten most common STEM positions that were not related to computers were sales representatives/wholesale and manufacturing/technical and scientific products (400,000 jobs), mechanical engineers (300,000 jobs), and civil engineers (300,000 jobs). Of all of the STEM occupations, 93% of them afford wages (average $87,570) that are above the National average (about $50,000 as of 2018). As far as STEM occupation growth, 817,260 new jobs (10.5%) in the STEM field were created between May 2009 and May 2015. This growth was double that of the percent net growth for non-STEM occupations (5.2%; Fayer, Lacey, & Watson, 2017).

3 Is STEM Really in Dire Need of Job Applicants?

While the past number of STEM jobs is quite clear, the future number of them seems to be murky. This is because there are two main conflicting schools of thought concerning the STEM occupations: (1) There is a shortage of STEM workers to fill the positions in the labor market; and (2) There is a current surplus of STEM workers. For example, according to the President's Council of Advisors on Science and Technology (2012), the number of STEM categorized positions are expected to increase an additional 1 million in the U.S. between 2012 and 2022, and this increase has been stated to be necessary in order for the U.S. to keep its leading hold on science and technology globally. On the other hand, Anft (2013) claimed the unemployment rates in STEM were higher than they have ever been in years and this signifies a shortage of those jobs rather than a shortage of qualified employees.

According to Xue and Larson (2015), both schools of thought are correct. They concluded this after conducting a literature review, interviewing 18 different talent recruiters, and finding there is wide variance in qualified professional demand across STEM disciplines. They found there were differences in supply and demand for particular disciplines (e.g. engineering, computer science, physics, etc.) as well as for different branches of employment (i.e. academia, government, and the private sector). Specifically, the findings showed there was a surplus of PhD's seeking tenure academic jobs; shortages in persons to fill government and government-related sector positions in STEM disciplines (e.g., electrical engineering, operations research, quantitative psychology, physics, computer science, electromagnetics) especially those candidates with advanced degrees in those areas. It is important to note that this was not to be considered a STEM shortage crisis at all but, rather, shortages of qualified U.S. citizens (required in order to work in many U.S. government positions) with advanced-degrees; and both shortages and surpluses in STEM workers in the private sector.

Some level of post-secondary education is often required for STEM employment. In fact 99% of STEM jobs require some kind of training after high school, unlike other occupations were only 36% of jobs require post-secondary training (BLS, 2017). The number of applicants needed at the different educational levels vary.

In general, the academic sector, consisting of 2-year and 4-year colleges and universities, is well-supplied because there are significantly more applicants than there are jobs (Xue & Larson, 2014). In fact, in 2010, less than 15 percent of new PhDs found tenure track jobs within three years after graduation in academic areas of science, engineering, and health-related fields (NSB, 2014). There is a surplus of PhD's in biomedical research and chemistry due pharmaceutical companies sending their jobs to countries that have cheaper labor (Xue & Larson, 2014).

Entry level STEM occupations that require a bachelor's degree are 73% of STEM employment (Fayer et al., 2017). While this is a large percentage of the STEM work force it only represents 21% of all occupations. The fastest growing jobs include statisticians, operations research analysts, cartographers and photometrist, forensic science technicians, biomedical engineers, mathematicians, computer systems analysts, actuaries, software developers, and information security analysts.

Because there is such a high need for skilled professionals at all educational levels there is a shortage of STEM workers who have either associate degrees or attended trade school (Xue & Larson, 2014). There are two career which usually require an associate degree, one is a computer user support specialist, which is expected to grow by 12.8% and the other is a web developer which is predicted to grow by 26.6% from 2014 to 2024 (Fayer et al., 2017).

The government and government-related sector has shortages in specific areas such as nuclear engineering, materials science, electrical engineering, cybersecurity, and intelligence. This is not necessarily because of lack of STEM professionals but due to the lack of STEM professionals who are U.S. citizens (Xue & Larson, 2014)). This is often criteria for employment that requires certain security clearances.

The private sector has shortages with petroleum engineers in certain geographic locations, data scientists, and software developers (Xue & Larson, 2014).

There is also demand for STEM skills below the bachelor's level. A 2011 survey of manufacturers found that as many as 600,000 jobs remain unfilled due to lack of qualified candidates for technical positions requiring STEM skills, primarily in production (machinists, operators, craft workers, distributors, and technicians) (The Manufacturing Institute and Deloitte, 2011). While the number of job openings in this sector has decreased since 2011, there were still 264,00 opening in 2014 (BLS, 2014).

4 What Really Is the STEM Career Projection?

In total, there were nearly 8.6 million STEM jobs in 2015. However, this only represented 6.2 percent of U.S. employment. Although STEM jobs only represented a small portion of the U.S. economy, employment in STEM occupations grew by 10.5 percent between 2009 and 2015, compared to only 5.2 percent in non-STEM occupations (Fayer, Lacey, & Watson, 2017). STEM workers also reported a lower rate of unemployment at 2.5% in 2015. During this same time period the non-STEM unemployment rate was 5.5 percent (Noonan, 2017).

Forty-five percent of all STEM employment were computer, and engineers made up an additional 19 percent of the STEM work force (Fayer, Lacey, & Watson, 2017). Applications software developer is the largest STEM occupation, followed by computer user support specialists and computer systems analysts. The largest STEM occupation not related to computers was wholesale and manufacturing sales representative of technical and scientific products.

Employment in computer occupations is projected to increase by 12.5 percent from 2014 to 2024. This growth would be nearly half a million jobs. These jobs would include computer programmers, computer support specialists, software engineers, database administrators, network systems and data communication analysts, and computer scientists and systems analysts. With the combination of new jobs and workers leaving occupations the computer occupational group is projected to yield over 1 million job openings from 2014 to 2024. The computer occupations group also has the fastest growing occupations that do not require a bachelor's degree with web developers and computer user support specialists (Fayer et al., 2017).

The STEM occupation group that is projected to grow fastest from 2014 to 2024 at 28.2 percent is the mathematical sciences occupation group, which includes statisticians and mathematicians (Fayer et al., 2017). However, this group has the lowest employment among the STEM groups, so the growth would result in about 42,900 new jobs. Although the mathematical sciences occupation group is projected to grow at the fastest pace, it will still have a small number of job openings (Vilorio, 2014). Another area of jobs is the engineering STEM occupation group, which is projected to yield 500,000 jobs when factoring in new jobs and workers leaving jobs (Fayer et al., 2017).

However, when discussing the current and projected labor markets, it is important to discuss industrial jobs including manufacturing and services occupations. Together, these two categories of employment retain 118 million workers in the United States. Specifically, 107 million Americans work in services, and 11 million Americans works in manufacturing (Delgado & Mills, 2017). Although less people are employed in the manufacturing type jobs, these jobs represent more STEM positions (9.5%) than do the service jobs (5.2%; Heckler, 2005). Additionally, the manufacturing jobs, on average, pay higher wages than do the service jobs ($54,200 versus $47,700, respectively; Delgado & Mills, 2017). Manufacturing positions also account for the majority of patents (Delgado & Mills, 2017), and, therefore, have been considered by many to be the driver of innovation (Pisano & Shih, 2009). With this stance in mind, many have argued the U.S. should focus on increasing manufacturing jobs that concentrate on new manufacturing processes and technologies (e.g. nanotechnology, smart production processes, and advanced materials). However, higher-wage manufacturing jobs in the United States have declined, and, overall, manufacturing employment decreased by more than 30%. This decline in an area of high focus has led many to observe the U.S. economy pessimistically. To correct this, policies have focused on initiatives to bring manufacturing back (Acemoglu et al., 2016).

However, Delgado and Mills (2018) argue this is perhaps misguided as the categorization of services versus manufacturing is not homogenous. That is, the "services" category includes engineering, cloud computing, retail, and restaurant occupations. This category lumps 90% of industry jobs together and leaves 10% of them for the manufacturing category (Delgado & Mills, 2017). In order to demonstrate the different kinds of services the U.S. economy offers, and reveal the potential for increasing the number of high-wage and high-technology occupations in some of the service subdivisions, Delgado and Mills (2018) restructured industry jobs into supply chain (SC) industries and business-to-consumer (B2C) industries categories. The SC category represents industries that sell, both services and goods, to governments or businesses; whereas, the B2C category represents industries that sell mainly to every-day people. Once the industry jobs were categorized using the Delgado and Mill's (2018) scheme, the data showed SC industries employed 44 million (37% of industry jobs) while the B2C industries employed 74 million (63% of industry jobs) Americans. Also shown, SC industries provided wages that are 57% higher than those in B2C industries ($61,700 versus $39,200, respectively), accounted for more STEM occupations than do the B2C industries (11% versus 2%, respectively), and involved the majority of patents.

Delgado and Mill (2018) then further broke down the data into sub-categories (i.e. SC manufacturing, SC services, B2C manufacturing, and B2C services). This revealed 35.5 million occupations supported by the SC service industry, 8.7 million jobs in the SC manufacturing industry, 2.6 million positions taken in the B2C manufacturing industry, and 71.5 million jobs in the B2C services industry. Again, the SC industries provided more STEM jobs than did the B2C industries (11.1% for SC manufacturing, 11.5% for SC services, 4.2% for B2C manufacturing, and 2% for B2C services). To examine the importance of supply chain services further, Delgado and Mills (2018) separated the data for industries that sell their goods and services across countries and regions, otherwise known as traded economy, into SC traded manufacturing, SC traded services,

B2C traded manufacturing, and B2C traded services. This showed the SC traded services category includes 18.7 million jobs with an average wage of $80,800 and a STEM intensity of 19.3%. The $80,800 wage was 70% higher than the average wage in the U.S. economy at the time. The other data showed 8.1 million jobs at an average wage of $57,400 and a STEM intensity of 11.7% for the SC traded manufacturing category; 2.4 million jobs at an average wage of $47,200 and a STEM intensity of 4.5% for the B2C traded manufacturing category; and 13.4 million jobs at an average wage of $57,800 and a STEM intensity of 6.1% for the B2C traded services category. This data ultimately showed the services industry, regardless of being categorized as SC or B2C, employed more people that the manufacturing industry. It also demonstrated there are higher wages associated with service jobs, on average, and there is a larger average STEM intensity in the services than in the manufacturing categories. Overall, Delgado and Mills (2018) showed the current importance of services, both traded and non-traded, to the U.S. economy as well as other countries' economies.

5   How Are We Going to Fill the Pipeline?

In the United States, we currently face a shortage of STEM majors and graduates (National Science Board, 2016; The Committee on STEM Education National Science and Technology Council, 2013). Approximately 16% of high school seniors proficient in mathematics are interested in a STEM career (U.S. Department of Education, n.d.) and only half of students who attend college wanting to pursue a STEM career actually work in a related STEM field.

In order to fill the pipeline of needed STEM professionals it is important that students are properly prepared to have the opportunity to purse a STEM career if they would want to. Research has shown that students who perform well in calculus, pre-calculus/trigonometry, physics, and chemistry in high school are more likely to be successful in college STEM gatekeeper courses (Redmond-Sanogo, Angle, & Davis, 2016). These gatekeeper courses can be STEM content courses that students are not able to be successful in and hinder them from completing a STEM degree. It is important that students are encouraged to take the above-mentioned courses in high school and also be provided adequate support in these courses. However, with this being said these courses are not being offered to all students at the high school level. According to the U.S. Department of Education, Office of Innovation, and Improvement (2016),

> … at the high school, both in rural communities and across the nation in more urban and suburban centers, many students are not even provided

with the courses they need to develop and deepen their mathematics and science interests, skills, and knowledge. (p. 2)

Some schools do not offer all of the core mathematics and science classes necessary for students to be successful in STEM careers (U.S. Department of Education, Office of Innovation, and Improvement, 2016). Table 2.1 shows the percentages of schools that do not offer core mathematics and science classes. According to the National Assessment of Educational Progress (NAEP), most students are not proficient in mathematics and science in the 8th grade (Table 2.2).

TABLE 2.1   Percentages of US schools that don't offer core mathematics and science classes

School demographics	Percentage of schools	Course missing from the curriculum
All high schools	20–25%	algebra I and II, geometry, biology, and chemistry
All high schools	50%	calculus
All high schools	37%	physics
High minority population	34%	chemistry
Low minority population	22%	chemistry
High minority population	24%	algebra II
Low minority population	17%	algebra II

SOURCE: U.S. DEPARTMENT OF EDUCATION, OFFICE OF INNOVATION, AND IMPROVEMENT (2016)

TABLE 2.2   Percentage of 8th grade students proficient in math and science according to the NAEP

	Ethnicity			
	White	Asian	Black	Hispanic
Mathematics	43%	61%	13%	19%
Science	45%	46%	Less than 20%	Less than 20%

SOURCE: U.S. DEPARTMENT OF EDUCATION, OFFICE OF INNOVATION, AND IMPROVEMENT (2016)

Even if advanced mathematics and science courses are offered at their high school they would not have access to them.

In general, in STEM careers there is an underrepresentation of women. Reigle-Crumb, Grodsky, and Muller (2012) suggest that academically qualified females' underrepresentation in STEM is a purposeful choice. Diekman et al. (2010) has noted females' tendencies to pursue communal careers such as teaching, health care, and law. These careers are seen as more focused on helping others in contrast to the stereotypical perception of STEM careers as being an isolated job. This indicates the need for increasing awareness of what different STEM careers entail and often the teamwork that is required. When women choose STEM degree programs they are just as successful as men in completing the programs (Griffith, 2010).

The lack of racial/ethnic diversity in the STEM pipeline is another focus area. For example, in 2014 there were 55.3 million Latina/os in the United States, comprising 17.3% of the total U.S. population. In 1980 Latina/os made up just 6.5% of the total U.S. population (Stepler & Brown, 2016). Around one-third of the Latina/o population is younger than 18 (Patten, 2016). Latina/os are the largest minority population in the U.S. and they are also the youngest. They are also the fastest growing population in U.S. K-12 schools, colleges, and universities.

Due to the growing youth Latina/o population it is imperative that they are provided with robust education that prepares them for careers and with valuable 21st century competencies that will help them in life and any career, which can be accomplished through quality STEM education. There is room for improvement as Latina/o students have the highest high-school dropout rate at 14%. This issue carries on to influence the number of Latina/o students with bachelor's degrees, who account for just 9% of young adults (ages 25 to 29) with a bachelor's degree compared to 69% of white young adults (Fry, 2014). In many STEM fields Latina/os are underrepresented, with less than 2% of Latina/os in engineering, computers, and mathematics fields (Stepler & Brown, 2016). Further research is needed on how to support minority populations to ensure that they have the opportunity and support to pursue STEM careers if they so choose.

In general, there is a need for an increase in the number of undergraduate STEM degrees. According to the President's Council of Advisors on Science and Technology (PCAST), the United States would need to increase yearly production of undergraduate STEM degrees by 34 percent over current rates to match the forecast demand for STEM professionals (PCAST, 2012). This is in part because of the percent of those with a bachelor's degree in STEM who are not employed in STEM occupations. The U.S. Census Bureau (2014) reported that

74 percent of those who have a bachelor's degree in STEM are not employed in STEM occupations. Because of this there are more workers with a STEM degree (11.9 million) than there are workers in STEM occupations. Though it should be noted some occupations considered "non-STEM" are healthcare practitioners or technicians and those who work in education (Noonan, 2017).

Although the majority of STEM bachelor's degree holders are not employed in STEM occupations, the acquired STEM skillset is widely applicable. The majority of workers with STEM degrees that are not employed in STEM occupations indicate that their job is related to their STEM education. The National Academies of Press (2014) state that, "Leaders in business, government, and academia assert that education in the STEM subjects is vital not only to sustaining the innovation capacity of the United States but also as a foundation for successful employment, including but not limited to work in STEM fields" (p.vii). In order to ensure a well-functioning STEM pipeline of diverse workers it is vital that students receive a quality K-12 STEM education that exposes students to a wide variety of careers and STEM careers and prepares them with 21st century competencies and STEM content knowledge.

6 Implications for K-12 Education

STEM education is not just an issue for employers in STEM fields. Not all students will go into a STEM major in college, but STEM literacy is vital for all students to be successful participants in society. As society becomes increasingly more dependent on STEM it will become important for those who do not plan on pursuing a career in a STEM field to receive an authentic STEM education (Surr, Loney, Goldston, Rasmussen, & Anderson, 2016). An authentic STEM educational experience should promote (a) teamwork and communication skills, (b) use student centered pedagogies (c) engage learners and have meaningful content, (d) engage students in the engineering design process, (e) involves problem solving skills that connect to the context of the problem and (f) include appropriate math and/or science content (Maiorca & Stohlmann, 2016). An authentic STEM learning experience is important for all students because it encourages students to use innovative thinking and encourages students to solve complex problems; a skill they need to be successful in life regardless of their chosen profession (NAS, 2014). Another advantage of the integrated approach used in STEM education is that it better prepares students to solve interdisciplinary problems that they will encounter in life regardless of their major or chosen career path (Guzey, Moore, & Hartwell, 2016).

Regardless of whether there is a STEM shortage or surplus, it is important for all students to receive an authentic STEM education. All students need to have some degree of STEM literacy in order to participate in society. Even if they do not pursue STEM careers students can benefit from an authentic STEM education because it promotes higher order thinking skills, makes students better problem solvers and lifelong learners.

All students should have access to STEM education. This can be accomplished by learning activities that invite students to explore, be creative and most importantly learn that failure is not bad. An authentic STEM education is student centered curriculum, culturally relevant and accessible to all students through the use of and developmentally appropriate practices (U.S. Department of Education, Office of Innovation, and Improvement, 2016). However, teachers will need professional development and support so that they can develop the skills necessary to provide students with these types of activities.

Not everyone is interested in pursuing a STEM degree or STEM career, but we must ensure that students are made aware of STEM career possibilities and are given the needed preparation and support to be successful in their pursuit of STEM careers if they choose to do so. There needs to be an increased focus on Career Technical Education in high school so that students are better prepared to enter the STEM workforce after graduation, if they choose to do so.

7 Future Research

Future research should include studying the effects of authentic STEM education on K-12 students when it is implemented in a school setting. Their beliefs about teaching and learning should be examined because they may influence their willingness to participate in the typical integrated STEM problems. The preconceived notions that teachers have regarding integrated STEM should be studied because they could influence how teachers implement integrated STEM in their classroom and who that think will be successful participating in it.

Specific research questions might include:
- How do CTE courses influence a student's desire to pursue a STEM career?
- How do students' beliefs about STEM disciplines influence their willingness and ability to learn the STEM disciplines in an integrated way?
- How do teachers' beliefs about the different STEM disciplines influence their willingness and ability to teach authentic integrated STEM lessons?

8 Conclusion

According to the National Science Board (2015), there are three roadblocks the STEM workforce needs to address. Everyone needs to be able to participate in quality STEM learning. There should be more access and opportunities for underrepresented student populations in STEM education. The workforce pathways need to be continually examined and monitored to ensure opportunity and access. In moving forward, it is important that clear messages about STEM are communicated along with data on the number of STEM jobs available and in what areas. The benefits of preparation in STEM education are clear, in that students are equipped with the necessary knowledge and skills to be productive citizens.

References

Acemoglu, D., Autor, D., Dorn, D., Hanson, G., & Prince, B. (2016). Import competition and the great U.S. employment Sag of the 2000s. *Journal of Labor Economics, 34*(1), 141–198.

Bureau of Labor Statistics. (2014). *Got skills? Think manufacturing.* Retrieved from https://www.bls.gov/careeroutlook/2014/article/manufacturing.htm

Bureau of Labor Statistics, U.S. Department of Labor. (2017, July). 8.8 million Science, Technology Engineering, and Mathematics (STEM) jobs in May 2016. *TED: The Economics Daily.* Retrieved from https://www.bls.gov/opub/ted/2017/8-point-8-million-science-technology-engineering-and-mathematics-stem-jobs-in-may-2016.htm

Diekman, A. B., Brown, E., Johnston, A., & Clark, E. (2010). Seeking congruity between goals and roles: A new look at why women opt out of STEM careers. *Psychological Science, 21,* 1051–1057.

Fayer, S., Lacey, A., & Watson, A. (2017). *BLS spotlight on statistics: STEM occupations-past, present, and future.* Washington, DC: U.S. Department of Labor, Bureau of Labor Statistics.

Fry, R. (2014). *U.S. high school dropout rate reaches record low, driven by improvements among Hispanics, blacks.* Washington, DC: Pew Research Center. Retrieved from http://www.pewresearch.org/fact-tank/2014/10/02/u-s-high-school-dropout-rate-reaches-record-low-driven-by-improvements-among-hispanics-blacks/

Griffith, A. (2010). Persistence of women and minorities in STEM field majors: Is it the school that matters? *Economics of Education Review, 29,* 911–922.

Guzey, S., Moore, T., & Harwell, M. (2016). Building up STEM: An analysis of teacher developed engineering design-based STEM integration curricular materials. *Journal of Pre-College Engineering Education Research, 6*(1), 11–29. doi:10.7771/2157-9288.1129

Heckler, D. (2005). High technology employment: A NAICS-based update. *Monthly Labor Review, 128*, 57–72.

Honey, M., Pearson, G., & Schweingruber, H. (Eds.). (2014). *STEM integration in K-12 education: Status, prospects, and an agenda for research.* Washington, DC: National Academies Press.

Maiorca, C., & Stohlmann, M. (2016). Inspiring students in integrated STEM education through modeling activities. In C. Hirsch & A. R. McDuffie (Eds.), *Annual perspectives in mathematics education 2016: Mathematical modeling and modeling mathematics* (pp.153–161). Reston, VA: NCTM.

National Academy of Sciences. (2014). *STEM integration in K-12 education: Status, prospects and an agenda for research.* Washington, DC: National Academies Press.

National Research Council (NRC). (2007). *Rising above the gathering storm: Energizing and employing America for a brighter economic future.* Committee on Prospering in the Global Economy of the 21st Century: An agenda for American science and technology; Committee on Science, Engineering, and Public Policy. Washington, DC: National Academies Press.

National Research Council (NRC). (2009). *Engineering in K-12 education: Understanding the status and improving prospects.* Committee on K-12 Engineering Education of the National Academy of Engineering and the National Research Council. Washington, DC: National Academies Press.

National Science Board (2003). *The science and engineering workforce: Realizing America's Potential.* Arlington, VA: National Science Foundation. Retrieved from https://www.nsf.gov/nsb/documents/2003/nsb0369/nsb0369.pdf

National Science Board. (2014). *Science and engineering indicators.* Arlington, VA: National Science Foundation. Retrieved from http://www.nsf.gov/statistics/seind14/

National Science Board (2015). *Revisiting the STEM workforce: A companion to science and engineering indicators 2014.* Arlington, VA: National Science Foundation.

Noonan, R. (2017). *STEM jobs: 2017 update.* Washington, DC: U.S. Department of Commerce.

Patten, E. (2016). *The Nation's Latino population is defined by its youth.* Washington, DC: Pew Research Center. Retrieved from http://www.pewhispanic.org/2016/04/20/the-nations-latino-population-is-defined-by-its-youth/

Pisano, G. P., & Shih, W. C. (2009). Restoring American competitiveness. *Harvard Business Review, 87*(7–8), 114–125.

President's Council of Advisors on Science and Technology. (2010). *Prepare and inspire: K-12 education in Science, Technology, Engineering, and Mathematics (STEM) education for America's future.* Retrieved from http://www.whitehouse.gov/administration/eop/ostp/pcast/docsreports

President's Council of Advisors on Science and Technology. (2012). *Engage to excel: Producing one million additional college graduates with degrees in science, technology, engineering, and mathematics.* Retrieved from https://obamawhitehouse.archives.gov/sites/default/files/microsites/ostp/pcast-engage-to-excel-final_2-25-12.pdf

Redmond-Sanogo, A., Angle, J., & Davis, E. (2016). Kinks in the STEM pipeline: Tracking STEM graduation rates using science and mathematics performance. *School Science and Mathematics, 116*(7), 378–388.

Riegle-Crumb, C., & King, B. (2010). Questioning a white male advantage in STEM: Examining disparities in college major by race/ethnicity. *Educational Researcher, 39*, 656–664.

Sargent Jr., J. F. (2017). *The US science and engineering workforce: Recent, current, and projected employment, wages, and unemployment.* Retrieved from https://www.everycrsreport.com/reports/R43061.html

Stepler, R., & Brown, A. (2016). *Statistical portrait of Hispanics in the United States.* Washington, DC: Pew Research Center. Retrieved from http://www.pewhispanic.org/2016/04/19/statistical-portrait-of-hispanics-in-the-united-states-key-charts/

Surr, W., Loney, E., Goldston, C., Rasmussen, J., & Anderson, K. (2016). *From career pipeline to STEM literacy for all: Exploring evolving notions of STEM.* Washington, DC: American Institutes for Research.

The Manufacturing Institute, and Deloitte. (2011). *Boiling point? The skills gap in U.S. manufacturing.* Retrieved from http://www.themanufacturinginstitute.org/~/media/A07730B2A798437D98501E798C2E13AA.ashx

U.S. Census Bureau. (2014). *Census bureau reports majority of STEM college graduates do not work in STEM occupations.* U.S. Census Bureau Newsroom. Retrieved from http://www.census.gov/newsroom/releases/archives/employment_occupations/cb14-130.html

U.S. Department of Education. (n.d.). *Science, technology, engineering, and math: Education for global leadership.* Retrieved from https://www.ed.gov/sites/default/files/stem-overview.pdf

U.S. Department of Education, Office of Innovation, and Improvement. (2016). *STEM 2026: A vision for innovation in STEM education.* Washington, DC: Author.

Vilorio, D. (2014). STEM 101: Intro to tomorrow's jobs. *Occupational Outlook Quarterly, 58*(1), 2–12.

Xue, Y., & Larson, R. (2014). *STEM crisis of STEM surplus? Yes and yes.* Engineering Systems Division Working Paper Series. Massachusetts Institute of Technology.

CHAPTER 3

Underrepresentation of Women and Students of Color in STEM

Dionne Cross Francis, Kerrie G. Wilkins-Yel, Kelli M. Paul and Adam V. Maltese

Abstract

The underrepresentation of women and racialized minorities in science, technology, engineering, and mathematics (STEM) disciplines endures. In response, research, policy initiatives, STEM Enrichment programs, school and teacher education programs, and after-school activities were implemented to address this need. Nonetheless, women and people of color remain markedly underrepresented in many aspects of STEM and studies often fail to discuss the interplay of the institutional, societal, and systemic contributors to this disparity. Our purpose with this chapter is to shed light on the ways in which these broader contextual factors influence girls' perceptions of, engagement with, and subsequent participation and persistence in STEM education. Identification and delineation of elements of the problem are important, but it is also critical for educators to consider ways to ameliorate these issues. For a girl to successfully pursue a STEM career path, she must first be able to see it as possible, then embrace it as an aspect of her identity and be provided with the resources to boldly traverse this pathway. Ensuring that girls receive support for STEM career development early, thereby building awareness of the range of possible STEM professions, how lucrative they are, and details on what the career pathway entails is essential for setting goals. Given the role identity development plays in shaping interest and engagement, it is imperative that minority girls have opportunities to connect with STEM professionals with whom they can identify. We further discuss promising practices for addressing inequities in the chapter.

1   Introduction

The underrepresentation of women and racialized minorities in science, technology, engineering, and mathematics (STEM) disciplines is an enduring concern. Numerous national reports, political speeches, presidential commissions, and research dating back to the 1970s have called upon the United States to address the lack of diversity in STEM fields (Cole & Cole, 1973; Epstein, 1970; Institute of Medicine, National Academy of Sciences, and National Academy of Engineering [IM/NAS/NAE], 2007; National Center for Women in Information Technology [NCWIT], 2011; President's Council of Advisors on Science and Technology [PCAST], 2010; The White House, 2009). In response to this clarion call, a burgeoning body of research, policy initiatives, STEM Enrichment programs, school and teacher education programs, and after-school activities were implemented to address this need. Nonetheless, women and people of color remain markedly underrepresented in many aspects of STEM, especially within engineering and computer science – two of the most elite and lucrative sectors of the U.S. labor force (Buchmann, 2009; National Research Council [NRC], 2011; National Science Foundation [NSF], 2017). In fact, over the past twenty-five years, women only marginally increased in their attainment of bachelor's degrees in engineering (from 14% to 17%) and demonstrated a striking *decline* in their share of bachelor's degrees in computer science (from 36% to 18%; NSF, 2017).

The stagnation and decline of diverse representation in STEM hints at a complex and multifaceted problem that begins as early as elementary school and occurs at the individual, institutional, and societal levels. However, extant research only focuses on girls' and students of color's limited interests or inclination toward STEM fields (Scott & White, 2013; Smyth & McArdle, 2004). Studies fail to discuss the interplay of the institutional, societal, and systemic contributors to this disparity. Our purpose with this chapter is to shed light on the ways in which these broader contextual factors influence girls' perceptions of, engagement with, and subsequent participation and persistence in STEM education. We also discuss promising practices for addressing inequities.

2 Persistent Underrepresentation and Its Global Impact

Recent evidence indicates that women in the United States are enrolling in college and universities at a much higher rate than men. From 1970 to 2015, women's enrollment increased from 42% to 56% (National Center for Education Statistics [NCES], 2017). Despite this increased enrollment, women in their first year of college are less likely to report intentions to major in science and engineering compared to their male counterparts. In 2014, only 5.8%, 2.1%, and 2.1% of women cited intentions to pursue engineering, computer science, or the physical sciences, compared to 19.1%, 7.8%, and 3.2% of incoming men, respectively (NSF, 2017). This underrepresentation continues at each educational level. Although women comprised 58% of the overall graduate student population, men accounted for the majority of recipients in engineering and the physical sciences. The gender gap in computer science and information technology is particularly concerning. During the 1980s, when the computing field reached an early peak in popularity, women earned 35% of the bachelor's degrees awarded in computer science. In 2014, women earned only 18% of computer science degrees (NSF, 2017). The computing workforce experienced a similar decline between 1990 and 2011. During this time, the share of women employed in computing decreased from 34% to 27% (United States Census Bureau, 2013).

The tendency of extant 'underrepresentation' research to focus primarily on the experiences of women and minorities creates the impression that these identities are distinct, with little difference among those who hold each identity. However, decades of work on "intersectionality" (e.g., Collins, 1991; Crenshaw, 1989) makes clear that a woman is never just her gender–she also is raced and classed, among other identities. This also is the case for a Black person, as they are simultaneously raced, gendered, classed, and much more. The extant literature on STEM diversity rarely acknowledges the experiences of those who reside at the intersections of both race and gender (i.e., women of color). Given their intersecting marginalized identities, some argue that women and girls of color are at a double disadvantage in STEM (Burning et al., 2015; Ong et al., 2011). For example, White and middle-class teachers, the majority of K-12 educators, tend to have lower expectations of Black girls than of White girls, and as a result give Black girls tasks that enhance their caretaking skills and White girls' tasks that promote their academic abilities (Brickhouse et al., 2000). These and other experiences contribute to the stark underrepresentation of women and girls of color in STEM. In 2014, White women numbered 10,508 (61.8%) of all engineering bachelor's degree recipients, compared to 1,954 (11.5%) Hispanic women, and 933 (8.8%) Black women (NSF, 2017).

The lack of diverse representation in STEM is more than just numbers. Numerous research studies highlight the importance of diverse classrooms and work environments in promoting creativity and increased problem-solving skills (Carnevale et al., 2011; Lehman, Sax, & Zimmerman, 2016). Thus, by not diversifying STEM, the quality of scientific output may be compromised (Lehman et al., 2016). For example, companies have developed voice recognition software that do not accurately recognize women's voices (Margolis & Fisher, 2002), and created personal assistant apps (e.g., Siri) that fail to respond to crises that disproportionately affect women (e.g., domestic abuse or sexual assault; Miner, 2016). When diverse groups are included in the development of new innovations, they can advance product designs that meet the needs of a wider cross-section of consumers as they draw from a more diverse set of experiences.

3 Why so Few Women and Girls in STEM?

The underrepresentation of women, racial/ethnic minorities, and especially women of color in STEM exists for a number of reasons. These can be categorized as barriers or challenges encountered at the individual, institutional, and societal levels. Due to the ways in which individuals are nested within institutions, which are in turn nested within society, these categories are by no means distinct; rather, they overlap greatly with some factors aligning with more than one category. In the sections below, we discuss some of the factors and how they impede progress along the STEM career pathway and support dissociation with STEM disciplines.

3.1 *Influential Factors at the Individual Level*

The pathways to entering STEM-related occupations begin early, often by elementary school or earlier (Maltese & Cooper, 2017). The most direct pathways are somewhat linear, and any deviations can make it significantly more challenging to reenter and persist (Elliott, Strenta, Adair, Matier, & Scott, 1996). This is due to the sequential and linear process of the prerequisite courses needed to prepare for and pursue a STEM major in college. We can see this more clearly if we consider the academic background a student needs to pursue an undergraduate STEM major successfully. For many university programs this would require significant conceptual knowledge accompanied by strong grades in high school science and math courses. Building such a strong academic record often necessitates having interest and access to meaningful and rigorous math and science experiences in the early grades. For example, to be able to take Advanced Placement (AP) mathematics in high school, a student needs to have robust algebraic reasoning skills and conceptions of the number system. These skills are strengthened with curricula taught in sixth through tenth grades that build on the number sense students developed in the elementary grades. Although it is possible to develop strong mathematical reasoning skills after having a shaky mathematical start, it is more challenging, especially for marginalized youth. For students of color, navigating this pathway is particularly difficult as schools with high populations of Black and Hispanic students tend to be plagued with issues of teaching inequality and instructional incoherence (Ladson-Billings, 1997; Lee, 2004).

It is evident that the nature and quality of students' identities (including interest and self-efficacy in STEM), early experiences with mathematics and science, and general academic preparedness are critical to influencing the pursuit of STEM career pathways. Girls tend to take fewer of certain STEM courses (e.g., calculus, physics) in high school compared to boys (Blickenstaff, 2005; Cheryan et al., 2017), making it less likely that girls will continue on into a STEM field after high school (Blickenstaff, 2005). For women, taking more high-level math courses (e.g., trigonometry, pre-calculus, calculus) and doing well, lead to greater probability of selecting a quantitative field of study and declaring a STEM major (Trusty, 2002). To develop these efficacy beliefs and academic competence requires early and robust experiences in STEM-related subjects. These early experiences are not only important to attaining the necessary prerequisites to pursue advanced STEM-related courses in high school, but they also support early enculturation into STEM thereby fostering the development of mathematics and science identities.

One's science identity is context dependent and is affected by one's gender and racial and ethnic identities. As such, identity is a "cultural production" and reflects how local practices and meanings of activities are enabled or constrained by the global, socio-historical contexts. Because individuals experience different contexts and bring their own unique backgrounds, not all individuals "are free to develop *any* kind of science identity" (Carlone & Johnson, 2007, p. 1192, italics in original). Rather, choices are shaped by the larger meaning of what it means to be a "science person." Two types of stereotypes exist that serve as barriers to girls adopting a STEM-identity and entering STEM fields, namely the perceptions of the kind of people who belong in STEM (including beliefs about one's own abilities) and perceptions of the type of work done in STEM (Master et al., 2014).

In the U.S. today, scientists are depicted as White and male, and sometimes quite odd (e.g., Einstein, Newton, Aristotle). Furthermore, professionals in computer science and information technology are portrayed as single-minded, lacking outside interests, and intelligent but deficient in interpersonal skills; in short, they are considered to be nerds, geeks, or hackers who sit in front of the computer all day (Cheryan, Plaut, Davies, & Steele, 2009; Margolis & Fisher, 2001). These images are consistently reinforced in the media, for example through popular television programs like *Silicon Valley*, *Scorpion*, and *The Big Bang Theory*. In all three programs, the engineers are portrayed as stereotypical nerds, underrepresented minorities are glaringly absent, and storylines perpetuate negative stereotypes about gender roles and STEM professionals overall. This negative and fairly non-representative image of engineers and computer scientists is highly problematic for encouraging Black females specifically, and underrepresented minorities in STEM more broadly. Sorge, Newson, and Haggerty (2000) found that although Hispanic students involved in a STEM program enjoyed participating in the program, most of these students struggled with perceiving themselves as scientists based on lack of exposure to role models and negative media stereotypes. Given this dominant portrayal, it is challenging for girls, students of color, and girls of color to envision themselves as scientists or future "STEM folk." These images, which in no way resemble themselves or anyone they know or would want to be, encourages disassociation with science and technology as they see these futures as being exclusive to White males, uninteresting, and not particularly conducive to familial identities (e.g., wife, mother) that they may also wish to embrace (Ladson-Billings, 1997; Riegle-Crumb, Moore, & Ramos-Wada, 2010). These outcomes often result in the false assumption that these young girls are intrinsically uninterested in or unable to do science and math.

Beliefs that women are less capable in math and science continue to be cited as an explanation of gender difference in STEM career aspirations, despite research to the contrary (Hyde, 2008; Riegle-Crumb et al., 2011). However, evidence suggests that achievement may be an obstacle for different race/ethnic groups whose test scores tend to trail those of their White peers. Math and science test scores provide information to students about how their skills match up to expectations of what is needed to obtain future success in these fields, such that lower scores may indicate a mismatch with a future STEM career. Additionally, lower levels of achievement may prompt teachers and counselors to discourage students from pursuing STEM careers despite expressed interest and enjoyment in science and math (Riegle-Crumb et al., 2010).

Academic aspirations and career choices also are shaped by the false and negative narratives held about a particular group. These false narratives coupled with negative stereotypes may lead to students experiencing stereotype threat – the negative thoughts and emotions that accompany feelings of risk to conforming to negative stereotypes about one's gender, racial, ethnic, or social group. Steele's (1997, 2010) research indicates that members of racial/ethnic minority groups for whom there is a negative perception of intellectual and academic inferiority tend to display greater psychological vulnerability to failure. Girls of color are at heightened risk of experiencing this threat given their increased likelihood of experiencing negative stereotypes about both their gender (i.e., girls are innately less capable at math and science than boys) and race/ethnicity (i.e., academically inferior to their White peers). Internalization of these negative societal perceptions often manifest in low efficacy and achievement which does not encourage persistence in STEM.

Gender role socialization and how females are characterized psychologically – as emotionally-driven, irrational, dependent followers – creates an image of females that seems incompatible with the STEM profession. Similar stereotypes often exist within the home and are supported in the parent-child relationship where girls are to be homemakers while boys get to build, manipulate, and invent things (Tindall & Hamil, 2004). As girls take up identities of caregiver and homemaker early on, trying to align these gender-based responsibilities with the full-time, challenging career of a STEM professional seems incompatible (van Langen & Dekkers, 2005) and thus deters pursuit of related careers. Social perceptions of Black and Hispanic females also serve as roadblocks to pursuing STEM career pathways. Society and familial messages, direct and indirect, about gender roles tend to be influential in shaping who girls of color think they can be and what they can do.

Stereotypes of the work done within STEM fields provide another barrier to entry onto STEM pathways. Women who are interested in helping or working with others tend to report less interest in STEM fields that are perceived as not involving these activities (e.g., computer science, engineering, math, and physical sciences) and greater interest in STEM fields that do (e.g., biological sciences; Cheryan et al., 2017). Specific values are associated with academic fields and specific occupations, especially in STEM. As early as adolescence, gender differences in occupational values begin to emerge (Eccles, 1994). Girls tend to value altruism, family, and developing knowledge, whereas boys tend to value achievement, power, and financial success. Stereotypes of STEM fields lead to the perception that STEM is antithetical to the more communal values held by girls, resulting in a perception by girls that they do not fit in STEM. As such, girls tend to move away from STEM (Dasgupta & Stout, 2014; Eccles, 2007, 2009).

3.2 *Institutional and Societal Influences*

Many of the individual psychological and cognitive barriers to entry and persistence in STEM career pathways are by-products of students' experiences within institutions and societal structures that generate and instill negative perceptions about women and students of color. Current K-12 schooling does not provide the instructional or curricular resources to adequately attend to the needs of women and students of color. Schools with large populations of students of color are more likely to have inexperienced/unqualified teachers, few to no college preparatory courses, more remedial courses, and higher teacher turnover (Lee, 2004; van Langen & Dekkers, 2005). These issues of access and equity are significant institutional barriers for girls of color. In particular, there is differential availability and access to rigorous math and science curricula (e.g., AP courses, Project Lead the Way courses) including well-equipped labs and other college preparatory and development courses. With respect to academic preparedness, studies found that Black women often have inadequate high school preparation for STEM college courses and that students felt underprepared academically while in STEM programs at predominantly White institutions (Joseph, 2012).

Even when students have access to the same resources, such as when they are enrolled in the same STEM classrooms, female students may have different experiences than their male counterparts (Deboer, 1984). For example, in high school physics classes, gender differences have been found in how students participate, such that females tend to focus more on reading the textbook and completing homework assignments while males spend more time designing projects and learning concepts (Hazari, Sonnert, Sadler, & Shanahan, 2010). Classroom activities that predicted having a physics identity in high school students included those that focused on a conceptual understanding of physics, labs that addressed beliefs about the world, discussions of currently relevant science topics, and discussions of the benefits of being a physicist. However, these activities were reported as occurring less often by female than male students (Hazari et al., 2010).

In addition to the different experiences girls and boys may have within the same STEM classrooms, the educational opportunities provided to students and assumptions made about students' interests and abilities within STEM classrooms also may differ (Hobbs, Jakab, Millar, Prain, Redman, Speldewinde, Tytler, & vanDriel, 2017). Teachers often dissuade girls from STEM, in essence reinforcing gender stereotypes of those in STEM fields, and hold lower expectations of girls compared to boys, such that girls' interest and competence in STEM often goes unrecognized (Carlone, 2004; Tan, Calabrese Barton, Kang, & O'Neill, 2013; Hobbs et al., 2017). Teaching strategies that emphasize performance mastery, an "only one answer is correct" approach, and that place less emphasis on individual student needs is especially discouraging for girls (Hobbs et al., 2017, p. 14).

School organization structures and policies, specifically standardized testing and tracking, also serve as mechanisms that marginalize girls and students of color within STEM. Black females who were initially interested in pursuing STEM careers have had "their hopes extinguished at an early age when excluded from many high-level math and science classes because of achievement tests and school tracking" (Farinde & Lewis, 2012, p. 423). Although in many cases these early failures are attributed to innate deficiencies and inferiority, this exclusion really stems from the disproportionately high number of Black female students, many from low socioeconomic statuses, who often receive inadequate academic preparation in STEM areas (Moses, Howe, & Niesz, 1999).

Given the increased focus on testing in middle and upper grades, there may be a shift away from pedagogical techniques that focus on increasing students' learning and enjoyment in science and math. As early as 4th grade, students across genders and race/ethnic groups show similar levels of enjoyment in science. However, by 8th grade, enjoyment of science has decreased across all groups, especially females who report significantly lower enjoyment compared to White males. These decreasing patterns of enjoyment suggest that the educational system is not able to maintain students' interest as they progress into adolescence (Riegle-Crumb et al., 2011).

4 Strategies to Address Underrepresentation

In this section, we present a variety of approaches that are shown to or have the potential to lead to improvements in the increased representation of women and minorities in STEM. We pay specific attention to teaching and instruction.

For a girl to successfully pursue a STEM career path, she must first be able to see it as a possibility, then embrace it as an aspect of her identity and be provided with the resources necessary to boldly traverse this pathway. As Black girls engage in authentic science practices, they begin to see themselves as scientists and mathematicians (Riedinger & Taylor, 2016), as "STEM folk." In our own work, we observed that following participation in a 40-hour curriculum where Black girls developed and used their knowledge of elementary statistics concepts to solve a problem in their school, the girls not only felt empowered as problem solvers, a significant number considered becoming statisticians (Cross, Hudson, Adefope, Lee, Rapacki, & Perez, 2012). Without these meaningful experiences in science and mathematics, activities where they are able to engage in the practices of scientists and engineers, there is a high likelihood that girls of color will disaffiliate with science-identified peers (Carlone, Haun-Frank, & Webb, 2011) or dissociate with the subject altogether. It is through these positive associations with science that interest will grow, deep knowledge will develop, and efficacy will be enhanced. Given the sequential nature of most STEM-related subjects, it is essential that meaningful engagement in science and math activities begin early and target the development of spatial and numeracy skills which has been shown to influence later STEM achievement (Uttal, Meadow, Tipton, Hand, Alden, Warren, & Newcombe, 2013). Research suggests that spatially and numerically enriched experiences, within formal (e.g., school), informal (e.g., extra-curricular school activities, museums) and recreational (e.g., playground, sports facilities) learning spaces, would greatly support increased participation in mathematics, science, and engineering (e.g., Zhang, Kopenen, Rasanen, Aunola, Lerkkanen, & Nurmi, 2014).

Opportunities to pursue STEM careers are greatly reduced if one does not have the prerequisite qualifications (namely, successful completion of relevant K-12 math and science courses). Ensuring that girls receive support for STEM career development early, thereby building awareness of the range of possible STEM professions, how lucrative they are, and details on what the career pathway entails is essential for setting goals (UNESCO, 2017). Arming girls with this knowledge empowers them with greater agency in decision-making around courses and opportunities that arise through K-12 schooling. In this regard, they can serve as their own advocates when needed. Also, educating other potential advocates similarly (e.g., teachers, parents, career counselors, community leaders) will first ensure that girls and students of color receive the appropriate, gender- and race-responsive guidance that they are well suited for STEM careers. Second, with this information, students can serve as informed support networks for other girls organized around similar goals. These networks can help increase the likelihood that girls will take more of the appropriate courses necessary to navigate successfully into a STEM profession (Harackiewicz, Rozek, Hulleman, & Hyde, 2012).

Access to quality instruction is also critically important, specifically exposure to culturally responsive teaching (CRT) practices. Culturally responsive teaching recognizes that culture is central to learning and essential in promoting effective communication and shaping the thinking process of groups and individuals (Ladson-Billings, 1994). It is an approach to instruction that acknowledges, responds to, and promotes information and knowledge as culturally bound. This approach to teaching offers a richer, more rigorous, and equitable access to education for students of color in particular (Gay, 2000; Nieto, 1999). Instructional approaches that fall outside of the parameters of CRT tend to be detrimental to the academic success of students, irrespective of whether we see short-term gains. For example, Carlone (2003, 2004) found that girls enrolled in a traditionally taught physics curriculum, where lectures and verification labs were prominent, experienced short-term knowledge gains but not sustained interest in science. She posited that the girls embraced the certainty of knowledge because it generally aligned with their good student identities (i.e., allowed them to earn good grades) but learning in this way did not develop science identities as the nature and quality of the tasks and experiences deemphasized scientific thinking (Carlone, 2003). Rather than having opportunities to develop a passion for math and science through exploration, investigations, and connections to their experiences, minority female students are often relegated to being taught only the fundamentals, and they tend not to receive opportunities to pursue rigor or be challenged academically – in regular and advanced math and science courses (van Langen & Dekkers, 2005). Fundamental for the CRT approach is ensuring equal standards for all and allowing students of color to pursue courses at the level of rigor of their White peers. Equal standards, accompanied by high-quality instruction and adequate support, will bode well for increasing efficacy related to STEM via setting high expectations with opportunities for mastery experiences, reduce learned helplessness by minimizing early failures, and develop a sense of belonging in STEM via multiple and varied opportunities for success in STEM (Farinde & Lewis, 2012).

Although CRT is framed as an approach that builds bridges between students' cultural referents and with their learning experiences, some (e.g., Carlone & Johnson, 2012; Seiler, 2013) are critical of the metaphor of "bridges" as they believe it suggests that academic cultures and the cultures of underrepresented minorities are two distinct and bounded entities. Rather, these researchers see academic cultural practices, in particular disciplines encapsulated under the umbrella of STEM, as permeable and intersecting. As such they recommended developing STEM learning environments that draw on "the array of cultural knowledge, skills, abilities, and contacts possessed by socially marginalized groups" (Yosso, 2005, p. 76) and "what youth know from everyday settings to support specific subject matter learning" (Lee, 2007, p. 15) that will serve as the launching pad for relevant and motivating STEM instruction (Moll et al., 1992; Barton & Tan, 2009). By situating learning within familiar contexts and artifacts, we create a space where students feel some sense of expertise and ownership to embrace the challenges that come with expanding their existing knowledge, learning about new phenomena, and designing solutions to relevant problems. One project that successfully drew on approaches aligned with CRT included Gutstein's (2003) work with students on mathematizing issues of fairness and social injustice. Through helping students better understand their life contexts and the sociopolitical conditions of their lives, thereby developing knowledge of "the forces and institutions that shape their world" (p. 40), they engender interest in math and use this newly found math knowledge as the basis for students being a part of the solution to injustice. In one project students read an article that described a developer's desire to transform a neighborhood park to create parking spaces for nearby lofts. The students drew on concepts of time, geometry and measurement to better understand the situation and how it related to gentrification. Over the course of two years and engagement in similar kinds of projects that supported students in using mathematics to address personal and community-relevant sociopolitical problems, students developed mathematical power (i.e., developed mathematical efficacy, expertise and achievement) and an orientation toward mathematics as a valuable tool to make sense of the world and issues that matter to them.

As Gutstein's study and others (e.g., Dimick, 2012; Grimberg & Gummer, 2012) show, having students work on projects that situate learning within personal and culturally meaningful contexts supports development of interest and engagement. In similar ways we can use our knowledge of what gets young girls and students of color interested in STEM (e.g., Maltese & Cooper, 2017) along with career intentions to make instruction in these areas more aligned with students' interests (e.g., medicine). A significant factor in creating these STEM-influencing learning environments is having access to qualified teachers, not only with strong knowledge of content and pedagogy but ones that desire to build personal knowledge of their students and their communities and the disposition to modify and/or develop curricula that provide opportunities for meaningful engagement in inquiry and problem solving.

Given the role identity development plays in shaping interest and engagement, it is imperative that minority girls have opportunities to connect with STEM professionals with whom they can identify. As such, there needs to be greater representation of women and minorities in public images of STEM at all levels, from comic books through television and movies. Research (e.g., Kearney, 2015) found that having female teachers can have a differential impact on female students' pursuit of STEM studies, and careers. Based on this some countries (e.g., Austria, Switzerland, Israel, Sweden, and the United Kingdom) foregrounded the recruitment of more female STEM teachers. In line with this, recruiting more Black STEM teachers may have a similar positive impact on Black females. The more opportunities girls have to experience STEM and see or interact with female STEM professionals, the greater the likelihood they will choose this career path (Liston, Peterson, & Ragan, 2007).

At the more macro-level, it is imperative that the "counter narrative" of girls of color as strong, competent and intelligent individuals become prevalent. Currently the master narrative depicts girls of color as "the stark contrast of what is deemed 'normal'" compared to the White and male narrative. Therefore, girls of color are deemed as inferior in intellect and capabilities (Farinde & Lewis, 2012, p. 422). Due to this perspective, whether overtly or tacitly, girls of color are often passed over, discouraged from entering STEM fields, and their contributions devalued. Such actions can be attributed to the fact that the unique perspectives of women of color are rarely seen or considered. Overcoming this bias is not an easy feat as these perspectives are rooted in a long history of inequity and marginalization. However, shifts in the images of who is portrayed as competent and capable to include women of color, in STEM and more broadly, can be a first step. Additionally, focused efforts to develop policy and legislation that seek to diversify the STEM workforce may create a demand for more women of color in the field (UNESCO, 2017). Greater opportunities to showcase girls and students of color positively may add more fuel to spreading this counter narrative.

5 Conclusion

Despite longstanding efforts to broaden participation among diverse groups in STEM, women, minorities, and women of color remain strikingly underrepresented in areas such as computer science, engineering, and the physical sciences. Continued stagnation and decline of a large untapped human resource (i.e., women and minorities) can have far reaching consequences in this technologically-driven global economy. If our nation is serious about narrowing this disparity, we must examine the complex socio-cultural factors that disadvantage certain groups more than others. In this chapter, we highlighted the crippling impact of factors such as inequity in access to foundational curriculum, inability to see one's self as a scientist given the stereotypical Whitemale representation in the media and combatting stereotype threat as well as gender role socialization. In addition to understanding the influential factors, we must also consider creative strategies to maximize access and engagement across all groups in STEM. One such strategy is adopting a culturally responsive teaching approach. In doing so, practitioners and educators can foreground the centrality of culture in learning and move away from outdated one-size-fits-all paradigms. Not only will this and other strategies broaden participation in STEM, but such diversification will have far-reaching implications on our nation's ability to retain its global standing as a leader in innovation and design.

References

Blickenstaff, J. C. (2005). Women and science careers: Leaky pipeline or gender filter? *Gender and Education, 14*(4), 369–386.

Brickhouse, N. W., Lowery, P., & Schultz, K. (2000). What kind of a girl does science? The construction of school science identities. *Journal of Research in Science Teaching, 37*(5), 441–458.

Bruning, M. J., Bystydzienski, J., & Eisenhart, M. (2015). Intersectionality as a framework for understanding diverse young women's commitment to engineering. *Journal of Women and Minorities in Science and Engineering, 21*(1), 1–26.

Buchmann, C. (2009). Gender inequalities in the transition to college, *Teachers College Record, 111*(10), 2320–234.

Carlone, H. B. (2003). (Re)producing good science students: Girls' participation in high school physics. *Journal of Women and Minorities in Science and Engineering, 9*(1), 17–34.

Carlone, H. B. (2004). The cultural production of science in reform-based physics: Girls' access, participation, and resistance. *Journal of Research in Science Teaching, 41*, 392–414.

Carlone, H. B., Haun-Frank, J., & Webb, A. (2011). Assessing equity beyond knowledge and skills-based outcomes. *Journal of Research in Science Teaching, 48*, 459–485.

Carlone, H. B., & Johnson, A. (2007). Understanding the science experiences of successful women of color: Science identity as an analytical lens. *Journal of Research in Science Teaching, 44*, 1187–1218.

Carr, P. B., & Steele, C. M. (2010). Stereotype threat affects financial decision making. *Psychological Science, 21*(10), 1411–1416.

Carnevale, A. P., Smith, N., & Melton, M. (2011). *STEM: Science Technology Engineering Mathematics*. Washington, DC: Georgetown University Center on Education and the Workforce.

Cheryan, S., Plaut, V. C., Davies, P. G., & Steele, C. M. (2009). Ambient belonging: How stereotypical cues impact gender participation in computer science. *Journal of Personality and Social Psychology, 97*, 1045–1060.

Cheryan, S., Ziegler, S. A., Montoya, A. K., & Jiang, L. (2017). Why are some STEM fields more gender balanced than others? *Psychological Bulletin, 143*(1), 1–35.

Cole, J. R., & Cole, S. (1973). *Social stratification in science.* Chicago, IL: University of Chicago Press.

Collins, P. H. (2000). *Black feminist thought: Knowledge, consciousness, and the politics of empowerment* (2nd ed.). New York, NY: Routledge.

Crenshaw, K. (1991). Mapping the margins: Intersectionality, identity politics, and violence against women of color. *Stanford Law Review, 43*(6), 1241–1299.

Cross, D. I., Hudson, R. A., Adefope, O., Lee, M. Y., Rapacki, L., & Perez, A. (2012). Success made probable: African-American girls' exploration in statistics through project-based learning. *Journal of Urban Mathematics Education, 5*(2), 55–86.

Dasgupta, N., & Stout, J. G. (2014). Girls and women in science, technology, engineering, and mathematics: STEMing the tide and broadening participation in STEM careers. *Policy Inisghts from the Behavioral and Brain Sciences, 1*(1), 21–29.

Deboer, G. E. (1984). Factors related to the decision of men and women to continue taking science courses in college. *Journal of Research in Science Teaching, 21*(3), 325–329.

Eccles, J. S. (1994). Understanding women's educational and occupational choices. Applying the Eccles et al. model of achievement-related choices. *Psychology of Women Quarterly, 18*, 585–609.

Eccles, J. S. (2007). Where are all the women? Gender differences in participation in physical science and engineering. In S. J. Ceci & W. M. Williams (Eds.), *Why aren't more women in science? Top researchers debate the evidence* (pp. 199–210). Washington, DC: American Psychological Association.

Eccles, J. S. (2009). Who am I and what am I going to do with my life? Personal and collective identities as motivators of action. *Educational Psychologist, 44*(2), 78–89.

Elliott, R., Strenta, A. C., Adair, R., Matier, M., & Scott, J. (1996). The role of ethnicity in choosing and leaving science in highly selective institutions. *Research in Higher Education, 37*(6), 681–709.

Epstein, C. F. (1970). *Woman's place: Options and limits in professional careers.* Berkeley, CA: University of California Press.

Farinde, A. A., & Lewis, C. W. (2012). The underrepresentation of African American female students in STEM fields: Implications for classroom teachers. *US-China Education Review B, 4*, 421–430. Retrieved from http://www.eric.ed.gov/contentdelivery/servlet/ERICServlet?accno=ED533550

Fisher, A., & Margolis, J. (2002). Unlocking the clubhouse: the Carnegie Mellon experience. *ACM SIGCSE Bulletin, 34*(2), 79–83.

Harackiewicz, J. M., Rozek, C. S., Hulleman, C. S., & Hyde, J. S. (2012). Helping parents to motivate adolescents in mathematics. An experimental test of a utility-value intervention. *Psychological Science, 23*(8), 899–906. doi:10.1177/0956797611435530

Hazari, Z., Sonnert, G., Sadler, P. M., & Shanahan, M.-C. (2010). Connecting high school physics experiences, outcome expectations, physics identity, and physics career choice: A gender study. *Journal of Research in Science Teaching, 47*(8), 978–1003.

Hobbs, L., Jakab, C., Millar, V., Prain, V., Redman, C., Speldewinde, C., Tytler, R., & van Driel, J. (2017). *Girls' future – Our future. The invergowrie foundation STEM report.* Melbourne: Invergowrie Foundation.

Institute of Medicine, National Academy of Sciences, and National Academy of Engineering (IM/NAS/NAE). (2007). *Rising above the gathering storm: Energizing and employing America for a brighter economic. Future.* Washington, DC: The National Academies Press.

Joseph, J. (2012). From one culture to another: Years one and two of graduate school for African American women in the STEM fields. *International Journal of Doctoral Studies, 7*, 125–142.

Ladson-Billings, G. (1997). It doesn't add up: African American students' mathematics achievement. *Journal for Research: Crossing Boundaries in Search of Understanding, 28*(6), 697–708.

Lee, C. (2004). *Racial segregation and educational outcomes in metropolitan Boston.* Cambridge, MA: The Civil Rights Project at Harvard University.

Lehman, K. J., Sax, L. J., & Zimmerman, H. B. (2016). Women planning to major in computer science: Who are they and what makes them unique? *Computer Science Education, 26*(4), 277–298.

Maltese, A. V., & Cooper, C. S. (2017). STEM pathways: Do men and women differ in why they enter and exit? *AERA Open, 3*(3), 1–16. doi:10.1177/2332858417727276

Master, A., Cheryan, S., & Meltzoff, A. (2014). Reducing adolescent girls' concerns about STEM stereotypes: When do female teachers matter? *Revue international de psychologie sociale, 27,* 79–102.

Moses, M. S., Howe, K. R., & Niesz, T. (1999). The pipeline and student perceptions of schooling: Good news and bad news. *Educational Policy, 13*(4), 573.

National Center for Women in Information Technology (NCWIT). (2011). *NCWIT scorecard: A report on the status of women in information technology.* Retrieved May 12, 2018, from http://ncwit.org/resources

National Research Council. (2011). *Expanding underrepresented minority participation: America's science and technology talent at the crossroads.* Washington, DC: The National Academies Press.

National Science Foundation, Division of Science Resources Statistics. (2017). *Women, minorities, and persons with disabilities in science and engineering: 2014.* Arlington, VA: Author.

President's Council of Advisors on Science and Technology (PCAST). (2010). *Prepare and inspire: K-12 education on Science, Technology, Engineering, and Mathematics (STEM) for American's future.* Washington, DC: Author.

Ong, M., Wright, C., Espinosa, L., & Orfield, G. (2011). Inside the double bind: A synthesis of empirical research on undergraduate and graduate women of color in science, technology, engineering, and mathematics. *Harvard Educational Review, 81*(2), 172–209.

Riegle-Crumb, C., Moore, C., & Ramos-Wada, A. (2010). Who wants to have a career in science or math? Exploring adolescents' future aspirations by gender and race/ethnicity. *Science Education, 95*(3), 458–476. doi:10.1002/sce.20431

Scott, K. A., & White, M. A. (2013). COMPUGIRLS' standpoint: Culturally responsive computing and its effect on girls of color. *Urban Education, 48*(5), 657–681.

Smyth, F. L., & McArdle, J. J. (2004). Ethnic and gender differences in science graduation at selective colleges with implications for admission policy and college choice. *Research in Higher Education, 45*(4), 353–381.

Sorge, C., Newsom, H. E., & Hagerty, J. J. (2000). Fun is not enough: Attitudes of Hispanic middle school students toward science and scientists. *Hispanic Journal of Behavioral Sciences, 22,* 332–345.

Steele, C. M. (1997). A threat in the air: How stereotypes shape intellectual identity and performance. *American Psychologist, 52*(6), 613.

Tan, E., Calabrese Barton, A., Kang, H., & O'Neill, T. (2013). Desiring a career in STEM-related fields: How middle school girls articulate and negotiate identities-in-practice in science. *Journal of Research in Science Teaching, 50*(10), 1143–1179.

The White House. (2009). *President Obama launches "educate to innovate" campaign for excellence in science, technology, engineering & math (stem) education* [Press release]. Retrieved May 12, 2018, from https://obamawhitehouse.archives.gov/the-press-office/president-obama-launches-educate-innovate-campaign-excellence-science-technology-en

Tindall, T., & Hamil, B. (2004). Gender disparity in science education: The cause, consequences, and solutions. *Education, 125*(2), 282–295.

Trusty, J. (2002). Effects of high school course-taking and other variables on choice of science and mathematics college majors. *Journal of Counseling & Development, 80*, 464–74.

United Nations Educational, Scientific and Cultural Organization. (2017). *Cracking the code: Girls' and womens' education in Science, Technology, Engineering and Mathematics (STEM)*. Retrieved from http://www.unesco.org/open-access/terms-use-ccbysa-en

Uttal, D. H., Meadow, N. G., Tipton, E., Hand, L. L., Alden, A., Warren, C., & Newcombe, N. (2013). The malleability of spatial skills: A meta-analysis of training studies. *Psychological Bulletin, 139*(2), 352–402. doi:10.1037/a0028446

van Langen, A., & Dekkers, H. (2005). Cross-national differences in participating in tertiary science, technology, engineering and mathematics education. *Comparative Education, 41*(3), 329–350.

Zhang, X., Koponen, T., Räsänen, P., Aunola, K., Lerkkanen, M. K., & Nurmi, J. E. (2014). Linguistic and spatial s etic development via counting sequence knowledge. *Child Development, 85*(3), 1091–1107. doi:10.1111/cdev.12173

CHAPTER 4

Teaching and Learning Integrated STEM: Using Andragogy to Foster an Entrepreneurial Mindset in the Age of Synthesis

Louis S. Nadelson and Anne L. Seifert

Abstract

In this chapter, we explore the notion that the Age of Synthesis requires the amalgamation of knowledge and skills to develop working solutions to complex problems. We promoted the idea of teaching of integrated STEM to facilitate students' development of the understanding of how to blend knowledge. Integrated STEM provides an opportunity to foster student awareness of how seemingly unrelated topics or domains are combined to solve problems. Further, through teaching integrated STEM, teachers can help students develop an array of 21st century skills, an entrepreneurial frame of mind, and a growth mindset. However, to effectively teach integrated STEM, teachers must shift their instruction to be student-centered and embrace the context of self-directed learning to allow the students to explore potential solutions as part of their learning. Therefore, in this chapter we argue for the shift from pedagogy to andragogy, the instructional approach in which students are provided with a problem to solve and are empowered to make the decisions about how to approach, develop, and test a solution. We close with some ideas for research and the implications of adopting integrated STEM on curricular and instructional decisions.

1   Introduction

Knowledge and information continues to grow and technology continues to advance new opportunities; ideas, processes, and possibilities begin to emerge (e.g. Cresswell, Worth, & Sheikh, 2010). For example, the access to the vast amounts of information through cellular networks that is accessible by portable devices seemed unimaginable two decades ago. The technology developments of the components that compose the phone, the cellular network, access, and software used by the phones and servers were developed simultaneously which led to a system that people can use to rapidly access and share information from around the world (e.g. Wang et al., 2014). The internet and technology advancements continues to grow to include self-driving vehicles, robots, remote sensors, and much more, further expanding the kinds of data available and information that can be shared and accessed (e.g. Ruttan, 2001).

The complexity and integration of technology requires the application of and attention toward science, technology, engineering, and mathematics (STEM) in ways that are very different than the traditional processes currently used to teach STEM concepts in many school's (Nadelson & Seifert, 2013b; Stohlmann, Moore, & Roehrig, 2012). Developments of complex solutions of grand challenges, such as technology advancements to increase human knowledge and communication, requires the integration of STEM; however, in schools, STEM is commonly segregated and frequently presented as being independent of multidisciplinary context and other disciplines (Tofel-Grehl, Callahan, & Nadelson, 2017; Wicklein & Schell, 1995). We argue that successful development of solutions to societal grand challenges that require STEM integration also requires the application of 21st century practices (e.g. working in teams, communication, etc.), entrepreneurial thinking (e.g. tolerating failure, connecting people and ideas, etc.), perception of life-long learning, and teaching aligned with preparing students for the age of synthesis (Nadelson & Seifert, 2017).

To prepare today's students for a future where they will be involved in developing solutions to grand challenges and using and contributing to the diverse and vast amount of information, requires them to be engaged in learning environments that parallel the activities taking place outside of school (Bybee & Fuchs, 2006; Nadelson, Seifert, & McKinney, 2014). In these type of learning environments, students are given assignments that are open-ended, allow for students to make choices, require students to develop novel solutions, support creative thinking, and necessitate integrating knowledge from multiple domains (Blumenfeld et al., 1991). We argue that one possible solution for achieving these goals is to engage students in problem and project-based learning activities that require the integration and application of knowledge and practices across STEM disciplines (Stohlmann, Moore, & Roehrig, 2012).

The integration of STEM which commonly occurs outside of K-20 learning environments may be useful for exposing students to models that prepare them for engaging in grand challenges such as developing reliable and clean energy solutions, accessible and effective medical care option, efficient transportation, or sufficient food for all. Preparing students to find solutions to grand challenges requires them to engage in experiences that require them to integrate knowledge, work in groups, apply 21st century practices, and combine seemingly unrelated information to synthesize new ideas and solutions. Students need to be prepared for work in the age of synthesis (Nadelson & Seifert, 2017).

2 Age of Synthesis

We have entered the age of synthesis – an age that is defined by the need for diverse groups of individuals to work collaboratively to share ideas and knowledge to synthesize solutions to complex problems (Cai, 2011). Such diversity of thought potentially brings forth new and creative innovations to solve the grand challenges of society such as disease control, clean water, food, transportation, cyber security, and energy, require consideration of multiple aspects of science, mathematics, technology, and engineering. Thus, by nature, many of the grand problems and challenges we face in society require a range of expertise and knowledge that is simultaneously specialized and broad. Further, the need for diverse groups of people to work together to come up with creative and innovative solutions reinforces the importance of well-developed 21st century skills such as communication, collaboration, cultural competency, creativity, and critical thinking (Nadelson & Seifert, 2017; Rotherham & Willingham, 2010).

Preparing students for the age of synthesis requires engaging them in activities and assignments that are authentic, open-ended, ill-structured, multidisciplinary, and require the students to work in teams. Open-ended projects that require the integration of STEM disciplines are ideal for preparing students for the age of synthesis because the assignments require students to apply 21st century skills to develop viable or reasonable solutions to the complex problems (Bell, 2010; Nadelson & Seifert, 2013b).

A potential project that would prepare students for the age of synthesis may be developing a scale model of an efficient parabolic solar hotdog cooker and testing the heat flow, air exchange, and light received. The project may also require students to determine the cost, materials, time to build, and other related parameters for building the cooker. The development of a plan for building the cooker, as well as, the identification and use of the required materials relies on technology and engineering knowledge. The efficiency testing of the model requires knowledge of science and technology. The interpretation of data, scaling to actual size, determination of actual cost, and reporting of data require understanding of mathematics and computing. Because the assignment requires the integration of STEM and team work, the designing, building, and testing of a parabolic solar hotdog cooker is aligned with the kinds of open ended problems that can be used to prepare students for the age of synthesis.

The reliance on one's ability to work with diverse groups of people to solve complex problems will continue to expand (Hudson, 1996), increasing the importance of preparing students to be productive in the age of synthesis. The lone researcher working to solve a problem is very rare and will continue to decline as the recognition and diversity of an array of information needs to be considered and applied to solve grand challenges continues to expand (Barlow et al., 2018). The nature of authentic integrated STEM group projects provides students with the opportunity to work on developing the skills and knowledge needed to be prepared for the age of synthesis.

3   Defining Integrated STEM

Defining integrated STEM is complicated, as integrated STEM can be configured in a multitude of ways which makes it challenging to create a definition that represents the concept comprehensively, consistently, and that is encompassing (Honey, Pearson, & Schweingruber, 2014; Kelley & Knowles, 2016). Because of the challenges with defining integrated STEM, we have shifted the focus from a definition that is integrated STEM as the amalgamation of the STEM domains to rather a focus on the STEM related learning goals, assignment structures, required tasks, instructional approaches, and essential knowledge necessary to successfully complete the assignments and/or projects (Nadelson & Seifert, 2017). Thus, we define integrated STEM based on learning that is student centered, problem or project-based, requires learning about and applying knowledge from multiple STEM disciplines, and involves application of 21st century skills such as collaboration, creativity, critical thinking, and communication (Nadelson & Seifert, 2017). We maintain by immersing students in integrated STEM, learning is more likely to take place when students engage in assignments or projects that have multiple possible solutions, require application of knowledge from multiple STEM disciplines, are aligned with authentic issues that we face in society, embrace diversity in thought, and require students to make choices to successfully complete the assignment.

A possible integrated STEM assignment is the creation and testing of a parabolic solar hotdog cooker. The assignment requires the application of knowledge from multiple STEM disciplines. Designing and building a parabolic solar hot-dog cooker requires the application of knowledge of engineering and design, mathematics to understand focal point and the equation for a parabola, science to understand solar radiation, heat, and cooking, and technology to understand the materials and process of building the cooker. The testing may involve the use of a thermocouple to determine the temperature of the cooker near or at the focal point; mathematics would be applied to graph and interpret the collected data; engineering would be applied in discussing the design and re-design of the cooker for increased efficiency and reduced costs; and technology would be relied upon through the testing of multiple materials to determine the most effective for reflecting and heating the sun for cooking. While the elements of STEM would be needed and applied in order for students to successfully engage in the project, the focus would be on the problem (need/relevance to the world) project and application of knowledge and practices from an array of STEM disciplines but also requires the 21st century skills, diversity of thought from the group, student decision making, and provides for a wide range of possible solutions. Thus, the project would involve the integration of STEM in ways that may be conceived as a toolbox, with the STEM knowledge and processes accessed, learned, and used as needed.

To further support providing a useful, comprehensive, and encompassing definition of integrated STEM, we have developed the STEM spectrum (Nadelson & Seifert, 2017) (see Figure 4.1). At the far left end of the spectrum, STEM is segregated and the teaching and learning is focused on discipline specific STEM disciplines. As one moves along the spectrum, the STEM disciplines become more integrated, knowledge becomes more applied, and the focus shifts from teacher-centered to student-centered learning. In integrated

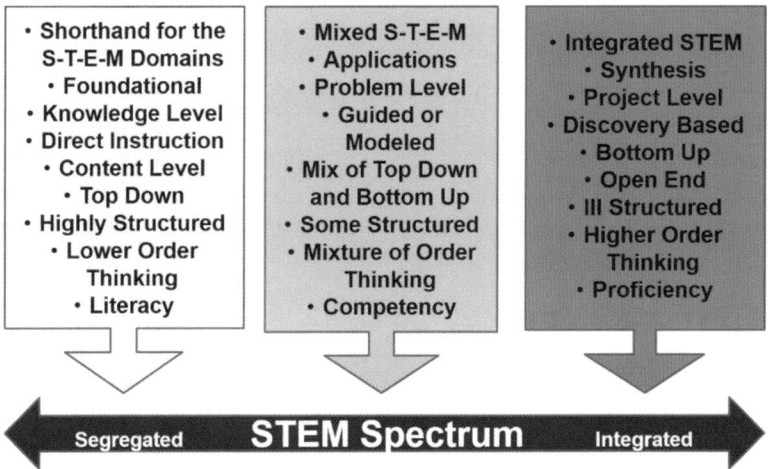

FIGURE 4.1   The STEM integration spectrum

STEM, the elements of STEM are applied seamlessly with the focus on solving complex, multidisciplinary problems or projects and are likely to include multiple STEM processes and an array of information and skills from each discipline.

In conjunction with our STEM spectrum, we designed a tool that can be used by teachers, informal educators, curriculum developers, or researchers to determine what specific STEM content might be needed or applied when teaching and using integrated STEM projects or assignments (see Figure 4.2). The purpose for the tool is to provide educators or curriculum developers a method for identifying appropriate integrated projects or assignments, planning for the implementation, and developing appropriate approaches for assessment. The tool could be used by researchers to examine the content or processes that promote learning or by teachers who are intending to teach with integrated STEM projects.

To use the tool, the educators, curriculum designers, and researcher, writes the project title in the center of the diagram as the focus of activities that will take place leading to STEM integration. The user then lists all the STEM concepts or processes that are needed, will be learned, or applied within the four quadrants in the tool, identifying STEM domains to complete the project. In Figure 4.3, we display a template that we completed using the parabolic hotdog cooker as the integrated STEM project (see Figure 4.3).

Thus, selecting the right integrated STEM project may be fundamental to assuring students are learning or applying the STEM knowledge that is desired

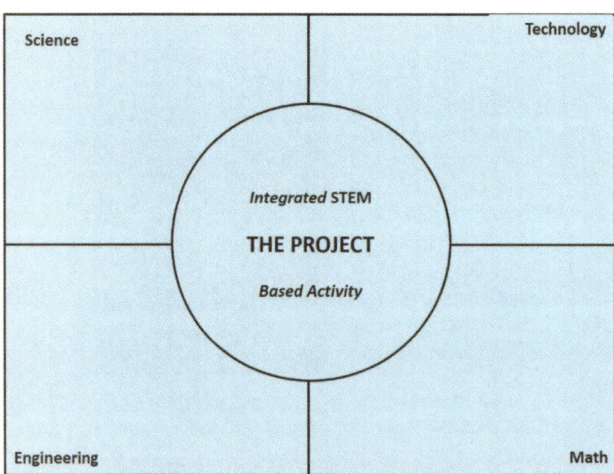

FIGURE 4.2   Tool for determining knowledge and process of STEM disciplines needed or used in integrated STEM projects

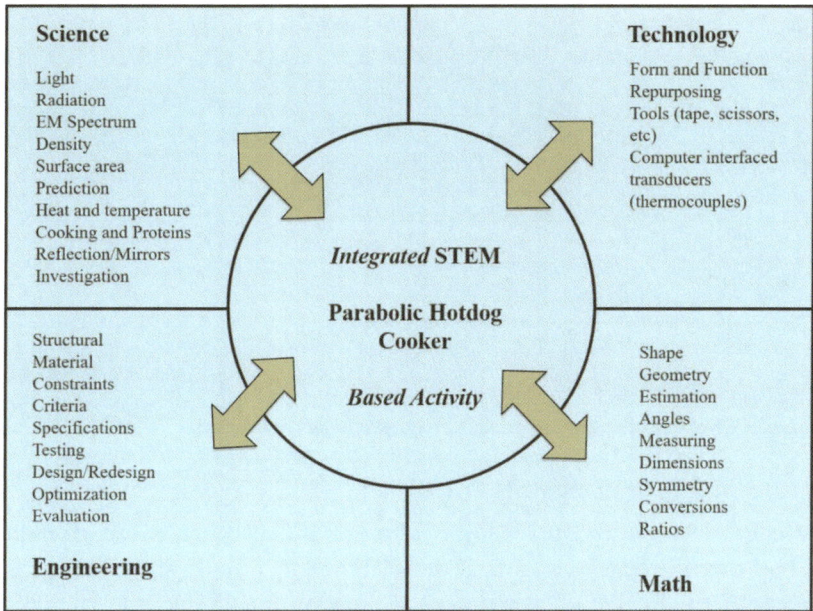

FIGURE 4.3 STEM processes and knowledge identification tool completed using the parabolic hotdog cooker as the project

through learning standards or curriculum expectations. However, using our tools researchers and teachers may be able to further determine which aspects of STEM students would engage in when they work on their integrated STEM projects or assignments. The challenge to researchers, curriculum developers, and teachers is to design or select project-based assignments that are aligned with the students' level of knowledge and abilities, and yet, meet education curriculum standards.

4   The Zone of Optimal Learning

Integrated STEM projects may be scaled in complexity to align with the students' level of knowledge and capacity in ways that optimize their learning. We maintain that assignments or projects are within the students zone of optimal learning (see Figure 4.4) when structured such that the level of complexity of the tasks is attainable by the students and, yet, requires them to put forth sufficient effort to complete the assignment while struggling, to some degree, to effectively carry out the related tasks (Nadelson & Seifert, 2017). We maintain that if integrated STEM assignments or projects are too sophisticated or complex for the students, exceeding their skills, capacity, and

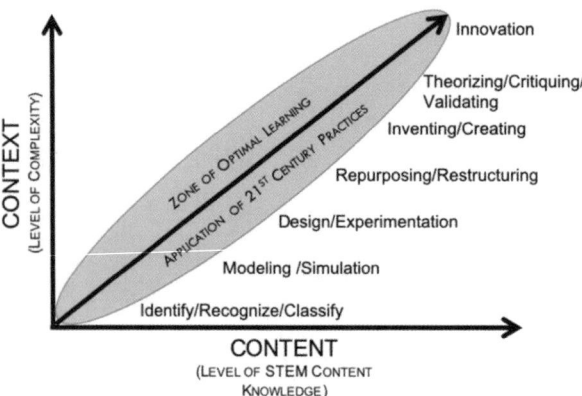

FIGURE 4.4   The zone of optimal learning where assignment complexity and students' capacity align

knowledge, then the students will get frustrated and discouraged, which then could potentially negatively impact their learning and motivation. Similarly, we argue if the integrated STEM assignments are too simple, the students could potentially lose interest, lack motivation to engage, and gain little from the experience.

With integrated STEM projects or assignments, there is opportunity to impact a range of student learning and development (Sias, Nadelson, Juth, & Seifert, 2017). Structuring assignments such that students have the opportunity to fail, make choices, and develop novel solutions is consistent with authentic work in the STEM workforce and, therefore, is a critical consideration when designing assignments or projects that are within the zone of optimal learning (McGrath & MacMillan, 2000). Further, by requiring students to make choices to complete a project, such as with the design and materials choices of a parabolic hotdog cooker, the teacher can reinforce the student's development of skills and experiences associated with learning to grapple with situations of ambiguity and develop deeper thinking skills (Cotabish, Dailey, Robinson, & Hughes, 2013). Similarly, by providing students with learning conditions in which they are safe to fail and have the opportunity to try again, the assignments or projects reinforces the skill of learning from failure and being resilient and the importance of a growth mindset (Hochanadel & Finamore, 2015). Many of the experiences that take place when working on integrated STEM projects reinforce skills and perceptions that are aligned with innovation and the entrepreneurial mindset.

4.1   *Entrepreneurial Mindset and Integrated STEM*

The entrepreneurial mindset is a way of thinking that affords people the opportunity to approach life in ways that allow them to thrive and achieve success,

particularly in situations of challenge or adversity (McGrath & MacMillan, 2000). There are multiple facets of the entrepreneurial mindset including:
- Resiliency and Tenacity
- Calculated Risk-taking
- Recognizing and Acting Upon Opportunity
- Connecting People and Ideas
- Respecting and Accepting Diversity of Thought
- Being Curious and Creative
- Explore and Develop Novel Solutions
- High Levels of Motivation
- High Level of Tolerance for Failure
- Consideration of Failure as an Opportunity for Learning

We embrace the notion that novel thinking is both trait and state dependent (Baron, 1998; Krueger, 2007), and therefore, is likely to be influential on the level of individual expression of the entrepreneurial mindset. For example, people may develop and act upon their entrepreneurial mindset continuously, which reflects a personality trait. However, people may also express and act in ways that are reflective of the entrepreneurial mindset elements based on the conditions of problem or task context, adjusting to the state they are experiencing.

In alignment with the work of Nadelson and colleagues (forthcoming), we also consider individual level of development and embrace of the entrepreneurial mindset to be defined as a spectrum. Similar to the STEM integration spectrum, the level of entrepreneurial thinking and engagement increases from one end of the spectrum to the other end of the spectrum. For example, individuals that are aligned at the end of spectrum that is representative of a cautious or conservative approach to situations would be considered to have a very limited entrepreneurial mindset because there are risk adverse, feel hopeless when they experience failure, avoid the unknown or unpredictable situations, tend to not think creatively, and are not motivated to connect people with ideas. Individuals at the opposite end of the spectrum tend to have an advanced entrepreneurial mindset accepting failure as part of learning, embracing novelty, thinking creatively, and taking risks.

Teaching and learning integrated STEM requires an entrepreneurial mindset (Nadelson, Seifert, & Sias, 2015). Because integrated STEM is a non-traditional way of thinking about teaching and learning STEM, content, concepts, and processes and require a different approach to thinking about the skills and mindset that is essential to effectively learn from integrated STEM assignments or projects (Nadelson & Seifert, 2016). For example, implementing integrated STEM activities requires tenacity and motivation to explore the unknown and is an unfamiliar way of teaching and learning, and requires creativity to develop novel assignments, projects, and solutions (Nadelson, Callahan, Pyke, Hay, Dance, & Pfiester, 2013).

Teachers with an entrepreneurial mindset would be better suited to teach integrated STEM because they would be more likely to experiment with assignments and curriculum, seek novel or creative approaches to teaching and learning, would be willing to take the risks with implementing the unknown, and would reflect on failure as part of the challenge of implementing an integrated STEM curriculum (Nadelson & Seifert, 2016). Teachers without an entrepreneurial mindset would be unlikely to embrace integrated STEM because they tend to be failure-adverse, choose not seek novel solutions, avoid taking risks, and lack resiliency. Thus, the entrepreneurial mindset is needed by teachers to explore and implement the unknown associated with integrated STEM teaching.

Similar to teachers, students with high entrepreneurial mindset are more likely to be successful with an integrated STEM curriculum because they would be willing to take initiative to self-direct their learning, embrace diversity of thought, persist in working on the project to completion, take risks, persist through failure, and develop innovative and novel solutions (Halverson & Sheridan, 2014). Students with a low entrepreneurial mindset would be less likely to take risks, would give up when faced with failure, discount diversity of thought, lack desire to develop creative solutions, and lack motivation to self-direct their learning.

The overlap and alignment between teacher and learner mindset in embracing STEM integrated assignments and projects suggests that in addition to teaching students the integrated STEM concepts, content and processes, there is also a need for teachers to help their students develop their mindset to be more entrepreneurial in thinking about and approaching problems (Buabeng-Andoh, 2012). Similarly, there are potential benefits to fostering the growth of teachers' entrepreneurial mindset through professional development, particularly when preparing teachers to take innovative approaches to teaching integrated STEM (Stein & Kim, 2011).

5   Integrated STEM and Life-Long Learning

Teaching and learning integrated STEM exposes both teachers and students to the unknown, and new possibilities. Unlike traditional approaches to teaching STEM in which concepts are established, answers appear to already be known, and STEM domains are segregated, integrated STEM activities have the potential to result in attainment of new knowledge and innovative thinking, provide potential solutions to conditions in which multiple answers to problems can be correct, and STEM knowledge is integrated. We argue that the potential of integrated STEM teaching and learning to lead to new knowledge, new solutions, and new perspectives of STEM, that working in integrated STEM environments reinforces the necessity to embrace and maintain the perspective of life-long learning (Shin, Haynes, Johnston, 1993; Shuman, Besterfield-Sacre, & McGourty, 2005).

Unlike the traditional approaches to teaching STEM that reinforce the perception that knowledge is fixed and STEM concepts are well established, integrated STEM reinforces perceptions that knowledge continues to develop, and there is a continual expansion of new concepts and ideas. Thus, with an integrated STEM approach to teaching, students have greater potential for developing views of knowledge as being dynamic and authentic, and, therefore, develop the perception that there is always more to learn (National Research Council, 2011). Further, teachers who embrace integrated STEM in their instruction also develop new course content, curricular arrangement, and novel instructional choices that require continual learning and reflection, which is reflective of life-long learning (Bryan, Moore, Johnson, & Roehrig, 2015).

There is a substantial overlap between an entrepreneurial mindset and life-long learning. Both entrepreneurial mindset and life-long learning may be learned through teaching and learning integrated STEM. However, we maintain that an entrepreneurial mindset is needed to engage in integrated STEM teaching and learning, and through the engagement, students and teachers develop a deeper awareness of life-long learning.

6 An Andragogic Approach Teaching and Learning Integrated STEM

Teaching and learning integrated STEM requires a different approach than traditional or domain segregated approaches to teaching STEM. In most domain segregated STEM instruction the focus is on the teacher or knowledge source (e.g. textbook, reference materials) who holds and delivers the content to the students. The students are passive in the process, absorbing the information. The direct instruction approach is in alignment with pedagogy (the education of the child) (Knowles, 1977). While the direct instruction approach may be effective for delivering content and introducing learners to certain new concepts or content, the approach is not effective for engaging students in learning more complex and novel concepts, thus. teaching students to be self-directed life-long learners, and helping them develop an entrepreneurial mindset (Neck & Corbett, 2018).

We argue that teaching and learning integrated STEM requires a student-centered approach to teaching and learning. Thus, the learning shifts from focus on the teacher directed learning to focus on student-centered learning and requires student to be more self-directed in their learning. A self-directed approach to learning and the associated instructional approach is defined as *andragogy* (Holton, Swanson, & Naquin, 2001; Knowles, 1977). Andragogy translates to education of the adult and refers to an instructional approach that involves providing contexts and support for learning in which learners are expected to explore, make decisions, determine direction for projects working within parameters of expectations, and self-monitoring their achievement of the associated learning expectations (Neck & Corbett, 2018). Most of the learning that takes place in the workplace is self-directed (Dirkx, 1996), and therefore taking an andragogical approach to teaching is needed to prepare students for the workplace and productivity outside of school.

An andragogical instructional approach is well aligned with problem and project-based active learning (Minhas, Ghosh, & Swanzy, 2012), in which students may be provided specific parameters and criteria but are expected to make choices to complete the project. Andragogy is an effective approach for teaching integrated STEM because of the options and decision making that assigned projects or assignments can require of learners. The open-end and student-centered nature of integrated STEM projects or assignments usually involve providing the learners with some parameters or goals, and then guiding the learners as they direct their learning through the choices they make as individuals or team members; their process and approaches decisions; their determination of the knowledge needed to complete the project; their reflection on their learning; and their exploration of additional potential approaches to solving the problem (Freeman et al., 2014).

Shifting to an andragogical instructional approach may be challenging for both students and teachers, particularly if both are conditioned for the traditional pedagogical approach to instruction (Delahaye, Limerick, & Hearn, 1994). Shifting to an andragogical approach requires teachers to take the role of facilitator or mentor, and the students to take more of an active role in their learning assuming higher levels of self-regulation, self-assessment, responsibility, accountability and choice making (Chan, 2010). Thus, there may be some scaffolding needed when shifting from teacher focused traditional pedagogy to student focused andrology to help all learn the new approach to teaching and learning so that an integrated STEM curriculum is fully integrated in a successful and authentic manner.

7 Integrated STEM Opportunities and Challenges

As we have shared, there are multiple benefits to taking an integrated problem-based approach to teaching STEM. Exposing students to the authentic learning experiences that parallel conditions of STEM professionals helps to prepare students for working in the age of synthesis. Providing students with ill-structured or open-ended problems or projects fosters the development of an entrepreneurial mindset and supports the development of a perception of learning as being a life-long endeavor and allows them opportunity to develop 21st century or rather employability skills needed in their careers.

There are multiple potential challenges that we have identified with adopting and implementing an integrated approach to teaching STEM. The potential lack of entrepreneurial mindset by teachers and students provides a barrier due to the lack of the necessary creativity, novel thinking, innovative spirit, respect for diversity of thought, risk taking, and handling failure that are all associated with teaching and learning integrated STEM. Thus, the adoption and implementation of integrated STEM should likely begin with professional development for the teachers to help them develop a strong entrepreneurial mindset and prepare them for teaching their students to think like entrepreneurs.

A lack of STEM knowledge or comfort with and motivation for learning STEM content may be a barrier that needs to be addressed in order to successfully engage in and complete integrated STEM projects or assignments. However, to have acquired the depth and breadth of STEM knowledge necessary to complete many integrated STEM project or assignments is substantial and therefore, learning the STEM content needed for the project should be considered as part of the project implementation process and desired outcomes. As with STEM professionals and researchers, when teachers and students engage in integrated STEM, they should engage in life-long learning and consider gaps in knowledge as an opportunity for learning.

The dearth of integrated STEM teaching and learning models may limit the ability for teachers and school administrators to understand how to prepare, adopt, and implement integrated STEM curriculum and activities. Similarly, there may be educational system constraints or barriers that may limit educator's capacity to implement integrated STEM, such as institutional support or the organizational culture. Thus, there may be a need for educators to work with their colleagues and engage with leadership in their organizations to increase overall understanding about the learning benefits and outcomes when students engage in integrated STEM teaching and learning.

Teachers' and students' perceptions of teaching and learning may provide a substantial challenge to implementing an integrated approach to STEM education. If teachers perceive teaching to be the conveying of knowledge rather than establishing context for students to learn and then facilitating and mentoring the students through projects, teaching intergrated STEM may feel very uncomfortable to them, especially with in context of student-centered learning. Similarly, if students have developed a perception of learning that involves finding a specific correct answer or the delivery of rote information from the teacher they may also be very uncomfortable with student centered learning. Thus, there may be a need to scaffold both the teachers and students toward a learning environment that is more self-directed (taking an approach of andragogy) deviating from the directed learning environment which takes more of a pedagogical approach.

8 Potential Lines of Teaching and Learning Integrated STEM Research

There is a foundation for teaching and learning integrated STEM in problem or project-based learning literature (e.g. Strobel & Van Barneveld, 2009). Similarly, the research on student-centered learning provides a foundation for investigation into integrated STEM teaching and learning (e.g. Hannifin & Land, 2012). There is also some research on the integration of curriculum that may be useful for informing integrated STEM education (e.g. Marshall, 2005). Thus, there are several extant lines of research that can be built upon for investigating integrated STEM teaching and learning.

New lines of integrated STEM education research include exploration of the focus, extent of authenticity, and structure of problems and projects that are effective for integrating STEM to enhance student STEM learning. There is a need to further explore the alignment between problems or projects and the zone of optimal learning. Thus, there is value in examining how teachers identify and select problems or projects that they perceive will be effective for engaging their students in learning integrated STEM concepts and processes.

There is a need to develop, validate, and scale the use of assessments to measure what and how students are learning when they engage in integrated STEM learning activities. Unlike segregated STEM, the knowledge gained through integrated STEM may be more difficult to assess with traditional assessment techniques. Integrated STEM learning may require the use of alternative assessment techniques such as student portfolios, rubrics, self-reflection, interviews, and observations. While a range of alternative assessment techniques have been documented in the literature as being effective for determining the levels of student learning, there may be additional aspects of the assessment process that may need to be explored in the context of integrated STEM.

An additional area of research is the association between organizational structure and culture and the adoption and implementation of integrated STEM. Similarly, a fruitful direction for research may involve the examination of organizational change when there is a commitment toward and resources for the adoption and support for integrated STEM. Again, there is a foundation of organizational change and educational innovation in the literature, however, there may be unique aspects of organizational change and innovation adoption in the context of integrated STEM (Nadelson & Seifert, 2016; Nadelson, Seifert, & Sias 2015).

There is a need for further assessment of how integrated STEM fosters the development of an entrepreneurial mindset, diversity of thought, perceptions of life-long learning, and preparation of the age of synthesis. The complexity of effectively assessing these association is may require investigators to explore and develop new approaches, tools, and techniques for research.

There are potentially many other lines for future research that may emerge as the adoption or resistance to integrated STEM continues to evolve. Further, the growth of contexts and opportunities such as makerspaces, robotics, and coding in formal and informal learning may catalyze additional promotion, adoption and implementation of integrated STEM which may further expand the research needs and opportunities. Thus, as integrated STEM education continues to evolve there is a need for an evolution and expansion of the associated research.

9 Conclusion

The complex overlap of multiple facets of content learning, psychological growth, and skill development associated with integrated STEM teaching and learning requires understanding of multiple aspects of student learning and development. Next steps may involve creating and testing a model that includes integrated STEM, entrepreneurial mindset, life-long learning, and preparation for the age of synthesis. In conjunction with shifts in teachers' instructional and content choices there is a need for shifts in organizational leadership structures to support the changes and foster the systemic transformations that are required for institutionalizing integrated STEM teaching and learning. There may addition aspects of societal influences and progressions and elements of student development that may need to be addressed as part of the model to assure a comprehensive and inclusive representation of student learning and teaching approaches of integrated STEM.

References

Barlow, J., Stephens, P. A., Bode, M., Cadotte, M. W., Lucas, K., Newton, E., ... Pettorelli, N. (2018). On the extinction of the single-authored paper: The causes and consequences of increasingly collaborative applied ecological research. *Journal of Applied Ecology*, 55(1), 1–4.

Baron, R. A. (1998). Cognitive mechanisms in entrepreneurship: Why and when entrepreneurs think differently than other people. *Journal of Business Venturing, 13*(4), 275–294.

Bell, S. (2010). Project-based learning for the 21st century: Skills for the future. *The Clearing House, 83*(2), 39–43.

Blumenfeld, P. C., Soloway, E., Marx, R. W., Krajcik, J. S., Guzdial, M., & Palincsar, A. (1991). Motivating project-based learning: Sustaining the doing, supporting the learning. *Educational Psychologist, 26*(3–4), 369–398.

Bryan, L. A., Moore, T. J., Johnson, C. C., & Roehrig, G. H. (2015). Integrated STEM education. In C. C. Johnson, E. E. Peters-Burton, & T. J. Moore (Eds.), *STEM road map: A framework for integrated STEM education* (pp. 23–37). New York, NY: Routledge.

Buabeng-Andoh, C. (2012). Factors influencing teachers' adoption and integration of information and communication technology into teaching: A review of the literature. *International Journal of Education and Development using Information and Communication Technology, 8*(1), 136–155.

Bybee, R. W., & Fuchs, B. (2006). Preparing the 21st century workforce: A new reform in science and technology education. *Journal of Research in Science Teaching, 43*(4), 349–352.

Cai, S. (2011). The age of synthesis: From cognitive science to converging technologies and hereafter. *Chinese Science Bulletin, 56*(6), 465–475.

Chan, S. (2010). Applications of andragogy in multi-disciplined teaching and learning. *Journal of Adult Education, 39*(2), 25–35.

Cotabish, A., Dailey, D., Robinson, A., & Hughes, G. (2013). The effects of a STEM intervention on elementary students' science knowledge and skills. *School Science and Mathematics, 113*(5), 215–226.

Cresswell, K. M., Worth, A., & Sheikh, A. (2010). Actor-network theory and its role in understanding the implementation of information technology developments in healthcare. *BMC Medical Informatics and Decision Making, 10*(1), 67.

Delahaye, B. L., Limerick, D. C., & Hearn, G. (1994). The relationship between andragogical and pedagogical orientations and the implications for adult learning. *Adult Education Quarterly, 44*(4), 187–200.

Dirkx, J. M. (1996). Human resource development as adult education: Fostering the educative workplace. *New Directions for Adult and Continuing Education, 1996*(72), 41–47.

Freeman, S., Eddy, S. L., McDonough, M., Smith, M. K., Okoroafor, N., Jordt, H., & Wenderoth, M. P. (2014). Active learning increases student performance in science, engineering, and mathematics. *Proceedings of the National Academy of Sciences, 111*(23), 8410–8415.

Halverson, E. R., & Sheridan, K. (2014). The maker movement in education. *Harvard Educational Review, 84*(4), 495–504.

Hannafin, M. J., & Land, S. M. (2012). Student-centered learning. In N. M. Seel (Ed.), *Encyclopedia of the sciences of learning* (pp. 3211–3214). Freiburg: Springer.

Hochanadel, A., & Finamore, D. (2015). Fixed and growth mindset in education and how grit helps students persist in the face of adversity. *Journal of International Education Research, 11*(1), 47–50.

Holton, E. F., Swanson, R. A., & Naquin, S. S. (2001). Andragogy in practice: Clarifying the andragogical model of adult learning. *Performance Improvement Quarterly, 14*(1), 118–143.

Honey, M., Pearson, G., & Schweingruber, H. (Eds.). (2014). *STEM integration in K-12 education: Status, prospects, and an agenda for research*. Washington, DC: National Academies Press.

Hudson, J. (1996). Trends in multi-authored papers in economics. *Journal of Economic Perspectives, 10*(3), 153–158.

Kelley, T. R., & Knowles, J. G. (2016). A conceptual framework for integrated STEM education. *International Journal of STEM Education, 3*(1), 1–11.

Knowles, M. (1977). Adult learning processes: Pedagogy and andragogy. *Religious Education, 72*(2), 202–211.

Krueger Jr., N. F. (2007). What lies beneath? The experiential essence of entrepreneurial thinking. *Entrepreneurship Theory and Practice, 31*(1), 123–138.

Marshall, J. (2005). Connecting art, learning, and creativity: A case for curriculum integration. *Studies in Art Education, 46*(3), 227–241.

McGrath, R. G., & MacMillan, I. C. (2000). *The entrepreneurial mindset: Strategies for continuously creating opportunity in an age of uncertainty*. Boston, MA: Harvard Business Press.

Minhas, P. S., Ghosh, A., & Swanzy, L. (2012). The effects of passive and active learning on student preference and performance in an undergraduate basic science course. *Anatomical Sciences Education, 5*(4), 200–207.

Nadelson, L. S., Callahan, J., Pyke, P., Hay, A., Dance, M., & Pfiester, J. (2013). Teacher STEM perception and preparation: Inquiry-based STEM professional development for elementary teachers. *The Journal of Educational Research, 106*(2), 157–168.

Nadelson, L. S., Jouflas, G., Basnet, R., Benton, T., Bissonnette, M., ... Palmer, A. (forthcoming). Developing next generation of innovators: Teaching entrepreneurial mindset elements across disciplines. *Entrepreneurship Education and Pedagogy*.

Nadelson, L. S., & Seifert, A. L. (2013a). Perceptions, engagement, and practices of teachers seeking professional development in place-based integrated STEM. *Teacher Education and Practice, 26*(2), 242–266.

Nadelson, L. S., & Seifert, A. L. (2013b). *i-STEM leads the way: Integrating STEM to develop the 21st century workforce*. Workshop presentation at the North Carolina New Schools, Scaling STEM: Strategies that Engage Minds, Durham NC.

Nadelson, L. S., & Seifert, A. L. (2016). Putting the pieces together: A model K-12 teachers' educational innovation implementation behaviors. *Journal of Research in Innovative Teaching, 9*(1), 47–67.

Nadelson, L. S., & Seifert, A. L. (2017). Integrated STEM defined: Contexts, challenges, and the future. *The Journal of Educational Research, 110*(3), 221–223.

Nadelson, L. S., Seifert, A. L., & McKinney, M. (2014). *Place based STEM: Leveraging local resources to engage K-12 teachers in teaching integrated STEM and for addressing the local STEM pipeline.* Proceedings of the American Society of Engineering Education Annual Conference and Exposition, Indianapolis, IN.

Nadelson, L. S., Seifert, A. L., & Sias, C. (2015). To change or not to change: Indicators of K-12 teacher engagement in innovative educational practices. *International Journal of Innovation in Education, 3*, 45–61.

National Research Council. (2011). *Successful K-12 STEM education: Identifying effective approaches in science, technology, engineering, and mathematics.* Washington, DC: National Academy Press.

Neck, H. M., & Corbett, A. C. (2018). The scholarship of teaching and learning entrepreneurship. *Entrepreneurship Education and Pedagogy, 1*(1), 8–41.

Rotherham, A. J., & Willingham, D. T. (2010). 21st-century skills. *American Educator, 17*, 17–20.

Ruttan, V. W. (2001). *Technology, growth, and development.* New York, NY: Oxford University Press.

Sais, C. M., Nadelson, L. S., Juth, S. M., & Seifert, A. L. (2017). The best laid plans: Education innovation in elementary teacher generated integrated STEM lesson plans. *The Journal of Educational Research, 110*(3), 227–238.

Shin, J. H., Haynes, R. B., & Johnston, M. E. (1993). Effect of problem-based, self-directed undergraduate education on life-long learning. *Canadian Medical Association Journal, 148*(6), 969–976.

Shuman, L. J., Besterfield-Sacre, M., & McGourty, J. (2005). The ABET "professional skills"—Can they be taught? Can they be assessed? *Journal of Engineering Education, 94*(1), 41–55.

Stein, M. K., & Kim, G. (2009). The role of mathematics curriculum materials in large-scale urban reform: An analysis of demands and opportunities for teacher learning. In J. T. Remillard, B. A. Herbel-Eisenmann, & G. M. Lloyd (Eds.), *Mathematics teachers at work: Connecting curriculum materials and classroom instruction* (pp. 37–55). New York, NY: Routledge.

Stohlmann, M., Moore, T. J., & Roehrig, G. H. (2012). Considerations for teaching integrated STEM education. *Journal of Pre-College Engineering Education Research, 2*(1), 28–34.

Strobel, J., & Van Barneveld, A. (2009). When is PBL more effective? A meta-synthesis of meta-analyses comparing PBL to conventional classrooms. *Interdisciplinary Journal of Problem-based Learning, 3*(1), 44–58.

Tofel-Grehl, C., Callahan, C. M., & Nadelson, L. S. (2017). Comparative analyses of discourse in specialized STEM school classes. *The Journal of Educational Research, 110*(3), 294–307.

Wicklein, R., & Schell, J. (1995). Case studies of multidisciplinary approaches to integrating mathematics, science and technology education. *Journal of Technology Education, 6*(2), 59–61.

CHAPTER 5

National Reports on STEM Education: What Are the Implications for K-12?

Sarah Bush

Abstract

While the acronym STEM has been around since the 1990s, STEM education was thrust into the spotlight in the early 2000s, with many reports that pre-dated it. Since then, there have been multitudes of national reports calling attention to the importance of STEM education and recommendations. This chapter will converge and highlight findings and recommendations across reports and consider implications for K-12 education. Specifically, this chapter identifies recent K-12 STEM reports in the United States that focus on integrated STEM which include recommendations to the field. A table of these reports and their key recommendations are presented. Four categories emerged from a synthesis of the key recommendations across national STEM reports: (1) K-12 STEM education curriculum, standards, programs, and literacy; (2) Increase number of students pursuing STEM majors and careers, with a focus on underrepresented populations; (3) Recruitment, retention, recognition, preparation, and professional development of STEM teachers; and (4) STEM education funding and research. After a discussion of these four categories, the author identifies limitations and considers implications for K-12 education. Directions for future research with suggested research questions are presented followed by concluding remarks.

1   Introduction

The acronym STEM (Science, Technology, Engineering, and Mathematics) was first coined in the 1990s by the National Science Foundation (NSF) to describe any policy, program, project, event, or practice related to one or more of the STEM disciplines (Bybee, 2010a, Bybee 2010b). The STEM issue in the United States is two-fold. The shortfall of STEM majors and graduates from universities (National Science Board, 2016) coupled with the projected growth of STEM occupations (Langdon, McKittrick, Beede, Khan, & Doms, 2011; U.S. Bureau of

Labor Statistics, 2008) perpetuates the need to focus on STEM education in the U.S. Specifically, prioritizing STEM education must begin in K-12. This chapter highlights and synthesizes findings from national reports on STEM education with a focus on K-12.

Since the early 2000s, national STEM reports have articulated goals, action steps needed, best practices, synthesized research, and have summarized key recommendations related to STEM education in the U.S. For example, the National Research Council (2011) established three goals for U.S. STEM education that include:

1. Expand the number of students who ultimately pursue advanced degrees and careers in STEM fields and broaden the participation of women and minorities in those fields.
2. Expand the STEM-capable workforce while broadening the participation of women and minorities.
3. Increase STEM literacy for all students, including those who do not pursue STEM-related careers or additional study in the STEM disciplines (pp. 4–5).

These three goals center on the idea that because of workforce demands and to be competitive in the global economy, the commitment to quality K-12 STEM teaching and learning has never been greater in the U.S., and we should recognize and be bold in our efforts to broaden the participation of women and minorities in STEM. In another example, stated in an executive report to President Obama, "STEM education will determine whether the United States will remain a leader among nations and whether we will be able to solve immense challenges in such areas as energy, health, environmental protection, and national security" (PCAST, p. 1). The state of K-12 STEM education in our country is at a critical juncture and we must work collaboratively to push the field forward in ways that foster opportunity, access, STEM literacy, and inclusion (Mohr-Schroeder, Bush, & Jackson, 2018). Therefore, the time to collectively review and synthesize national reports on STEM education is now.

2 National Reports on STEM Education

An abundance of U.S. policy documents and reports on STEM speak to the notion that preparing future STEM professionals is a key ingredient to the sustained success of the U.S. Table 5.1 provides the lists of recommendations from a sample of national reports on STEM education, organized by year, 2010–2017. This table is meant to provide an overview ($n = 8$) and is not exhaustive. Inclusion criteria included reports that were focused on the U.S. rather than one state (such as in State Superintendent of Public Instruction Tom Torlakson's STEM Task Force, 2014) or other countries (such as in Chapman & Vivian, 2017); were specific to STEM rather than focused on a specific subset of the STEM disciplines (such as in National Science Board, 2016); did not target a specific subgroup (such as in Hill, Corbett, & St. Rose, 2010); was the most recent version of report (such as in American College Testing, 2017); and include a list of overall recommendations or similar (for example, National Research Council, 2013 did not have a list of recommendations). Next, findings portrayed in these recommendations are synthesized.

A synthesis review of the recommendations listed in Table 5.1 revealed the following four key categories: 1) K-12 STEM education curriculum, standards, programs, and literacy; 2) Increase number of students pursuing STEM majors and careers, with a focus on underrepresented populations; 3) Recruitment, retention, recognition, preparation, and professional development of STEM teachers; and 4) STEM education funding and research. Each category will now be discussed.

2.1 K-12 STEM Education Curriculum, Standards, Programs, and Literacy

Recommendations regarding K-12 STEM education curriculum, standards, programs, and literacy were present across all national reports in Table 5.1. Two overarching recommendations related to this category include ensuring our K-12 students leave high school with the skills (in addition to content knowledge) to pursue STEM college and career paths and that the U.S. prepares our students to become and remain competitive globally.

Several suggestions for accomplishing these two recommendations are presented such as including educational experiences for K-12 students that engage them in interdisciplinary work to solve authentic problems while keeping a clear focus on the content of the STEM disciplines. Such educational experiences are important because they provide students with the opportunity to work in collaborative teams, witness first-hand how STEM content and skills can be used to solve problems to improve our world, and introduce them to STEM college majors and careers paths. Further, this type of learning environment models for students that in today's society those with different expertise must collaborate to solve interdisciplinary problems. Another suggestion is to create learning opportunities in flexible environments that invite play, innovation, and risk-taking. Such opportunities foster students' natural curiosity and help students learn that failure and perseverance are all part of successful learning experiences. A third suggestion is to create opportunities to inspire students in STEM both during the school day and through informal learning opportunities.

TABLE 5.1 Recommendations of national reports on STEM education

Title	Authors	Year	Recommendations
Refueling the U.S. Innovation Economy: Fresh Approaches to Science, Technology, Engineering and Mathematics (STEM) Education	Robert D. Atkinson, Merrilea Mayo (The Information Technology and Innovation Foundation)	2010	1. Shift accountability measures for high schools from a content-based to a skills-based paradigm. 2. Substantially pare the breadth requirements/mandatory course lists required for high school graduation. 3. Provide funding to the department of education to create 300 new specialty STEM high schools over the next decade. 4. Establish a national STEM talent recruiting system. 5. Provide substantially more research opportunities for freshmen STEM students. 6. Create new kinds of STEM colleges and programs. 7. Require all colleges and universities receiving federal money to report results from the National Survey of Student Engagement. 8. Offer prizes up to $35 million to colleges and universities that have dramatically increased student STEM degrees and maintained those increases over 5 years. 9. Significantly increase industry co-funded academic research and graduate student fellowships. 10. Develop an industry-ranked list of the best STEM departments. (pp. 11–12, identified as the 10 most important and transformative recommendations of the report)

(cont.)

TABLE 5.1 Recommendations of national reports on STEM education (*cont.*)

Title	Authors	Year	Recommendations
Rising Above the Gathering Storm, Revisited: Rapidly Approaching Category 5	National Academy of Sciences, National Academy of Engineering, & Institute of Medicine	2010	1. Move the United States K-12 education system in science and mathematics to a leading position *by global standards*. 2. Double the real federal investment in basic research in mathematics, the physical sciences, and engineering over the next seven years (while, *at a minimum*, maintaining the recently doubled *real* spending levels in the biosciences). 3. Encourage more United States citizens to pursue careers in mathematics, science, and engineering. 4. Rebuild the competitive ecosystem by introducing reforms in the nation's tax, patent, immigration and litigation policies.
Prepare and Inspire: K-12 Education in Science, Technology, Engineering, and Math (STEM) for America's Future	President's Council of Advisors on Science and Technology	2010	(pp. 19–20, overarching recommendations) 1. Standards: Support the current state-led movement for shared standards in math and science 2. Teachers: Recruit and train 100,000 great STEM teachers over the next decade who are able to prepare and inspire students 3. Teachers: Recognize and reward the top 5 percent of the nation's STEM teachers, by creating a STEM master teachers corps 4. Educational Technology: Use technology to drive innovation, by creating an advanced research projects agency for education

(*cont.*)

TABLE 5.1 Recommendations of national reports on STEM education (cont.)

Title	Authors	Year	Recommendations
			5. Students: Create opportunities for inspiration through individual and group experiences outside the classroom
			6. Schools: Create 1,000 new STEM-focused schools over the next decade
			7. Ensure strong and strategic national leadership
			(p. viii-x, recommendations)
Successful K-12 STEM Education: Identifying Effective Approaches in Science, Technology, Engineering, and Mathematics	National Research Council	2011	1. Expand the number of students who ultimately pursue advanced degrees and careers in STEM fields and broaden the participation of women and minorities in those fields.
			2. Expand the STEM-capable workforce and broaden the participation of women and minorities in that workforce.
			3. Increase STEM literacy for all students, including those who do not pursue STEM-related careers or additional study in the STEM disciplines.
			(pp. 4–5, goals identified)

(cont.)

TABLE 5.1 Recommendations of national reports on STEM education (*cont.*)

Title	Authors	Year	Recommendations
STEM Integration in K-12 Education: Status, Prospects, and an Agenda for Research	Margaret Honey, Greg Pearson, and Heidi Schweingruber (Committee on Integrated STEM Education; National Academy of Engineering; National Research Council)	2014	1. In future studies of integrated STEM education, researchers need to document the curriculum, program, or other intervention in greater detail, with particular attention to the nature of the integration and how it was supported. When reporting on outcomes, researchers should be explicit about the nature of the integration, the types of scaffolds and instructional designs used, and the type of evidence collected to demonstrate whether the goals of the intervention were achieved. Specific learning mechanisms should be articulated and supporting evidence provided for them. 2. Researchers, program designers, and practitioners focused on integrated STEM education, and the professional organizations that represent them, need to develop a common language to describe their work. This report can serve as a starting point. 3. Study outcomes should be identified from the outset based on clearly articulated hypotheses about the mechanisms by which integrated STEM education supports learning, thinking, interest, identity, and persistence. Measures should be selected or developed based on these outcomes. 4. Research on integrated STEM education that is focused on interest and identity should include more longitudinal studies, use multiple methods, including design experiments, and address diversity and equity.

(*cont.*)

TABLE 5.1 Recommendations of national reports on STEM education (cont.)

Title	Authors	Year	Recommendations
			5. Designers of integrated STEM education initiatives need to be explicit about the goals they aim to achieve and design the integrated STEM experience purposefully to achieve these goals. They also need to better articulate their hypotheses about why and how a particular integrated STEM experience will lead to particular outcomes and how those outcomes should be measured.
			6. Designers of integrated STEM education initiatives need to build in opportunities that make STEM connections explicit to students and educators (e.g., through appropriate scaffolding and sufficient opportunities to engage in activities that address connected ideas).
			7. Designers of integrated STEM experiences need to attend to the learning goals and learning progressions in the individual STEM subjects so as not to inadvertently undermine student learning in those subjects.
			8. Programs that prepare people to deliver integrated STEM instruction need to provide experiences that help these educators identify and make explicit to their students connections among the disciplines. These educators will also need opportunities and training to work collaboratively with their colleagues, and in some cases administrators or curriculum coordinators will need to play a role in creating these opportunities. Finally, some forms of professional development may need to be designed as partnerships among educators, STEM professionals, and researchers.

(cont.)

TABLE 5.1 Recommendations of national reports on STEM education (cont.)

Title	Authors	Year	Recommendations
			9. Organizations with expertise in assessment research and development should create assessments appropriate to measuring the various learning and affective outcomes of integrated STEM education. This work should involve not only the modification of existing tools and techniques but also exploration of novel approaches. Federal agencies with a major role in supporting STEM education in the United States, such as the Department of Education and the National Science Foundation, should consider supporting these efforts.
			10. To allow for continuous and meaningful improvement, designers of integrated STEM education initiatives, those charged with implementing such efforts, and organizations that fund the interventions should explicitly ground their efforts in an iterative model of educational improvement.
			(pp. 137–151, recommendations)
Revisiting the STEM Workforce: A Companion to Science and Engineering Indicators in 2014	National Science Board (National Science Foundation)	2015	1. The "STEM workforce" is extensive and critical to innovation and competitiveness. It is also defined in various ways and is made up of many sub-workforces.
			2. STEM knowledge and skills enable multiple, dynamic pathways to STEM and non-STEM occupations alike.
			3. Assessing, enabling, and strengthening workforce pathways is essential to the mutually reinforcing goals of individual and national prosperity and competitiveness.
			(pp. 1–2, primary insights)

(cont.)

TABLE 5.1 Recommendations of national reports on STEM education (cont.)

Title	Authors	Year	Recommendations
STEM 2026: A Vision for Innovation in STEM Education	U.S. Department of Education	2016	1. Engaged and networked communities of practice 2. Accessible learning activities that invite intentional play and risk 3. Educational experiences that include interdisciplinary approaches to solving "grand challenges" 4. Flexible and inclusive learning spaces 5. Innovative and accessible measures of learning 6. Societal and cultural images and environments that promote diversity and opportunity in STEM (p. ii–iii, six interconnected components)
STEM Education in the U.S.: Where We Are and What We Can Do, 2017	American College Testing	2017	1. Ensure that state graduation requirements emphasize the importance of rigorous science and math courses for all students. 2. Pay teachers more. 3. Establish a loan forgiveness program for STEM teachers. 4. Provide equitable access to both high-quality math and science courses and real-world work experiences for all students via dual enrollment programs. (pp. 20–23, policy recommendations)

Providing students with transformative learning experiences may not only spark interest but also help students see themselves as a "STEM person."

These national reports also speak to the development of infrastructures focused on K-12 STEM education. Such infrastructure might include creating high schools that specifically focus on STEM, intentionally recruiting talent, providing more STEM research opportunities for high school students, providing dual enrollment opportunities for each and every student, and ensuring that graduation from high school emphasizes coursework in mathematics and science. By prioritizing STEM in these ways, more students will have access to learn about STEM and future college and career options.

Finally, several reports focused on the notion of STEM literacy (as in Bybee, 2010a; Zollman, 2012). STEM Literacy advocates for each and every student to develop foundational knowledge in STEM needed for both STEM or non-STEM careers. Such knowledge is what prepares our students as well as empowers them to become adults who are informed consumers and citizens. For example, STEM literate adults are able to understand and interpret financial documents such as mortgages and investments; comparison shop; read and understand basic medical documents; consider ways to protect the environment; understand, interpret, and determine the reasonableness of news reports; measure, calculate, and create a materials list for a home improvement project; and much more.

2.2 Increase Number of Students Pursuing STEM Majors and Careers, with a Focus on Underrepresented Populations

A critical component of the recommendations from the national STEM education reports is increasing the number of high school graduates pursuing STEM majors and careers. While the ideas in the previous paragraph provide suggestions as to how this might be accomplished, intentionally stating it as a recommendation highlights the urgency. Some reports encourage more participation in STEM in general terms, while other reports draw much-needed attention to broadening participation of underrepresented groups in STEM.

The importance of providing access and strengthening those pathways for each and every student is an essential component of STEM equity. This is closely related to building infrastructures in schools in such a way that each and every student will participate in STEM learning and opportunities. Such infrastructures importantly consider how to make such learning equitable by monitoring the progress of students, and when needed, differentiating instruction, making accommodations, and presenting additional challenges (similar to NCTM, 2014). Such infrastructures are needed in all schools across the country to make STEM learning opportunities the norm, rather than an add-on, special area, elective, or enrichment only.

Furthermore, one report notes that in addition, we must also work to shift society and cultural norms and purposefully promote and encourage diversity in STEM through intentional opportunities. This includes programs that target student populations that are underrepresented in STEM. Such programs prioritize reaching and providing inspiring opportunities to students that might not otherwise have access to such opportunities. This might include special recruiting efforts. We also must promote and encourage diversity by ensuring that each and every student can see themselves in STEM. This includes intentionally learning about diverse scientists, mathematicians, and engineers; showcasing the diverse landscape of successful STEM professionals and innovators through posters, media, books, and more; guest speakers in STEM that our students can see themselves in; and providing opportunities that enable students to see what STEM possibilities exist that they may not have had access to otherwise.

2.3 Recruitment, Retention, Recognition, Preparation, and Professional Development of STEM Teachers

The success and breadth of STEM in K-12 education will be determined in large part by K-12 teachers. Rightfully so, many of the STEM education reports in Table 5.1 provide recommendations specific to teachers. These recommendations focus on retention and recruitment and also on preparation and professional development of STEM teachers.

The recommendations that focus on recruitment and retention of teachers suggest strategies such as paying teachers more, providing student loan forgiveness for STEM teachers, and rewarding and recognizing the top STEM teachers in our nation. With regards to low teacher pay, for example, the average high school science or mathematics teacher starting pay in the U.S. is only $39,000 but the average starting pay for an electrical engineer is $62,500 (ACT, 2017). Furthermore, while many states offer some type of teacher loan forgiveness especially for high-needs areas such as mathematics and science, many of these programs target only those teachers working in a low-income schools or districts and are often only available after a number of years of teaching. Some of these reports suggest beginning to offer this benefit after the first full year of teaching and not just to teachers in low-income schools or districts (ACT, 2017).

The recommendations specific to teacher preparation and professional development include providing professional development opportunities where connections to and across disciplines in STEM are made explicit and that professional development should iteratively focus on improving STEM education. STEM teacher preparation and professional development is different because the content spans multiple disciplines and by nature, the problems posed are ill-defined and complex with varied solution paths and strategies (Bush, Mohr-Schroeder, Cook, Rakes, Ronau, & Saderholm, under review). Teachers of K-12 STEM must have professional opportunities to practice making such connections across disciplines for themselves as learners and to their students and have time to work collaboratively with fellow teachers and administrators (Honey, Pearson, & Schweingruber, 2014). Furthermore, PD specifically focused on STEM should provide time for teachers to practice identifying key science and mathematics content and practices to be integrated (Asunda & Mativo, 2015).

2.4 STEM Education Funding and Research

Another key idea that emerged from these reports include the notion of increased funding for STEM Education as well as recommendations for research. Funding recommendations include providing funding incentives to colleges and universities that create innovative STEM programs as well as to those that substantially increase their number of STEM graduates.

Related to research, recommendations were made regarding funded research reporting which include more explicit documentation of interventions used, outcomes measured, evidence collected, and whether goals were met. Teaching STEM in ways that are integrated is not new, but as a field, we still lack a comprehensive body of literature that paints a complete picture of best practice and most effectives strategies.

Furthermore, there is an emphasis on more intentional alignment between outcomes and measures as well as common language used across the field. It was also recommended that research on STEM education include more longitudinal studies, employ multiple methodologies, and address issues of access and equity. These suggestions focus on the quality of STEM education research studies and align to criticisms of educational research in general. For example, in their systematic study of mathematics education technology dissertation scope and quality, Ronau, Rakes, Bush, Driskell, Niess, and Pugalee (2014) stated that "The mathematics education technology research community must in turn begin to demand greater quality in its published studies, through both how researchers write about their own studies and how they review the works of others" (p. 29). As a field, more research on STEM education that is valid and reliable and reported in ways that are replicable is needed.

3 Limitations

There are several key limitations when considering the reports discussed in this chapter as well as generally speaking, K-12 STEM education in the U.S.

First, this chapter only reports on a limited number of national reports related to K-12 STEM education. It is likely that other reports that did not meet the inclusion criteria previously outlined or that were inadvertently omitted would have shed light on additional recommendations. Second, a key limitation to the advancement of STEM Education in the United States is limited funding – in essentially every way imaginable. Funding to recruit and retain highly qualified K-12 STEM teachers is limited and teachers do not get paid enough. The budgets of many K-12 schools and districts are in crisis which greatly inhibits their ability to enact the infrastructure recommendations set forth in many of the national reports. Further, higher education in general as well as for STEM related research is grossly underfunded. Third, to remain globally competitive, we have to adapt to innovations in STEM at lighting speed – and the world of K-12 and higher education struggles to keep up with the speed at which advances in industry often moves, which presents challenges.

4 Implications for K-12 Education

Despite the limitations outlined above, it is imperative that the U.S. position themselves as a global competitor in STEM (PCAST, 2010), and that work starts in K-12. The national reports summarized above highlight the essential components that must be implemented in K-12. Some recommendations may be much simpler to enact than others. For example, schools and districts can identify current successful STEM initiatives and then intentionally provide access to those opportunities to more learners, creating more equitable STEM infrastructures. Instructional shifts such as implementing curriculum that engages each and every student in authentic learning that crosses the STEM disciplines and establishing more open and flexible learning environment all in efforts to promote STEM literacy for all learners are changes that can be enacted directly at the school or district level. Other recommendations, such as ensuring rigorous mathematics and science coursework is emphasized as a key ingredient to a high school diploma and the encouragement of STEM focused schools are calls to action that must be addressed at the state level.

Operationalizing STEM education recommendations in K-12 will only be as effective as our K-12 STEM teachers. Strengthening the knowledge and skills of STEM teachers is greatly recognized as a need in national reports (National Academy of Sciences, National Academy of Engineering, & Institute of Medicine, 2007, 2010). The National Research Council (2011) articulated that effective professional development in the STEM disciplines focuses on teachers' subject matter content and pedagogical knowledge, is set in the context of teachers' classrooms, and provides iterative learning through sustained professional development. Further, integrated STEM professional development is inherently more complex, as noted in in the literature (as described in part by Asunda & Mativo, 2015; Baker & Galani, 2017; Brophy, Klein, Portsmore, & Rogers, 2008; Bybee, 2010a; Driskell, Bush, Ronau, Niess, Rakes, & Pugalee, 2016; Estapa & Tank, 2017; Honey, Pearson, & Schweingruber, 2014; Hsu, Purzer, & Cardella, 2011; Moore, Mathis, Guzey, Glancy, & Siverling, 2014; Olster, 2012; Shernoff, Sinha, Bressler, & Ginsburg, 2017). Effective integrated STEM programs provide students the opportunity to spend time on inquiries that would hone their 21st century learning skills (Bybee, 2010b) and see connections between the STEM subjects and improve interest in STEM careers (Moore et al., 2014). Lee and Nason (2012) state that when STEM programs engage in inquiries that authentically task students in addressing the needs of the school, community, and industry, the resulting curriculum has an increased focus on design and problem solving and there is a natural flow both within and between the different disciplines of STEM. To that end, schools, districts, and states should work together to build long-term sustained teacher professional development opportunities that address the recommendations set forth in the national reports and uniquely prepares teachers to effectively teach STEM, including the integration of the STEM disciplines.

5 Future Research

The U.S. STEM challenge spans beyond measuring students' academic achievement as students also lack interest in pursuing STEM majors and careers (PCAST, 2010). Therefore, future research on STEM education must address both how to increase academic achievement while at the same time increasing students' interest in STEM – which necessitates the implementation of large-scale longitudinal studies. While research on effective instruction on the individual STEM disciplines (e.g. mathematics or science) is plentiful, additional research on integrated STEM learning is still needed. Honey and colleagues (2014) clearly articulate that through research on integrated STEM learning we must, as a field, identify the methods of content integration that are most effective, what instructional strategies are used, and how to best collect evidence.

As the recommendations set forth in these national reports are implemented, evaluation and research on their effectiveness is needed to advance the field. What strategies work best and which ones fall short? In order to determine this, practitioners and researchers need to work together as their shared expertise is essential for both strong implementation and high quality research on improving the field of STEM education. Further, we must work to use more common language to describe our work (Honey et al., 2014) and use a comprehensive framework to guide STEM education professional development (as in Bush, Cook, Ronau, Rakes, Mohr-Schroeder, & Saderholm, 2018; Bush et al., under review; Saderholm, Ronau, Rakes, Bush, & Mohr-Schroeder, 2017).

6 Future Research Questions

As we consider future research directions in K-12 STEM education, the following are a few broad research questions that we as a field must answer:
- What are the most effective instructional strategies for integrated STEM learning that leads to the greatest gains in (a) students' academic achievement; (b) interest in STEM majors and careers; (c) broadening underrepresented students' participation in STEM; and (d) developing STEM literacy over time?
- In what ways can we best provide future and current K-12 STEM teachers with the professional development, resources, and support needed so they can confidently and effectively implement integrated STEM instruction?
- What strategies for STEM teacher recruitment, retention, and recognition are most effective?

7 Conclusions

This is a pivotal time in K-12 STEM education. National reports on STEM education highlight key recommendations regarding K-12 student STEM experiences, K-12 STEM teachers, increasing the number of students pursuing STEM majors and careers, including broadening underrepresented students' participation in STEM, and STEM education funding and research. This chapter is not exhaustive of all literature on K-12 STEM education and focuses specifically on a sample of national STEM reports. This chapter is intended to be a synthesis resource and conversation starter. Consider your role in STEM education and how you can work collaboratively to do your part to enact the ideas set forth in these national reports. What recommendations shared in this chapter can you put into action? Working together, we can positively shape the landscape of K-12 STEM education and the future global competitiveness of the U.S.

References

American College Testing. (2017). *STEM education in the U.S.: Where we are and what we can do, 2017*. Retrieved from http://www.act.org/content/dam/act/unsecured/documents/STEM/2017/STEM-Education-in-the-US-2017.pdf

Asunda, P. A., & Mativo, J. (2015). Integrated STEM: A new primer for teaching technology education. *Technology and Engineering Teacher, 76*(5), 8–13.

Atkinson, R. D., & Mayo, M. (2010). *Refueling the U.S. innovation economy: Fresh approaches to Science, Technology, Engineering and Mathematics (STEM) Education*. Information Technology and Innovation Foundation. Retrieved from https://www.itif.org/files/2010-refueling-innovation-economy.pdf

Baker, C. K., & Galani, T. M. (2017). Integrated STEM in elementary classrooms using model-eliciting activities: Responsive professional development for mathematics coaches and teachers. *International Journal of STEM Education, 4*(10), 1–15. doi:10.1186/s40594-017-0066-3

Brophy, S., Klein, S., Portsmore, M., & Rogers, C. (2008). Advancing engineering education in P-12 classrooms. *Journal of Engineering Education, 97*(3), 369–387. doi:10.1002/j.2168-9830.2008.tb00985.x

Bush, S. B., Cook, K. L., Ronau, R. N., Rakes, C. R., Mohr-Schroeder, M. J., & Saderholm, J. (2018). A highly structured collaborative STEAM program: Enacting a professional development framework. *Journal of Research in STEM Education, 2*(2), 106–125.

Bush, S. B., Mohr-Schroeder, M. J., Cook, K. L., Rakes, C. R., Ronau, R. N., & Saderholm, J. (under review). Structuring STEM professional development: Challenges revealed and insights gained from a comparative case analysis.

Bybee, R. W. (2010a). Advancing STEM education: A 2020 vision. *Technology and Engineering Teacher, 70*(1), 30–35.

Bybee, R. W. (2010b). What is STEM education? *Science, 329*(5995), 996. doi:10.1126/science.1194998

Chapman, S., & Vivian, R. (2017). *Engaging the future of STEM: A study of international best practice for promoting the participation of young people, particularly girls, in Science, Technology, Engineering and Mathamatics (STEM)*. Chief Executive Women. Retrieved from https://cew.org.au/wp-content/uploads/2017/03/Engaging-the-future-of-STEM.pdf

Driskell, S. O., Bush, S. B., Ronau, R. N., Niess, M. L., Rakes, C. R., & Pugalee, D. (2016). Mathematics education technology professional development: Changes over several decades. In M. L. Niess, S. O. Driskell, & K. F. Hollebrands (Eds.), *Handbook of research on transforming mathematics teacher education in the digital age* (pp. 107–136). Hershey, PA: IGI Global.

Estapa, A. T., & Tank, K. M. (2017). Supporting integrated STEM in the elementary classroom: A professional development approach centered on an engineering design challenge. *International Journal of STEM Education, 4*(6), 1–16. doi:10.1186/s40594-017-0058-3

Hill, C., Corbett, C., & St. Rose, A. (2010). *Why so few? Women in science, technology, engineering, and mathematics.* Washington, DC: American Association of University Women. Retrieved from http://files.eric.ed.gov/fulltext/ED509653.pdf

Honey, M., Pearson, G., & Schweingruber, H. (2014). *STEM integration in K-12 education: Status, prospects, and an agenda for research.* Washington, DC: National Academies Press.

Hsu, M.-C., Purzer, S., & Cardella, M. (2011). Elementary teachers' views about teaching design, engineering, and technology. *Journal of Pre-College Engineering Education Research, 1*(2), 31–39.

Langdon, D., McKittrick, G., Beede, D., Khan, B., & Doms, M. (2011). *STEM: Good jobs now and for the future* (Report #03-11). Washington, DC: US Department of Commerce.

Lee, K. T., & Nason, R. A. (2012). Reforming the preparation of future STEM teachers. In Y. Shengquan (Ed.), *Proceedings from the 2nd International STEM in Education Conference.* Beijing, China.

Mohr-Schroeder, M., Bush, S. B., & Jackson, C. (2018). K12 STEM education: Why does it matter and where are we now? *Teachers College Record* (ID Number: 22288).

Moore, T. J., Mathis, C. A., Guzey, S. S., Glancy, A. W., & Siverling, E. A. (2014, October). *STEM integration in the middle grades: A case study of teacher implementation.* Proceedings of the Frontiers in Education Conference (FIE), Madrid, Spain. doi:10.1109/FIE.2014.7044312

National Academy of Engineering & National Research Council. (2014). *Integration in K-12 STEM education: Status, prospects, and an agenda for research.* Washington, DC: National Academies Press.

National Academy of Sciences, National Academy of Engineering, & Institute of Medicine. (2007). *Rising above the gathering storm: Energizing and employing America for a brighter economic future.* Washington, DC: National Academies Press.

National Academy of Sciences, National Academy of Engineering, & Institute of Medicine. (2010). *Rising above the gathering storm, revisited: Rapidly approaching category 5.* Washington, DC: National Academies Press.

National Council of Teachers of Mathematics. (2014). *Access and equity in mathematics education.* A position of the National Council of Teachers of Mathematics. Reston, VA: Author.

National Research Council. (2011). *Successful K-12 STEM education: Identifying effective approaches in science, technology, engineering, and mathematics.* Washington, DC: National Academies Press.

National Research Council. (2013). *Monitoring progress toward successful K-12 STEM education: A nation advancing?* Washington, DC: National Academies Press.

National Science Board. (2015). *Revisiting the STEM workforce: A companion to science and engineering indicators 2014.* Washington, DC: National Science Foundation.

National Science Board. (2016). *Science and engineering indicators 2016* (Report No. NSB-2016-1). Washington, DC: National Science Foundation.

Ostler, E. (2012). 21st century STEM education: A tactical model for long-range success. *International Journal of Applied Science and Technology, 2*(1), 28–33.

President's Council of Advisors on Science and Technology (PCAST). (2010). *Prepare and Inspire: K-12 education in Science, Technology, Engineering, and Mathematics (STEM) for America's future.* Executive Office of the President of the U.S.A. Retrieved from https://nsf.gov/attachments/117803/public/2a--Prepare_and_Inspire--PCAST.pdf

Ronau, R., Rakes, C., Bush, S. B., Driskell, S., Niess, M., & Pugalee, D. (2014). A survey of mathematics education technology dissertation scope and quality: 1968–2009. *American Education Research Journal, 51*(5), 974–1006. doi:10.3102/0002831214531813

Saderholm, J., Ronau, R. N., Rakes, C. R., Bush, S. B., & Mohr-Schroeder, M. (2017). The critical role of a well-articulated conceptual framework to guide professional development: An evaluation of a state-wide two-week program for mathematics and science teachers. *Professional Development in Education, 43*(5), 789–818. doi:10.1080/19415257.2016.1251485

Shernoff, D. J., Sinha, S., Bressler, D. M., & Ginsburg, L. (2017). Assessing teacher education and professional development needs for implementation of integrated approaches to STEM education. *International Journal of STEM Education, 4*(13), 1–16. doi:10.1186/s40594-017-0068-1

State Superintendent of Public Instruction Tom Torlakson's STEM Task Force. (2014). *Innovate: A Blueprint for science technology, engineering, and mathematics in California public education.* Dublin, CA: Californians Dedicated to Education Foundation.

U.S. Bureau of Labor Statistics. (2008). *State occupational employment and wage estimates.* Retrieved from https://www.bls.gov/oes/2011/may/oessrcst.htm

U.S. Department of Education. (2016). *STEM 2026: A vision for innovation in STEM education.* Washington, DC: U.S. Department of Education.

Zollman, A. (2012). Learning for STEM literacy: STEM literacy for learning. *School Science and Mathematics, 112*(1), 12–19. doi:10.1111/j.1949-8594.2012.00101.x

PART 2

How to Provide Rigorous STEM Education

CHAPTER 6

The Role of Interdisciplinary Project-Based Learning in Integrated STEM Education

Alpaslan Sahin

Abstract

The pivotal role of STEM education has been recognized and become a priority for countries to maintain their economic stability and improve their innovation capacity. To be able to achieve these goals, providing quality STEM education has been the ultimate goal for education systems in the two last decade. For example, the United States of America has invested hundreds of millions of dollars to increase the rigor, relevance, and content of STEM education for their K-12 students so they can continue as a world leader in all the areas they have been so far such as economy, innovation, and technology. Part of this campaign, many educational organizations in the U.S. has adopted or developed an instructional model named *Project-based learning* (PBL) that is engaging, rigorous, teacher-facilitated, student-centered, standards-based, and relevant. This chapter summarizes research findings on PBL; its definitions, benefits, challenges, and future research questions for the audiences of this book.

1   Introduction

Due to the rapid change in information dissemination with the invention of internet and technological advancement, the world has become a global village. This let science, technology, engineering, and mathematics (STEM) education become a priority for almost each and every country because it plays a pivotal role in countries' economic well-being, global leadership, innovation capacity, and creation of strong workforce. Numerous reports have noted the impact of a well-rounded K-12 STEM education on preparing the next generation's scientists, leaders and innovators (National Academy of Sciences, National Academy of Engineering, and Institute of Medicine, 2007; President's Council of Advisors on Science and Technology, 2010). However, policymakers, educators, and researchers have been concerned for years about whether students leave high school with the skills and qualities they need to be successful in college and workplace where science and technology are prevalent (Epstein & Miller, 2011). In addition, U.S. students have historically not performed well on international tests such as TIMSS and PISA (National Research Council, 2011). For example, U.S. students performed below the Organization for Economic Co-operation and Development (OECD) average in mathematics (ranking 31st out of the 35 OECD countries) and near average in science (ranking 19th out of 35) on the 2015 PISA (OECD, 2015). What is more pressing is that there is shortage of qualified graduates in STEM field needed for STEM occupations, especially data sciences, electrical engineering, material science, and cybersecurity (Xue & Larson, 2014). Contemporary and more rigorous teaching methods where students have opportunities to engage in integrated, complex, and relevant learning activities, such as project-based learning, have been shown to increase student interest in the STEM fields (e.g., Asghar, Ellington, Rice, Johnson, & Prime, 2012; Edmunds, Arshavsky, Glennie, Charles, & Rice, 2017; Larmer & Mergendoller, 2010). Indeed, proponents of project-based learning often argue that it is an ideal teaching method in preparing students for academic, personal, and career success especially in STEM fields (Buck Institute for Education, 2018; Harada, Kirio, & Yamamoto, 2008).

2   Project-Based Learning

Integrated STEM education is a great way to apply cross-curricular content (White, n.d.). One of the best ways to accomplish an effective integrated STEM education is through a project-based approach. While there is no singular definition of project-based learning (PBL), three of them that are widely accepted by researchers and educators. First, Tal, Krajcik, and Blumenfeld (2006) provided one of the most comprehensive and earliest definitions of PBL as:

> The design principles of PBS [project-based science] include a context that engages students in extended authentic investigations through a driving question, collaborative work that allows students to communicate their ideas, learning technologies to find and communicate solutions, and the creation of artifacts that demonstrate student understanding and serve as the basis for discussion, feedback, and revision. (p. 724)

Second, Holm (2011) defined PBL as "student-centered instruction that occurs over an extended time period, during which students select, plan, investigate and produce a product, presentation or performance that answers a real-world question or responds to an authentic challenge" (p. 1). Finally, the Buck Institute for Education's (2018) – a leading organization in PBL implementation – provides a more formal definition of PBL as "a teaching method in which students gain knowledge and skills by working for an extended period of time to investigate and respond to an authentic, engaging, and complex question, problem, or challenge" (p. 3). The institute also emphasizes the importance of students' voice and choice in project selection and completion in a PBL environment.

From the three definitions, there are several common themes that emerge. First, all definitions imply that the completion of PBL will not be limited to 45 or 90 minutes classroom time only. They all use the term 'extended time period' which means longer than one class period. Indeed, the Buck Institute (2018) explains this by indicating that students might work on their projects over the course of days, weeks, and even months. This might mean working after school hours and on weekends on school property or neighboring medical school or university labs throughout the project. The second theme that emerges is the student voice and choice, which is also one of the requirements of Buck Institute's Gold Standard PBL design elements. The third theme relates to how students take ownership of their projects and responsibility of their learning by addressing the importance of collaboration, communicating ideas, and presenting their questions and works. The fourth point is about the type of project or challenge or question students work on. Regardless of the naming, the project students work on should be authentic, real world-related, complex, engaging, and challenging. Last but not least, a culminating product should result from the project and be made public. This product can be in the form of an oral presentation, an artifact, a digital video, and/or a sketch of a city they are going to build. The main idea here is to make project work public and be ready to answer any questions they might receive from a variety of audiences.

In addition, in a PBL environment, since projects used in schools are authentic and involve real life problem solving and 21st century skills, students have opportunities to see the connections between what they learn in the classroom and the real-world application of such knowledge (LaForce, Noble, & Blackwell, 2017; Sahin, 2015). This will help students understand the relationships between subjects and the world. In addition, students will be able to answer questions like *"Why do I have to learn this?"* and *"Where will I use this ever?"* because the projects are real, engaging, and contextualized.

3   Previous Research on Project-Based Learning

Project-based learning prepares students for academic, personal, and career success, and equips them to tackle the challenges the world they live in will inherit (Buck Institute, 2018.). One of the earliest research studies on PBL revealed that projects increase student interest in STEM because they involve students in solving authentic real life problems, working with others, and producing artifacts (Fortus, Krajcikb, Dershimerb, Marx, & Mamlok-Naamand, 2005). Later, Berk et al. (2014) found that students who were taught with PBL had more positive attitudes toward STEM and were more likely to choose a STEM-related career. In a recent study, LaForce, Noble, and Balckwell (2017) investigated the relationship between inclusive STEM high school students' perceptions of PBL and their interest in STEM subjects and career. They found that students' higher ratings of PBL are associated with higher interest in STEM subjects and careers. In other words, PBL as an instructional tool provide critical learning opportunities for K-12 grade students to develop and cultivate STEM interest (LaForce, Noble, & Blackwell, 2017).

Moreover, studies have shown that the use of PBL as an instructional tool can improve students' 21st century skills such as creativity (Bell, 2010; Sahin, 2015), critical thinking and problem-solving skills (Albanese & Mitchell, 1993; Ertmer, Schlosser, Clase, &Adedokun, 2014; Sahin, 2015), reflective thinking (Dominguez & Jamie, 2010), communication and collaboration skills (Bell, 2010; Dominguez & Jamie, 2010; Lou, Shih, Ray Diez, & Tseng, 2010; Sahin, 2015) and ability to self-direct learning (Albanese & Mitchell, 1993; Bell, 2010; Norman & Schmidt, 2000). Because preparing today's students for 21st century workplace and college has become an important priority for educators, teachers started incorporating more 21st century skills into their teaching and assessment when teaching with PBL (Ravitz, Hixson, English, & Mergendoller, 2012). In addition, utilizing PBL may be an effective way to both engage students in STEM learning and provide them with the necessary foundational skills to pursue STEM professions (LaForce, Noble, & Blackwell, 2017). Researchers agree that PBL might be one of the most successful ways – if not the most successful one – to help students develop a set of core areas including mastery of key content areas (Fortus, Krajcikb, Dershimerb, Marx, & Mamlok-Naamand, 2005) and 21st century skills as long as PBL units are taught with rigor, relevance, and relationships (Buck Institute for Education, 2015; Edmunds, Arshavsky, Glennie, Charles, & Rice, 2017; Harada, Kirio, & Yamamoto, 2008). Therefore, it seems STEM education and PBL approach go hand in hand to accomplish what policymakers, researchers, and educators have been championing.

PBL has also been shown to increase students' engagement and cultivate positive attitudes towards STEM subjects, especially for girls and underrepresented students (Baran & Maskan, 2010; Mergendoller, Maxwell, & Bellisimo, 2006). Research has found that students' engagement in PBL is associated with increased self-efficacy and confidence in STEM subjects (Baran & Maskan, 2010; Cerezo, 2015). Masa et al. (2009) revealed that project-based learning increased students' learning of STEM subjects, intrinsic motivation, and self-efficacy. This was especially true for middle grade female students where their STEM self-efficacy increased after they were taught with PBL (Cerezo, 2015).

Unsurprisingly, research on PBL has indicated promising findings in regards to closing the achievement gap by engaging lower-achieving students (Boaler, 2002; Penuel & Means, 2000). In a qualitative dissertation study, Berry (2013) studied the contextual pedagogical influences of PBL on school engagement, creativity, and problem solving for 14 African American male students in a suburban Minnesota high school. He found that PBL teaching approach engaged several African American male students. The students did not show any disruptive behavior during PBL sessions. In addition, SRI Education researchers collaborated with researchers from University of Colorado Boulder and Michigan State University to study science curriculum materials in light of the Next Generation Science Standards (Harris, Penuel, DeBarger, & Gallagher, 2014). They conducted a randomized controlled trial involving sixth grade science classroom across 42 schools in one large urban school district. They randomly assigned each school to either a project-based middle school science curriculum or a comparison group that used a standard textbook for Grade 6. They found that the students who participated in the project-based science curriculum outperformed students in the comparison groups. They also found that project-based curriculum materials showed positive effects for all students regardless of their demographic background. Overall, these findings suggest that project-based curriculum that include opportunities for all students to engage in science practices may close the achievement gaps among students of all backgrounds.

In terms of STEM content preparation, a number of research studies have found PBL to be more effective than traditional methods of teaching mathematics and science. For example, Ravitz et al. (2012) found that PBL creates opportunities for students to learn rigorous content knowledge. More specifically, they showed that teachers who use PBL and attend regular professional development trainings do more teaching and students of those teachers also spend more time learning about important STEM contents through activities that emphasize collaboration, critical thinking, creativity, and communication. Moreover, Laforce, Noble, and Blackwell (2017) found that students from inclusive STEM high schools, where PBL is the main instructional method, indicated higher rates of interest in STEM disciplines and intrinsic motivation for science, and students' ability beliefs for both mathematics and science. In another study, the researchers randomly assigned second-grade social studies teachers in high-poverty schools that had low performance on state tests to two groups (Duke et al., 2017). One taught social studies using the PBL approach while the other taught it as they normally did (e.g., a traditional, lecture-based approach). Both teachers taught 80 social studies lessons over the course of the year. They found statistically significant gains overall favoring the PBL students over the control group in social studies.

4 Challenges in Implementing Effective Project-Based Learning

Although many schools in today's educational arena claim to have PBL incorporated in their school models, findings do not support what they claim. For example, in one study, researchers indicated that there was a gap between schools' claim of utilizing PBL as an instructional method and quality of PBL they implement (LaForce, Noble, & Blackwell, (2017)). Some researchers have already shown that project-based learning is not always accompanied by academic rigor – conceptualized as students engaging with rich, complex content using higher level of thinking and communication (Cook & Weaver, 2015; Lee & Bae, 2008). This implies that schools still need more time and effort to get to the bottom of how to benefit most from PBL teaching method.

Another study summarized challenges to implementing PBL effectively based on their review of prior research. They reported that high quality PBL requires teachers to have a deep understanding of the topic being taught and the skills to make the content understandable to their students (Kanter & Konstantopolous, 2010; Schneider, Krajcik, & Blujmenfeld, 2005). Teachers also need to be comfortable in changing their roles from the sole authority of their classrooms to a facilitator role where they share the authority with their students (Goodnough & Cashion, 2006; Han et al., 2015; Lee & Bae, 2008). This transition may cause a problem for some teachers in implementations of PBL (Cook & Weaver, 2015; Tamim & Grant, 2013). For example, in a case study of high school STEM teachers, Cook and Weaver (2015) found that student's choice, as one of the components of a high quality PBL, was not observed across teaching of all teachers who participated in PBL professional development training. They also noted that although all teachers created collaborative group works and students completed their PBL projects, some teachers could not integrate content expectations well to the projects students completed.

Another common challenge in implementing PBL is about teachers' bias because some teachers do not believe implementing PBL will help their students pass the end-of-year exam most states require (Han et al., 2015). Still another common challenge teachers encounter when teaching through PBL is the lack of proper training for teachers about what PBL is and its implementation. This yields teachers to implement it as they saw or read or heard (Tamim & Grant, 2013). Sahin and Top (2015) uncovered another challenge teachers face in PBL implementation: teachers do not have the support they need during teaching. The authors claim that this is a problem for teachers because teachers do not feel comfortable due to their lack of sufficient and regular trainings. Therefore, they need continuing professional development trainings to make a smooth transition from teacher-centered approach to teacher-guided and student-centered approach. They also need administrational support to have extra planning time so they can design their own PBL lessons. It is important to provide teachers with these support because preparing an interdisciplinary PBL lesson plan requires common planning, communication across participating teachers, and the help of a PBL expert. Because of these implementation challenges, some teachers may prefer doing "short duration and intellectually light weight activities and projects" (Larmer & Mergendoller, 2010, p. 1). Yet, they may still consider themselves doing PBL without meeting the standards of high quality and rigor of PBL (Edmunds et al., 2017).

5 Future Research Directions

In a recent brief, Sally Kingston (2018) summarized the weaknesses of PBL research stating there was a "lack of experimental studies, varying fidelity of PBL, implementation challenges, and lack of validity and reliability of measures" (p. 3). My research on the effectiveness and role of PBL in providing quality STEM education yielded similar results. Therefore, it is not difficult to say "until more rigorous research is conducted, the effects of PBL are" (Kingston, 2018, p. 3) "promising, but not proven" (Condliffe et al., 2017, p. iii).

Even though there have been number of newspaper and magazine articles published on the importance and necessity of PBL as a student-centered, active, engaging, and collaborative teaching method, there remains a critical need for continued scholarly studies showing the impact of effective PBL on the learning of mathematics, science, reading, and other subjects, in K-12 classrooms. In reviewing the prior literature and implications from prior research studies, the following questions should be studied in depth in order to continue moving the PBL field forward:

1. Do students in PBL classrooms/schools learn better mathematics, science, and reading than their counterparts in non-PBL classrooms?
2. To what extent does use of PBL close the achievement gap between female and male students in mathematics and science?

3. To what extent does use of PBL close the achievement gap in underrepresented populations (e.g., Latino/a, Black, Students with Special Needs) in mathematics and science?
4. What types of life and career skills do students develop in an environment where PBL is the main instructional method?
5. Do students taught with PBL develop better mathematics and science self-efficacy compared to those who are taught with traditional instructional methods?
6. Do students taught with PBL feel more prepared for college or careers than their counterparts taught with traditional instructional methods?
7. What are the impacts of STEM schools – adopted PBL as a school-wide teaching approach – on students' achievements, college-readiness, and development of 21st century skills?

6 Conclusion

This chapter addressed the roles of PBL instructional approach in realizing the benefits that are expected from K-12 STEM education. It was shown that PBL is a successful active engaging method that gives students freedom to investigate daily life problems of their own or curriculum associated with state standards with the guidance of their subject matter teachers. Although we presented how crucial the adoption of PBL to K-12 teaching in implementation of STEM, research suggests that providing ongoing professional development training for teachers is critical to its success. These trainings should include visiting other PBL schools who successfully use PBL in their teaching and learning so they have a chance to compare and improve their teachings in addition to having access to well-prepared and rigorous PBL curriculum. Overall, then, the change and quality STEM education for all might happen.

References

Albanese, M. A., & Mitchell, S. (1993). Problem-based learning: A review of literature on its outcomes and implementation issues. *Academic Medicine, 68*, 52–81.

Allen, D. E., Duch, B. J., & Groh, S. E. (1996). *The power of problem-based learning in teaching introductory science courses.* In L. Wilkerson & W. H. Gijselaers (Eds.), *Bringing problem-based learning to higher education: Theory and practice* (New Directions for Teaching and Learning Series, No. 68). San Francisco: Jossey-Bass.

Asghar, A., Ellington, R., Rice, E., Johnson, F., & Prime, G. M. (2012). Supporting STEM education in secondary science contexts. *Interdisciplinary Journal of Problem-Based Learning, 6*(2), 85–125.

Baran, M., & Maskan, A. (2010). The effect of project-based learning on pre-service physics teachers electrostatic achievements. *Cypriot Journal of Educational Science, 5,* 243–257.

Bell, S. (2010). Project-based learning for the 21st century: Skills for the future. *The Clearing House, 83,* 39–43.

Berk, L. J., Muret-Wagstaff, S. L., Goyal, R., Joyal, J. A., Gordon, J. A., Faux, R., & Oriol, N. E. (2014). Inspiring careers in stem and healthcare fields through medical simulation embedded in high school science education. *Advances in Physiology Education, 38,* 210–215.

Berry, T. A. (2013). *Contextual pedagogy: A praxis for engaging black male high school students toward eliminating the achievement gap* (Unpublished doctoral dissertation). Minnesota State University, Mankato, MN.

Buck Institute of Education. (2015). *Gold standard PBL: Essential project design elements.* Retrieved from https://www.pblworks.org/research/research-project-based-learning-literature-review

Buck Institute for Education. (2018). *What is PBL?* Retrieved from https://www.bie.org/about/what_pbl

Cerezo, N. (2015). Problem-based learning in the middle school: A research case study of the perceptions of at-risk females. *RMLE Online, 27,* 1–13.

Condliffe, B., Quint, J. Visher, M. G., Bangser, M. R., Drohojowska, S. Saco, L., & Nelson, E. (2017). *Project based learning: A literature review – Working paper.* Oakland, CA: MDRC. Retrieved from https://www.mdrc.org/sites/default/files/Project-Based_Learning-LitRev_Final.pdf

Dominguez, C., & Jamie, A. (2010). Database design learning: A project-based approach organized through a course management system. *Computer Education, 55,* 1312–1320.

Duke, N. K., Halvorsen, A., Strachan, S., Konstantopoulo, S., & Kim, J. (2017). *Putting PBL to the test: The impact of project-based learning or second-grade students' social studies and literacy learning and motivation.* Retrieved from https://docs.google.com/viewer?a=v&pid=sites&srcid=dW1pY2guZWR1fG5rZHVrZXxneDpkNGE5OGZiMGZiMGE0ZGI

Edmunds, J., Arshavsky, N., Glennie, E., Charles, K., & Rice, O. (2017). The relationship between project-based learning and rigor in STEM-focused high schools. *Interdisciplinary Journal of Problem-Based Learning, 11*(1), 1–23. doi:10.7771/1541-5015.1618

Epstein, D., & Miller, R. T. (2011). *Slow off the mark: Elementary school teachers and the crisis in science, technology, engineering and math education.* Washington, DC: Center for American Progress.

Ertmer, P. A., Schlosser, S., Clase, K., & Adedokun, O. (2014). The grand challenge: Helping teachers learn/teach cutting-edge science via a PBL approach. *Interdisciplinary Journal of Problem Based Learning, 8*(1), 1–20.

Fortus, D., Krajcikb, J., Dershimerb, R. C., Marx, R. W., & Mamlok-Naamand, R. (2005). Design-based science and real-world problem solving. *International Journal of Science Education, 27*(7), 855–879.

Han, S., Yalvac, B., Capraro, M. M., & Capraro, R. M. (2015). Inservice teachers' implementation and understanding of STEM project based learning. *Eurasia Journal of Mathematics, Science and Technology Education, 11*(1), 63–76.

Harada, V. H., Kirio, C., & Yamamoto, S. (2008). Project-based learning: Rigor and relevance in high schools. Library *Media Connection, 26*(6), 14–16.

Harris, C. J., Penuel, W. R., DeBarger, A., D'Angelo, C., & Gallagher, L. P. (2014). *Curriculum materials make a difference for next generation science learning: Results from year 1 of a randomized controlled trial*. Menlo Park, CA: SRI International.

Geier, R., Blumenfeld, P. C., Marx, R. W., Krajcik, J. S., Fishman, B., Soloway, E., & Clay-Chambers, J. (2008). Standardized test outcomes for students engaged in inquiry-based science curricula in the context of urban reform. *Journal of Research in Science Teaching, 45*(8), 922–939.

LaForce, M. Noble, E., & Blackwell, C. (2017). Problem-Based Learning (PBL) and student interest in STEM careers: The roles of motivation and ability beliefs. *Education Sciences, 7*(92), 1–22.

Larmer, J., & Mergendoller, J. R. (2010). *The main course, not dessert: How are students reaching 21st century goals? With 21st century learning*. Novato, CA: Buck Institute of Education. Retrieved from https://www.bie.org/?ACT=160&file_id=158&filename=-FreeBIEs_Main_Course.pdf

Lou, S. J., Shih, R. C., Ray Diez, C., & Tseng, K. H. (2010).The impact of problem-based learning strategies on stem knowledge integration and attitudes: An exploratory study among female Taiwanese senior high school students. *International Journal of Technology Design Education, 21*, 195–215.

Kanter, D. E., & Konstantopolous, S. (2010). The impact of a project-based science curriculum on minority student achievement, attitudes, and careers: The Effects of teacher content and pedagogical content knowledge and inquiry-based practices. *Science Education, 94*(5), 855–887.

Kingston, S. (2018). Project based learning & student achievement: What does the research tell us? *PBL Evidence Matters, 1*(1), 1–11.

Massa, N., Dischino, M., Donnelly, J., & Hanes, F. (2009). *Problem-based learning in photonics technology education: Assessing student learning*. Proceedings of the 11th Education and Training in Optics and Photonics, Wales, UK.

Mergendoller, J. R. (2014). *The importance of project based teaching*. Retrieved from http://www.bie.org/blog/the_importance_of_project_based_teaching

Mergendoller, J. R., Maxwell, N. L., & Bellisimo, Y. (2006). The effectiveness of problem-based instruction: A comparative study of instructional methods and student characteristics. *Interdisciplinary Journal of Problem Based Learning, 1*(2), 49–69.

National Academy of Sciences, National Academy of Engineering, and Institute of Medicine. (2007). *Rising above the gathering storm: Energizing and employing America for a brighter economic future*. Washington, DC: The National Academies Press.

National Research Council. (2011). *Successful K-12 STEM education: Identifying effective approaches in science, technology, engineering, and mathematics*. Washington, DC: NAP.

Norman, G. R., & Schmidt, H. G. (2000). Effectiveness of problem-based learning curricula: Theory, practice and paper darts. *Medical Education, 34,* 721–728.

Organization for Economic Co-Operation and Development (OECD). (2015). *Country note: Key findings from PISA 2015 for the United States*. Paris: OECD Publishing. Retrieved from https://www.oecd.org/pisa/PISA-2015-United-States.pdf

President's Council of Advisors on Science and Technology. (2010). *Prepare and inspire: K-12 education in Science, Technology, Engineering, and Mathematics (STEM) for America's future*. Washington, DC. Retrieved from http://www.whitehouse.gov/sites/default/files/microsites/ostp/pcast-stem- ed-final.pdf

Ravitz, J. Hixson, N., English, M., & Mergendoller, J. (2012). *Using project based learning to teach 21st century skills: Findings from a statewide initiative*. American Educational Research Association Conference, Vancouver, BC, Canada. Retrieved from https://www.pblworks.org/blog/top-pbl-news-stories

Sahin, A. (2015). How does the STEM S.O.S. model help students acquire and develop 21st century skills? In A. Sahin (Ed.), *A practice-based model of STEM teaching: STEM Students on the Stage (SOS)* (pp. 171–186). Rotterdam, The Netherlands: Sense Publishers.

Sahin, A., & Top, N. (2015). STEM Students on the Stage (SOS): Promoting student voice and choice in STEM education through an interdisciplinary, standards-focused project-based learning approach. *Journal of STEM Education: Innovation and Research, 16*(3), 24–33.

White, T. (n.d.). *How project-based learning enhances STEM education*. Retrieved from https://www.studica.com/project-based-learning-enhances-stem-education-tom-white

CHAPTER 7

Inclusive STEM Schools: Origins, Exemplars, and Promise

Sharon J. Lynch

Abstract

This chapter discusses the origins of STEM high schools in the U.S. and how they were stimulated by grassroots level economic needs, funding from private foundations, and state-level governmental initiatives. Policy makers offered policy supports for STEM schools that emphasized the twin goals of preparing more students for STEM careers, and better access to high quality STEM education for all students, especially those have been underrepresented in STEM fields. Inclusive STEM high schools (ISHSs), have sprung up across the U.S., although no umbrella organization nor an agreed upon model had been established for these educational experiments. Although some researchers documented instances where inclusive STEM schools failed to live up to their promises, others went on to show how successful ISHSs were organized. Carefully done quantitative research demonstrated their promise, especially as an alternative to traditional comprehensive high schools. Research showed that successful ISHSs had in place a strong STEM focused curricula, highly qualified, collaborative teaching staffs, supports for students underrepresented in STEM, and administrative structures open to innovation and flexibility. ISHSs also created positive school cultures, becoming STEM learning communities where diverse learners felt supported and valued as they developed the academic background and interpersonal skills to be successful in STEM in college and in careers.

1   Introduction

Change was in the air.

 In the first decades of the 21st century, the U.S. federal government and individual states were searching for ways to improve educational systems and make them better attuned to current economic and workforce demands. The business and technology communities and policymakers saw STEM education as having the potential to reinforce the economy and to shrink stubborn socio-economic, social mobility, and opportunity gaps. Some educators saw STEM education as a way to breathe new life into an education system locked into 20th century practices and that left behind too many students (Locke, 2011; NRC, 2010, 2011; PCAST, 2010, 2012; Robelen, 2013). The term "STEM education", an acronym for "Science, Technology, Engineering and Mathematics", had been around for some time. But in the real world, these disciplines were increasingly integrated in research, development, and production. Their artificial separation in schools seemed both archaic and unrealistic. Moreover, traditional science and mathematics classes were often perceived as boring, and failed to inspire students to STEM college majors and careers (Obama, 2010).

Education leaders, policy makers, teachers, and researchers began to see the promise of STEM education as something more than shorthand for the four disciplines. Some began to use it as a vehicle for education reform, adapting the engineering design process to create new schools, or to spur new approaches to curriculum and instruction such as project-based learning.

Progressive definitions of STEM education emerged and gained acceptance. For instance, STEM education was defined as "an interdisciplinary approach to learning where rigorous academic concepts are coupled with real-world lessons as students apply science, technology, engineering, and mathematics in contexts that make connections between school, community, work, and the global enterprise enabling the development of STEM literacy and with it the ability to compete in the new economy" (Tsupros, Kohler, & Hallinen, 2009). State governments, colleges and universities, and business and industry leaders introduced policies, laws, and incentives to improve traditional science and mathematics education in their respective states, as well as to create a more innovative and responsive set of education opportunities that better matched the definition of STEM education offered above (Lynch, Peters-Burton, & Ford, 2014).

The National Academy of Sciences points out that the U.S. science and engineering workforce stands at 5 million and will be the fastest growing sector of the economy in the years ahead. Students from underrepresented groups comprise 39% of the K-12 public school enrollment but just 18% of bachelor's degrees and 5% of doctorates in science and engineering fields (National Academy of Sciences, National Academy of Engineering, & Institute of Medicine, 2011). The United States has been overtaken by other countries in the development of STEM expertise (ranking 29th out of 109 countries in the percentage of 24-year-olds with a mathematics or science degree), and the fastest growing ethnic groups in our population are underrepresented in STEM degree programs (National Academy of Sciences et al., 2011). Until recently, American industry made up for the shortfall in STEM degree holders schooled in the U.S. by hiring scientists and engineers from abroad. But this strategy has become less tenable as other economies generate more opportunities for STEM professionals within their own borders (Atkinson, Hugo, Lundgren, Shapiro, & Thomas, 2007). Moreover, the strategy does nothing to lessen the inequality of educational and economic opportunity within the United States (Carnegie Corporation, 2009).

The problem of underrepresentation in STEM fields has a long history not only as an equity issue, but also as an economic issue. A PCAST report (2010) claimed that the success of the U.S. in the 21st century, its wealth and welfare, depends on the ideas and skills of its population. To meet immense challenges in energy, health, the environment and national security, a greater portion of populace needs to be better prepared in STEM, and generally more STEM literate. In addition, proficiency in STEM increases an individual's capacity to earn more money and have more flexibility in jobs and careers (Rothwell, 2013; U.S. Department of Commerce and Economics and Statistics Administration, 2011).

2 The Evolution of Specialized STEM Schools for Underserved Students

In the last hundred years, increasing interest in preparing students for careers as scientists, mathematicians and engineers led to experiments with new kinds of high schools with specialized curricula and instruction. Some students and families sought more STEM opportunities than neighborhood comprehensive high schools could offer. One solution was to create new high schools focused primarily on STEM education. These specialized high schools arose from a variety of perceived needs by families, communities, and state governments as a result of both economic and social justice goals. What they have in common was their focus on "STEM education."

2.1 *Highly Selective STEM High Schools*
Historically, the first specialized STEM high schools were designed for highly motivated and high achieving students who were seen as needing more advanced science and mathematics classes than comprehensive high schools could offer. These schools have an extensive record of successful outcomes, and offer specialized coursework emphasizing one or more STEM disciplines to students chosen through competitive examination or record of past academic achievement (Finn & Hockett, 2012; Hanford, 1997; Subotnik, Rayhack, & Edmiston, 2006). Highly selective STEM high schools are scattered throughout

the U.S., but are primarily located in cities where they can be accessible to large numbers of students. They have a strong track record for preparing students for STEM careers, and often provide research opportunities that help students select academic and career paths. Moreover, they have proven to be sustainable over time.

Despite their successes, highly selective STEM high schools alone cannot solve 21st century issues in STEM education because they only serve a narrow band of the student population. They have not been able to recruit and prepare representative numbers of students from underserved groups in STEM, including women, or African American or Hispanic students, or low income students (Badia & Chapman, 2014; Means, Confrey, House, & Bhanot, 2008). Selective STEM schools simply have not sufficiently expanded the pipeline of students motivated and prepared for STEM majors in college. Consequently, other approaches were needed to address the issues of economic competitiveness, educational quality, and equity (Means, Young, & Wang, 2014).

2.2 *Magnet Schools*

After calls for ending the practice of unequal, racially segregated schools in the middle of the 20th century, there were many forces at work that sought to solve problems of unequal opportunity to learn and to achieve. In the 1970s, a response to court-ordered racial desegregation of schools resulted in the creation of new "magnet schools." These schools were often designed around unifying themes and new organizational structures. They brought African American, Hispanic, and white children together in the same building. This was achieved through voluntary bussing and schools that attracted students from areas outside of their neighborhoods, resulting in racially and socioeconomically integrated magnet schools. Many of these new magnet schools had science and mathematics themes, and that emphasis continues to present, as they especially embrace STEM education. These schools offer expanded and enhanced STEM educational programs and facilities to attract students and families.

Early versions of magnet schools were often criticized because they simply brought diverse students into a school building, only to re-segregate them through tracking systems that placed students in racially segregated classes. However, second generation magnet schools have changed; they provide advanced educational programs that integrate students at the classroom level and provide opportunities for students to learn and work together regardless of ethnicity/race. The Magnet Schools Assistance Program (U.S. Department of Education, n.d.) is a long-standing, federally funded program that has strongly encourages its member schools to focus on STEM education to create integrated schools. For instance, the New London School High School for Science and Technology in New London, Connecticut is one such example. Students are bussed in from throughout the region to take advanced STEM courses. They also attend the adjacent comprehensive high school for the rest of their classes (Lucadamo, 2016). In addition to such federally funded magnet high schools, states and school districts have also created their own versions of magnet schools. These schools have similar goals. They are specialized STEM high schools that are not highly selective, but provide a high quality STEM programs to their widely diverse student populations.

2.3   Charter Schools and School Choice Systems

By the end of the 20th century, there was increasing dissatisfaction with the pace of school reform in the U.S. by both parents and policy makers. Traditional public comprehensive high schools were failing to reach too many students, especially in low-income urban areas. Some states and school districts embraced the idea of experiments with new kinds of schools. The charter school movement was seen as a means to propel innovation in education and to offer school alternatives to families and to students. Many large cities began to develop systems where students might choose to attend either traditional public schools or public charter schools.

As the number of these charter schools increased in some cities (e.g., Milwaukee, Washington, DC, and New Orleans), entire public school systems responded by developing a means to offer more choice in schools, especially at the high school level when students could handle the associated transportation challenges. Large cities established choice systems that allow all students to make selections of high schools among existing public charter schools, traditional comprehensive high schools, and themed high schools with defined and specialized offerings. The idea of a theme-based school has always existed in the U.S. to some extent, as well as in other countries that did not embrace the comprehensive high school model. At present, themed high schools include public charter schools, public high schools, and school-within-a-school academies. As had happened with the magnet school movement, STEM-themed schools are popular and are designed to draw students and families.

There are probably hundreds of STEM-themed schools in existence in the U.S., some long established and others brand new. The U.S. federal government does not keep track of the number of STEM-themed schools, and has no system in place to recognize them, certify them, or evaluate them. Nationally, their precise number is impossible to know, although Means et al. (2008) used a large survey and located about 300 such STEM schools.

2.4 Turnaround Schools

Turnaround schools are another category of U.S. schools that have been nudged to experiment with STEM themes as a school reform strategy. Turnaround schools ("Turn Around Schools", 2018) are struggling schools that consistently failed to make Annual Yearly Progress under federally required benchmarks, or have a similar designation under state-level guidelines. Such schools are either in danger of being taken over by their state, or they have already met such a fate. Some have been reconstituted as STEM-themed schools. These schools are often located in neighborhoods where there are high proportions of students from families that are poor; where school resources are in short supply; and, where teacher turnover and student mobility are high. Providing turnaround schools with a new STEM focus has been seen as a means of improving student achievement and providing more a more engaging school program (Carnegie Corporation, 2009).

The challenge of creating successful STEM-themed turnaround schools seems far greater at the high school level than in the early grades due to the demands for qualified STEM secondary teachers and adequate STEM facilities. Unlike selective, magnet, and charter schools, turnaround high schools are neighborhood schools; they remain "local" and are usually not schools of choice. It unclear that struggling, turnaround schools can provide the heavy lift required to become an innovative STEM high school unless they are well funded, designed, and administered (discussed later in this chapter). (Eisenhart et al., 2015; Weis et al., 2015) documented the problems of struggling new STEM high schools in Buffalo, New York and Denver, Colorado; these new STEM-themed schools simply could not deliver on their promises to students as specialized STEM learning environments. These study schools failed to offer as strong a STEM program of studies as the nearby comprehensive high schools. The lesson that can be taken from turnaround high schools is that to merely call a school a "STEM high school" without investing in careful planning, design, and evaluation, accompanied by start-up funding and school district-level ongoing support, is likely not a solution for struggling high schools nor STEM equity issues.

2.5 STEM Career and Technical Education (CTE)

Specialized vocational schools for career and technical education (CTE) high schools have been in existence for decades in the U.S. and other countries. In the U.S., they are often thought of alternative schools for students who are more interested in trades and vocations than academics and careers. They have a federal funding stream to supplement state and local funds. CTE high schools are schools of choice. They may have some admissions criteria, although the criteria may be more behavioral than academic. They are not highly selective. Technology- and engineering-focused CTE high schools have emerged over the last two decades. These schools can be seen as STEM schools; some are college preparatory, and some provide both college preparation and vocational training in technology, engineering, or STEM-related fields such as agriculture or health and medical sciences. These new STEM high schools have tapped into CTE federal funding, and use these funds to support innovative programs in STEM (NRC, 2011).

2.6 Summary

Most U.S. public high schools are traditional comprehensive high schools that draw students from local neighborhoods. They are designed to serve all students, no matter their needs, interests, and goals. In such schools, students must cobble together their STEM programs based on an array of choices. While comprehensive high schools may offer breadth in STEM classes the classes may lack depth, or classes are specialized to serve well a small proportion of the student body that are well-prepared to take STEM classes. In other words, high quality "honors" or advanced placement STEM classes are provided only to some students. They often do not reach out or serve underrepresented students well (Lynch, 2000; Oakes, 2011). Consequently, the U.S. has seen the development of a variety of specialized STEM-focused schools to solve problems related to economic and equity concerns.

Dissatisfaction with the status quo in the public school system has resulted in increased experimentation with different types of schools, including the creation of STEM-focused high schools. At the same time, there was increasing interest in school choice, i.e., allowing students and families to choose from a range of publically funded schools. This was first applied to the charter school movement, but used increasingly in public school systems more generally, and has spurred the development of alternative school models. These movements for improved STEM education at the school level for both economic and social justice reasons, coupled with an enthusiasm for new types of school models in the 21st century, were fertile ground for the evolution of inclusive STEM high schools.

3 Inclusive STEM High Schools (ISHSs)

Though their roots can be found in decades of policy and practice described in the section above, inclusive STEM-focused high schools (ISHSs) are a relatively new feature in the U.S. education landscape. They have implications for school reform, STEM curriculum and instruction, and improving both opportunities to learn and learning outcomes for underrepresented groups of students. Recognizing their potential, in 2010 President Obama issued a challenge to the U.S. educational system to create more than 1000 new STEM-focused schools, including 200 high schools (Obama, 2010; PCAST, 2012). Federal legislation made it easier to use funds for STEM schools, although at the discretion and direction the states.

Recent research shows that some ISHSS are making strides to build opportunity structures for students who are underrepresented in STEM (Lynch, Peters-Burton, et al., 2017; Means, Wang, Young, Peters, & Lynch, 2016; Means et al., 2017; Scott, 2009). ISHSS intentionally reach out to youth from groups underrepresented in STEM who may be underserved by comprehensive high schools due to institutionalized practice of tracking, i.e., the sorting of students through ability grouping. In other cases, students may attend poor schools with few resources, lacking qualified STEM teachers, and with few connections to STEM in the community. In traditional high schools and often in middle schools, students from low-income families, or African American and Hispanic students, or students new to U.S. cannot access to the kind of coursework required for STEM college preparation (Lynch, 2000; Oakes, 2005).

There is no common umbrella organization to define, certify, govern, or organize ISHSS across the U.S., and in most states. Moreover, there are varied definitions of STEM and STEM education, as well as for what might be called a "STEM high school." Many ISHSS were created in the first decade of the 21st century, some haphazardly in response to local demands or sudden infusions of funding designated for STEM. Others were more thoughtfully designed with careful attention to mission and organization within a larger educational school district or state system. Some have developed vibrant, innovative education programs focused on STEM (Lynch, Peters-Burton, et al., 2017).

A substantial body of new research encompassing ISHSS (Lynch, Peters-Burton, et al., 2017; Means, Wang, Young, Peters, & Lynch, 2016; Means et al., 2017) has adopted a consistent definition of Inclusive STEM high schools (ISHSS): ISHSS can be a school, or a school within a school, that accepts students primarily on the basis of their *interest in* STEM *rather than aptitude or prior achievement*. Admission is often by lottery. Equally important, ISHSS provide all students who attend them with the college preparatory mathematics and science course of study that they will likely need to succeed in a STEM college major. They also require specialized STEM coursework in engineering, technology or STEM-related career technical education (CTE), depending on an ISHS's theme or specific goals. ISHSS can be grassroots organizations developed in response to community and economic needs and opportunities for future work. Or, they can be stimulated at the state level to provide a network of ISHSs (Young, Lynch, House, & Peters, 2017). ISHSs provide students with a STEM program of greater depth and breadth than their states require for high school graduation (Carnegie Corporation, 2009; Lynch, Peters-Burton, et al., 2017; Means, Confrey, House, & Bhanot, 2008; Means et al., 2017). In addition, many ISHSs seek to integrate STEM subjects in cross-disciplinary coursework (Lynch, Spillane, et al., 2017) or offer ambitious experiences that integrate the humanities or STEM subjects in long-term, carefully planned projects (Spillane, Lynch, & Ford, 2016).

ISHSs are intentionally designed to provide students with access to a 21st century opportunity culture that promises more social mobility and a better chance to achieve personal goals. They seek to challenge common school organizational practices that too readily constrain and exclude students and families that lack capital, both real and social (Lynch & Ross, 2014). ISHSs may offer at least a partial solution to problems of opportunity to learn STEM because they are schools of choice in which all students take a college preparatory STEM sequence. If students are not "prepared" for high school STEM classes such as algebra 1 or biology, ISHSs find ways to bring students up to speed, rather than relegating them to tracked low-level classes. ISHSs' missions focus on developing the STEM expertise and talent of underrepresented students, and to open STEM fields to a wide range of students from demographic groups underrepresented in STEM (including race/ethnicity, socioeconomic status, or gender). In addition, some ISHSs aim at recruiting students who are the first generation in their families to attend college, or who are new to the U.S. (Lynch & Ross, 2014; Lynch, Spillane, et al., 2017). ISHSs are also interesting educational experiments that may show the way toward more inspiring and effective education for STEM learners in the 21st century (Bryk & Gomez, 2008; Carnegie Corporation, 2009).

In summary, in the first decade of the 21st century, philanthropists, the business community, and policy makers began to discuss the possibility of new types of schools with education leaders; small schools, early college high schools, and inclusive STEM high schools were planned and implemented. An important study, commissioned by the Gates Foundation, identified ISHSs that were springing up and their prevalence across the U.S. (Means et al., 2008). The National Research Council followed with a series of publications on STEM education and STEM schools (NRC, 2011), and the President's Office of Science and Technology Policy promoted ISHSs (2010, 2012). However, counting or advocating for ISHSs was not enough. The policy and research communities needed to know more about how they were organized, if they were effective, or whether each was a unique creation or if they had characteristics in common.

4 The OSPrI Study

Spurred by formal inquiries from the U.S. Congress, the National Science Foundation began fund new studies on ISHSs. One of these studies, Opportunity Structures for Preparation and Inspiration (OSPrI), was designed to open up the black box of ISHSs (Lynch, Peters-Burton, et al., 2017), i.e., describe ISHSs and explain how they work. Reasoning that the best way to do this was to study ISHSs with strong track records of success, OSPrI researchers identified eight "exemplar" ISHSs located across the U.S. They used an expert nomination process, followed by vetting each school's public records of academic success, especially for groups of students underserved in STEM. The final eight ISHSs selected for study were not affiliated with one another, and were located in seven states. OSPrI researchers set out to capture the organizational, curricular, and instructional features of each ISHS, and how these features worked together to produce strong student outcomes (high graduation and college admission rates, high test scores, awards, and recognition in their communities).

5 OSPrI's Theoretical Framework, Research Questions, and Methods

OSPrI researchers hypothesized that successful ISHSs would do more than focus on STEM or employ new technologies. Rather, they would need to view high school as a means to create ways of learning STEM and developing new opportunity structures related to STEM. Kenneth Roberts (1968, 1984) used the term *opportunity structure* to describe the social and economic structures that influence who careers are selected. Roberts noted "momentum and direction of school leavers' careers are derived from the way in which their job opportunities become cumulatively structured and young people are placed in varying degrees of social proximity, with different ease of access to different types of employment" (p. 179). In other words, according to Roberts, psychological choice did not govern student success so much as the actual physical and social affordances that are found in some geographic locations or communities, but not others. Determinants of occupational paths include the home; the environment; the school; peer groups; and job opportunities. Roberts (1984) later expanded his opportunity structure model to include factors such as distance to work (or school); job qualifications; informal contacts in business; ethnicity; gender; and, cyclical and structural factors operating within the economy that result in a demand for labor with high skill levels.

Roberts' notion of opportunity structures was a useful way to think about what it would take for students underrepresented in STEM – often less affluent, students of color who may live in neighborhoods with less direct access to STEM professionals – to move into rewarding STEM fields with the goal of a STEM college preparation for all. ISHSs, either deliberately or intuitively, must create opportunity structures designed to guide and support interested and motivated students for STEM jobs, college majors and careers.

The OSPrI study asked:

1. Is there a core set of likely critical components (listed in Table 7.1) shared by 8 well-established, promising ISHSs? Do other components emerge from the study?
2. How are the critical components implemented in each ISHS?
3. What are the most prominent critical components in evidence in the 8 ISHSs, and can they be used to create a theory of action that explains the success of ISHSs for students underrepresented in STEM?

The OSPrI study sought to find out whether the eight successful ISHSs had common features in design, implementation and organizational supports outside of the school in the district or community. The literature on both selective and ISHSs suggested an initial set of ten critical components (Peters-Burton, Lynch, Behrend, & Means, 2014) that may work together to form new opportunity structures for students, listed in Table 7.1. It was anticipated that there would be additional components not on this list or emerging themes.

The study's multiple instrumental case study design (Stake, 2006; Yin, 2003) allowed systematic comparisons across eight ISHSs. Table 7.1 shows the first ten critical components found through a review of the literature. The approach provided a means for cross case analyses, but also was open-ended to allow new empirical evidence and interpretation. Over a period of three years, the OSPrI research team created a 75–100 page long case study for each of the eight schools; the case studies have a similar organizational structure and were built around the discussion of the critical components as designed and implemented in each school. The full case studies can be found on the OSPrI website (https://ospri.research.gwu.edu/) and shorter versions or the cases and summaries have also been published (Lynch, Spillane, et al., 2017; Peters-Burton et al., 2014; Spillane, Lynch, & Ford, 2016).

As each case study neared completion, the OSPrI study team conducted an ongoing, omnibus cross-case analysis for each critical component in Table 7.1, and systematically identified and discussed emergent themes. The final list of critical components included 14 found in Table 7.1 (Lynch, Peter-Burton, et al., 2017). All 14 critical components were in evidence in all eight schools, but in different combinations. However, across all schools, four critical components were consistently prominent and can be viewed as foundational; it is hard to imagine a high-functioning STEM high school without them.

TABLE 7.1 Critical components for Inclusive Stem High Schools ISHSs in Opportunity Structures for Preparation and Inspiration in STEM (OSPrI Study) developed deductively and inductively

Critical components developed through literature review	Definition of critical components
CC1. College-Prep, STEM Focused Curriculum for All	Rigorous courses in all four STEM disciplines, or, engineering and technology are explicitly, intentionally integrated into STEM subjects and non-STEM subjects in preparation for college.
CC2. Reform Instructional Strategies and Project-Based Learning	STEM classes emphasize instructional practices informed by research for active teaching and learning, immersing students in STEM content, processes, habits of mind and skills. Opportunities for project-based learning are encouraged and measured by performance-based assessment practices that have an authentic fit with STEM disciplines.
CC3. Integrated, Innovative Technology Use	The school's use of technology connects students with information systems, models, databases and research; teachers; mentors; and, STEM-related social networking resources.
CC4. STEM-rich, Informal Experiences	Learning spills into areas regarded as "informal STEM education" and includes apprenticeships, mentoring, social networks, and engaging in STEM activities outside of school. As a result, the relationships between students, teachers, and knowledge are altered and hierarchies flatten.
CC5. Connections with Business, Industry, and the World of Work	The school boundaries extend beyond the school by creating partnerships with business and industry. The school environment intentionally reflects the workplace; students have the opportunity to think like professionals.
CC6. College Level Coursework	The school schedule is flexible, providing opportunities for students to take classes at institutions of higher education or online.
CC7. Well-Prepared STEM Teachers and Professionalized Teaching Staff	Teachers are highly qualified and have advanced STEM content knowledge and/or practical experience in STEM careers. There are in-house opportunities for professional development, collaboration, and interactions with STEM professionals in the field.

(cont.)

TABLE 7.1 Critical components for Inclusive Stem High Schools ISHSs in Opportunity Structures for Preparation and Inspiration in STEM (OSPrI Study) developed deductively and inductively (*cont.*)

Critical components developed through literature review	Definition of critical components
CC8. Inclusive STEM Mission	The school's stated goals are to prepare students for STEM, with emphasis on recruiting students from underrepresented groups.
CC9. Flexible and Autonomous Administration	The school has autonomy from the school district to address the goals of its innovative STEM program The school may have partnerships with charter networks and non-governmental organizations that provide leverage, expertise, leadership, and resources for the school.
CC10. Supports for Underrepresented Students	The school provides supports (tutoring, advisories, and special classes during and outside of school hours) for students to strengthen their STEM content and skills and to prepare them for STEM college majors.
Critical components developed as emergent themes	
CC11. Data Driven Decision Making for Continuous Improvement	The school community supports continuous improvement through data systems that inform future learning, teaching strategies, student supports, professional development, and resource allocation.
CC12. Innovative and Responsive Leadership	The school leadership is proactive and continuously addresses the needs of teachers, students, and the greater community through innovative solutions, open communication, and uplifting leadership.
CC13. Positive School Community and Culture of High Expectations for All	ISHSs have culture of high expectations for students and staff, and a school environment where students and staff feel a sense of personal, intellectual, and socio-emotional safety.
CC14. Agency and Choice	Students choose to attend a STEM-focused high school and understand the challenges that will be involved and develop a sense of purpose coherent with the school mission, committed to a different approach to high school due to its STEM focus.

6 Findings: Four Consistently Prominent Critical Components

6.1 STEM-Focused Curriculum

The OSPrI case studies showed that all eight ISHSs had focused on developing strong STEM curricula with a "traditional" college preparatory core sequence in science and mathematics and with additional classes in technology, engineering or CTE. The ISHSs had more rigorous STEM requirements for graduation than those designated by their respective states; students took at least one more science and one more mathematics class to graduate. The ISHSs also required engineering courses for graduation. The exceptions were the two CTE high schools where students took extensive STEM-related CTE courses rather than engineering. Within the ISHSs, tracking of students into high and low level courses was avoided. Rather, the ISHSs used a curricular strategy that had all students progressing through a rigorous course sequence. Some students moved through the required STEM core more quickly than others. Students were expected to master course material before moving on to the next course; this required some students to re-take a course if they did not achieve mastery in their first attempt. However, there was little stigma placed on repeating a class at ISHSs; rather, the emphasis was on mastery of course material. This signaled to students and families that learning STEM was important and that students should understand relationships between STEM ideas and disciplines, as they progressed toward college readiness for STEM. All students in the ISHSs were prepared for STEM for some form of post-secondary education or for a career path. Students had the choice after graduation to pursue STEM majors or choose another direction, but reported they valued attending their STEM-focused high schools, and the conceptual and soft skills that they had learned (Lynch, Peters-Burton, et al., 2017; Lynch, Ford, & Matray, 2017).

Textbooks were not widely used in the ISHSs. Rather the teachers taught students to be consumers of reliable digital resources. Although this may have created some challenges when transitioning to traditional lecture and recitation in colleges where success is dependent on note-taking skills from lecture and textbooks, alumni of ISHSs reported that the lessons they learned in high school allowed them to adapt to traditional settings in college. Their high level of student responsibility for learning at ISHSs, taught them to seek help and learn independently (House & Peters-Burton, 2014).

The ISHSs had non-traditional school schedules and consequently were able create new opportunities for learning (House & Peters-Burton, 2014). This allowed access to college coursework, integration of subjects across content, systematic faculty-wide professional development, common planning for team teaching, internships and project based learning experiences. For example, one school did not have classes on Monday mornings, providing time for teachers to plan together and evaluate each other's work in a "critical friends" workshop (Lynch, Spillane, et al., 2017). Students pursued internships or worked on their projects during this time. Another school moved to a semester schedule with double-blocked classes so students could take courses at the local community college.

The STEM curricula were developed by ISHS teachers and often created in accordance with state standards. Schools had a range of cross-disciplinary courses and experiences for students. Sometimes, the curricula and integration of topics could change from year to year, given the inclinations of the teachers and facilitated by collaborative time for planning. Often, STEM and humanities classes were integrated, as well as STEM disciplines. For example, at one school, biology and engineering students produced prototypes of museum displays for biological processes (e.g., distribution of mold spores), one of which was selected by museum professionals for permanent display (House & Peters-Burton, 2014).

A notable exception to the integration of subject matter was in mathematics for all eight ISHSs in the study (House & Peters-Burton, 2014). Mathematics appeared to be the subject in which students showed the widest range of pre-high school preparation. In other words, some students came to an ISHS ready for geometry, while others needed to catch up on pre-algebra mathematics skills before taking algebra I. School administrators and teachers worked together to support students who needed help with mathematics, often by scheduling two mathematics classes at a time to help students catch up to their peers or offering intense summer sessions. Schools all reported that mathematics was the driver of the curriculum; the mathematics level of the students determined all other levels of STEM coursework. By the time students graduated, the highest level of mathematics for graduation was usually pre-calculus or calculus.

6.2 *Well-Prepared STEM Teaching Staff*

ISHS school leaders had the autonomy to hire teachers whom they described as a good fit for the job (Spillane, 2014) seeking teachers who were philosophically aligned and willing to buy in to the ISHS's mission, vision, and beliefs. They sought teachers who had strong content knowledge in subjects they would teach; who were flexible and open-minded; and, who were willing and able to collaborate in an active STEM learning community. Leaders at one school described "a wall-to-wall program" where all of the adults were constantly learning. At another school, the leaders sought teachers who were willing to challenge prior ideas of standard teaching practices. An ISHS that relied on problem based learning as the primary means of instruction looked for teachers with real-world experience in their STEM fields. Another ISHS that focused on getting students into college wanted teachers who themselves were high achievers and high performers in college, and who displayed the grit to push through challenging circumstances. That school often hired Teach For America veterans who already had succeeded in their in their first assignments as well as teachers with a range of preparatory experiences. A rural ISHS cultivated a caring, supportive environment and hired teachers who were willing and able to differentiate instruction while maintaining academically high standards.

According to teachers' responses on OSPrI surveys, virtually all STEM teachers had earned undergraduate degrees in their STEM content areas and many held masters degrees. Engineering and technology teachers were somewhat more likely to have had previous business or industry experience than those who taught only a science or mathematics subject. Teachers' prior experience in the classroom ranged from 6 to 9 years, although every school had novice teachers and highly experienced teachers (Spillane, 2014).

Professional development was an integral component of the teacher experience across the ISHSs (Spillane, 2014). Professional development was embedded during the school day, often on a weekly basis. Teachers reported that it was thoughtfully constructed (often by the teaching staff), well used, and appreciated. It could range from formal sustained summer STEM education programs to regular weekly meetings during the school year to ad hoc gatherings to address an immediate concern. It could serve purposes as varied as new teacher induction, experiences with new methods for teaching or assessment, group tuning processes to refine lessons, designing integrated curriculum to meet standards in different subject areas, or using student data to find the root of a problem. Professional development experiences were planned and facilitated by school administrators, by experts from within the school network, and the teachers themselves. Importantly, professional development was aligned with the current needs of teachers and the goals of the school.

Teacher professionalism was a trait at all ISHSs, manifest in the form of teacher autonomy in the classroom and sense of empowerment within the school (Spillane, 2014). Teachers collaboratively developed curriculum and assessments, and took on leadership roles in coordinating across STEM and often humanities courses. Teachers were also involved in a number of formal STEM programs to advance their learning; some worked their way through programs to become its leaders. Teachers were committed to carrying out the ISHSs' missions. They participated in administrative or decision-making responsibilities within their roles as teachers, and moved in and out of positions of greater responsibility, as called for by the occasion. Teachers had opportunities for continued professional learning and pathways for advancement.

What appeared to be different about the STEM teachers at ISHSs (compared to comprehensive high schools) was not so much their STEM teacher qualifications – many high schools have highly qualified STEM teachers who work hard – but more about the high level of dedication, collaboration, and commitment to innovative ISHS programs. They worked long hours, creating strategies for innovative learning opportunities for students. Teachers used their interpersonal skills to connect deeply with students and families and members of the STEM community to provide personalized learning experiences in small, mission-driven, STEM-focused schools.

6.3 *Administrative Structure*

Given the number of school-level educational experiments over the last decade, it seems important to understand how "exemplar" ISHSs were administratively structured, as this could be crucial in understanding scale-up and sustainability. The OSPrI study found that a sound, well-organized, mission-driven administrative structure was crucial to launching a STEM school, and to keeping it functioning in an innovative and flexible fashion (Ford, Behrend, & Peters-Burton, 2014; Ford, 2017; Lynch, Peters-Burton, et al., 2017). Administrative structure was defined to include: the strength and organization of school leadership; its relationships with district, community, and funding organizations; the architecture and use of school facilities; the process of using data to make decisions about instruction; planning for school leadership and administration in times of transition (transfer of leadership); funding considerations; and, school size (Ford et al., 2014).

The organizational structures of the eight ISHSs in the OSPrI study varied according to whether the school was a part of the public school district system with a somewhat traditional structure, or a magnet school, or a member of a charter school network (Ford et al., 2014). Overall, these ISHSs tended to have relatively small administrative staffs led by the principal as the instructional leader, but with strong leadership contributions from teachers, especially STEM teachers who were the subject matter specialists who could guide the direction of an ISHS. Teachers often took on dual roles as administrators or counselors for a portion of their day. The ISHSs did not have large student bodies – the largest had about 600 students. The size of administrative staff was also relatively small, but nonetheless created school environments that were highly organized and responsive to individual student interest and learning needs. Two of the schools in the OSPrI study were charter schools was led by charter network Chief Executive Officer (CEO) who was supported by a school-level Chief Academic Officer (CAO) and other administrative personnel. The CEO was responsible for the overarching vision for its schools, and the CAO was responsible for implementation of academic programs and professional development. The charter ISHSs' administrative staffs were especially lean, e.g., at one school it consisted of a school director, college counselor, and an IT specialist. A strength of the charter management organization was that it more readily connected the ISHS to a larger network of people and organizations outside the school that enhanced the STEM program of studies. Overall, however, all the successful ISHSs in the study had flexible administrative structures that allowed individual school autonomy, and were dynamic enough to incorporate new resources and expand their offerings to students without losing common purpose.

Each ISHS had strong relationships with public, non-profit, and for-profit organizations to assist in expansion, curriculum development, professional development, or informal learning opportunities for students. Partnerships were valuable for maintaining the vibrant and innovative character of the ISHS. Partnerships also facilitated growth by enhancing the ISHS legitimacy, reputation, and access to a wide variety of resources – financial, political and organizational.

Each school had a principal or CEO whose leadership style might be categorized as transformational (Ford et al., 2014; Ford, 2017). ISHS leaders were active, hands-on leaders in curriculum and instruction. They did not hide in their offices. It appeared that each principal knew every student and family by name and the student's goals and accomplishments. This was especially important for ISHSs where the missions were to prepare every student for college. Moreover, in some ISHSs, about half of the students were first generation in their family to attend college, so the goal was not only about college readiness in STEM, but also admission to a four-year college. This required that students have the soft skills for college success (Lynch, Ford, & Matray, 2017). Families were guided through financial aid and admissions process.

The ISHSs in this study were all relatively new schools and considered "education experiments" in their regions. Given the scrutiny that an educational experiment attracts, most school leaders used data for decision-making and for closely monitoring school progress. Databased decision-making also helped teachers to improve instruction and to keep watch for students who might need extra supports. For instance, ISHSs had strong systems for monitoring student data both for incoming students and for those who had graduated. They could justify the successes of their programs because incoming students did not differ much from the population of students in the district, and because student outcomes at graduation were superior to those of other schools. Year after year, the schools' compelling data sets were used to defend a school's project-based learning environment, or early college program or CTE approach. Some schools have gained national attention as existence proofs of a successful ISHS.

6.4 Supports for Students in Under-Represented Groups

Creating schools that function as opportunity structures for students underrepresented in STEM required that ISHSs have student support systems as intrinsic components (Lynch & Ross, 2014; Lynch, Peters-Burton, et al., 2017). Tutoring was the most elaborately developed support system at every ISHS and was offered during, before, and after school, and most often by teachers. Moreover, tutoring was seen as a normal aspect of school, rather than as "remedial." Some students went to tutoring out of interest in a project or idea rather than need. Classroom teachers carefully followed the progress of each student and could recommend tutoring, or students might decide they wanted tutoring on their own. There were a variety of tutoring systems available, ranging from teachers as tutors, to university student tutors, to peer tutoring, and crowdsourced online tutoring and collaboration.

ISHSs deliberately developed close connections between students and teachers in order to realize academic goals (Lynch & Ross, 2014). Most schools had advisories, which were meetings with teachers and a small group of students that occurred weekly or more often. The advisories helped hold students accountable and kept advisors aware of problems that a student might face, academic or otherwise. Over time, students, families and advisors came to know one another well and built trusting relationships. This was in part due to small school size and frequent communications. The advisories also allowed time for character education, team building activities, discussions about what was happening in the students' world, and how to improve the school. Advisories were used in ninth and tenth grades to orient students to the ISHS's expectations, new ways of learning, and opportunities outside of school. In eleventh and twelfth grades, the emphasis was on college admissions.

College advising at ISHSs began when students entered ISHSs and continued for four years. Often, alumni of the ISHS dropped in for visits, or were invited to give presentations about their experiences in college. This was the schools' way of signaling to students that the focal mission was about college preparation and STEM. For instance, at one ISHS, the advisory program began with a home visit during ninth grade to meet the family and better understand the goals of the family for the student. At school, advisors worked with their students on a weekly basis to orient them to college life and help them find internships at universities, apply for summer programs, edit their resumes and college cover letters, or offer support on an individual basis. Students traveled with advisory groups to visit colleges and universities. Special attention was given to students from families that never had another family member attend college. Students and families had to learn the terminology and the mindset of college admissions. The ISHSs were an opportunity structure leading to college admissions and beyond (Lynch, Peters-Burton, et al., 2017).

Other supports included special education services that often were available to any student who needed specialized assistance as well as students with IEPs; summer bridge programs; extended school day or year; and special mathematics courses to "catch students up" who entered the school behind grade level. The ISHSs were innovative and flexible enough to be able to develop supports to solve specific problems, and personalized enough to address individual student needs.

6.5   ISHS School Types and Critical Components

The OSPrI study began with a list of ten critical components that the literature suggested might operate in successful inclusive STEM high schools. The four most consistently prominent critical components found across all of these successful schools were discussed in the section above. These critical components can be seen as foundational.

However, the ISHSs in this study had different contexts; they were urban, suburban or rural; located in different states; and, had different governing structures. Consequently, each ISHS had a unique profile of critical components that was based upon context. For instance, the ISHS located in Chicago with the agriculture CTE theme was organized differently than the rural North Carolina ISHS that focused on early college STEM by partnering with a local community college. These schools were designed by selecting and making use of different combinations resources in context and the critical components in Table 7.1 (Lynch, Peters-Burton, et al., 2017). Some ISHSs were newer and still developing components, such as better use of technology for teaching and learning.

In another example, two of the ISHSs had adopted project-based learning as the primary mode of instruction, and were having tremendous success with students in STEM. They focused on innovative instruction aimed at problems and projects relevant to students in their school communities, and used student performances to evaluate learning. In one ISHS, each project was designed to carefully align with state curriculum standards (Lynch, Spillane, et al., 2017). Teachers were expected to teach cross-disciplinary courses in double time blocks, and the schedule provided co-planning time for teachers and in-house professional development. In these schools, critical component 2,

Reform Instructional Strategies and Project Based Learning, in Table 7.1 was most prominent. In addition, these schools made extensive use of digital technologies (critical component 3) for organizing student and teacher work.

Two other ISHSs in the OSPrI study focused on early college opportunities for their students as way to create STEM opportunity structures. Critical components 5 and 6 in Table 7.1 were highly prominent. Those ISHSs chose to capitalize on college resources to enrich STEM programs. The school schedules were built around college schedules, and students moved in and out of the high school building to access learning opportunities at the colleges or related opportunities in their communities (Lynch, Peters-Burton, et al., 2017).

Emphasis on career technical education (CTE) required that the schools bring STEM resources in the community into the school during school, and had students interacting with members of STEM community after school and during the summer. The experiential learning in STEM was crucial to the success of these schools, so they formed STEM advisory boards, partnerships, and opportunities for students to job shadow and be mentored. They also relied on national and international organizations such as Future Farmers of America to engage students and help them meet students from all over the U.S. and sometimes beyond. University contacts in CTE helped develop CTE coursework. Consequently, critical components 4, 5, and 6 were highly prominent (Lynch, Peters-Burton et al., 2017).

6.6 *Emergent Themes*

The OSPrI project started from the premise that successful ISHSs would have a certain set of critical components in common; these features were developed from a review of the research on effective schools, best practices in science and mathematics education for students from underrepresented groups, and published articles on STEM schools that were not research-based but available c. 2011. The OSPrI case studies were able to confirm their presence and prominence, but other components emerged from cross-case analyses (Lynch, Peters-Burton, et al., 2017).

Student, parent, and educator interviews and focus groups expressed in discussions the ISHSs' Positive School Climates and High Expectations for All, Critical Component 13 (see Table 7.1). Each ISHS was viewed as a unique place, family, home, house, and learning community. This was deliberately designed into ISHSs' formal mission statements or stated vision for a new kind of school. For instance, one ISHS publically announced and practiced core values that included respect, integrity, responsibility, perseverance, and trust. Students and teachers discussed how the relatively small size of the ISHSs created opportunities for caring communication about students and how to support them. ISHS teachers and administrators sought personalized approaches to teaching and advising. Students were seen as individuals and had coursework, outside of school activities, and advising tailored to their academic and socio-emotional progress and needs. The positive school climate coupled with the STEM theme was crucial to developing STEM opportunity structures. Students described teachers as friends or like trusted family members. Students reported that teachers and administration cared about them academically and personally, and that they could talk to anyone when they needed academic or personal support.

Agency and Choice, critical component 14, was an inherent feature of ISHSs, beginning with the high school admissions process when students (and families) had to choose to apply in order to attend the ISHS. Their first experiences required understanding of how the ISHSs' expectations would be different from those of other schools. Students knew that they would commit to doing more work, learn in different ways, and take more STEM coursework. New STEM learning opportunities would occur during and outside of school, and that would require trust and responsibility, as well as parent understanding of the commitment to STEM. There was a culture of respect and caring that permeated the schools. Students learned not only to take responsibility for their own actions but also to collaborate and support each other. While the ISHSs fostered these important interactions, students had to choose to adopt the values and express them in unique ways.

The two remaining emergent critical components 11 and 12, Data Driven Decision Making for Continuous Improvement and Innovative and Responsive Leadership, respectively, were discussed earlier in this chapter and woven into discussions of Administrative Structure and STEM focused Curriculum. Frequently in the literature on school change, success is attributed to the presence of charismatic principal. The downside of this feature, however, is that when the principal leaves, the innovative school is not sustained. The ISHSs in this study often had charismatic, transformational leaders, but the leaders deliberately planning for succession so that the ISHS would continue, evolve, and grow. This was most often accomplished through a flattened hierarchy, with teachers integral to leadership in curriculum and instruction in STEM, and as participants in the day-to-day administration. Several of the ISHSs also worked with charter management agencies or with university staff; and as a consequence, STEM leadership was not in the hands of one person, but distributed.

Perhaps because ISHSs were experimental and needed to prove themselves to school districts and communities, or because ISHSs were based on engineering and design principles, the schools in this study also placed more reliance on data-driven decision-making. The use of technology made this possible at the student level (mastery learning systems and recommendations for tutoring, or outside-of- school opportunities in STEM) to classroom level (management of project-based learning or aligning projects with state standards and tests) to school level (measuring the rate that students entered 4-year colleges). Data-driven decision-making seemed to fit well into the organization of ISHSs.

7 Final Thoughts: The Role of Inclusive STEM High Schools in STEM Education

The introductory sections of this chapter documented school change efforts, born in the 20th century, with a goal to improve schools, STEM education, and increase opportunity to learn for underrepresented students in STEM. The efforts ranged from highly selective schools for preparing scientists and mathematicians, to magnet schools, charter schools, CTE-themed high schools, and turnaround schools. Each effort illuminated new ways for schools to help students achieve the twin goals of a high quality STEM education and equity. Each idea has contributed to the field's understanding how to improve STEM education for underrepresented students *at the school level*.

A purpose of this chapter was to introduce and discuss a 21st century approach – inclusive STEM high schools – to a 21st century phenomenon – STEM education – in the service of solving a problem that has stretched for decades – the need to produce more students prepared for college science, mathematics, and engineering majors. This has been especially troublesome for students from groups underrepresented in STEM who are not progressing under the current system. The failure to create more equitable school outcomes in STEM has important policy implications for national, state, regional and local economies. Moreover, the families and students who are part of those economies want to have a better chance at achieving the American Dream.

The argument made here is that traditional comprehensive high schools have failed to meet the needs of too many students who are interested and able to pursue STEM, and that new experiments at the school level are required. The research community and school leaders can learn from these innovative ISHSs. For underrepresented students, *access* to a high quality secondary STEM curriculum is not enough – attending a traditional school in a building that offers advanced STEM classes to a small proportion of the student body, while excluding large numbers of students, is inequitable and bad policy. Underrepresented students display may interest and ability, but can lack the preparation or exposure to STEM that should have occurred before high school. The ISHS model developed by the OSPrI study (Lynch et al., 2015) provides evidence that such students can penetrate the STEM pipeline, or simply become more STEM literate and better able to navigate education, jobs, and careers in the 21st century.

The inclusive STEM high schools in the OSPrI study each stand as existence proofs of what can be done if the old patterns of school organization are torn down and replaced by a system that aims to get students up to speed in STEM as they enter high school, then supports them as they progress. STEM policy makers and leaders need to re-think who can succeed in STEM, and how to re-organize STEM education systems so that more actually do so.

A strong STEM education can be viewed as an opportunity structure. Remarkably, ISHSs have intuitively or intentionally adopted approaches that do more than teach STEM concepts and skills. They are schools that help students build additional opportunity structures. This includes learning with a like-minded, supportive peer group that has similar STEM interests, goals, and aspirations. ISHSs promote contacts with professionals in the STEM community in school and outside of school through internships, mentorships and other experiences in the STEM workplace, providing students with a STEM social capital (Lynch, Peters-Burton, et al., 2017). Students develop a more sophisticated understanding of STEM jobs and careers, the concomitant required preparation in college. They learn about STEM career opportunities within their communities, and travel to college campuses to see what college life is like and what it requires; this is especially important for students who are the first generation in their families to attend college. These activities can occur during school time for credit, or they may occur after school or during the summer. The point is that that these experiences are as much a part of the STEM curriculum as are pre-calculus, chemistry or introductory engineering. Changes to STEM teacher education, professional development, curriculum and instruction, and standards all aim to improve STEM teaching and learning. School-level change in the form of ISHSs aimed at underrepresented students in STEM show immediate promise for research and policy, as well as for students and families who desire a more direct and immediate route to success in STEM.

Acknowledgements

This work was conducted by the OSPrI research project, with Sharon Lynch, Tara Behrend, Erin Peters-Burton, and Barbara Means as principal investigators. Funding for OSPrI was provided by the National Science Foundation (DRL 1118851). Any opinions, findings, conclusions, or recommendations are those of the authors and do not necessarily reflect the position or policy of endorsement of the funding agency.

References

Atkinson, R. D., Hugo, J., Lundgren, D., Shapiro, M. J., & Thomas, J. (2007). *Addressing the STEM challenge by expanding specialty math and science high schools*. Washington, DC: The Information Technology & Innovation Foundation.

Badia, E., & Chapman, B. (2014, September 19). Officials request changes to city's test specialized high schools in effort to stem racial bias. *New York Daily News*. Retrieved from http://www.nydailynews.com/new-york/education/officials-request-city-testspecialized-high-schools-article-1.1945368

Behrend, T. S., Ford, M. R., Ross, K. M., Han, E. M., Peters Burton, E., & Spillane, N. K. (2014). Gary and Jerri-Ann Jacobs high tech high: A case study of an inclusive STEM-focused high school in San Diego, California. *OSPrI Report, 3*.

Bryk, A. S., & Gomez, L. M. (2008). Ruminations on reinventing an R&D capacity for educational improvement. In F. M. Hess (Ed.), *The future of educational entrepreneurship: Possibilities for school reform*. Cambridge, MA: Harvard Education Press.

Carnegie Corporation of New York. (2009). *The opportunity equation: Transforming mathematics and science education for citizenship and the global economy*. New York, NY: Carnegie Corporation of New York and Institute for Advanced Study Commission on Mathematics and Science Education.

Eisenhart, M., Weis, L., Allen, C. D., Cipollone, K., Stich, A., & Dominguez, R. (2015). High school opportunities for STEM: Comparing inclusive STEM-focused and comprehensive high schools in two US cities. *Journal of Research in Science Teaching, 52*(6), 763–789.

Ford, M. R. (2017). *Approaches to school leadership in inclusive STEM high schools: A cross-case analysis* (Doctoral dissertation). The George Washington University, Washington, DC.

Ford, M. R., Behrend, T. S., & Peters-Burton, E. (2014). *A cross-case analysis of four exemplar STEM-focused high schools: Administrative structure*. Paper presented at Critical components of inclusive STEM-focused high schools: A cross-case analysis (E. Peters-Burton, Chair). Symposium conducted at the 2014 Annual Meeting of the American Education Research Association, Philadelphia, PA.

Finn, C. E., & Hockett, J. A. (2012). *Exam schools: Inside America's most selective public high schools*. Princeton, NJ: Princeton University Press.

Hanford, S. (1997). *An examination of specialized schools as agents of educational change*. New York, NY: Columbia University.

House, A., & Peters-Burton, E. (2014). *STEM-focused curriculum in inclusive STEM high schools: A cross-case analysis*. Paper presented at Critical components of inclusive STEM-focused high schools: A cross-case analysis (E. Peters-Burton, Chair). Symposium conducted at the 2014 Annual Meeting of the American Education Research Association, Philadelphia, PA.

Locke, G. (2011, July 14). *STEM jobs help America win the future* [Blog post]. Retrieved from http://whitehouse.gov/blog/2011/07/14/stem-jobs-help-America-win-future

Lucadamo, K. (2016, September 26). Can all students succeed at science and tech high schools? *US News and World Report* Retrieved from http://www.usnews.com/news/articles/2016-09-26/can-all-students-succeed-at-science-and-technology-high-schools

Lynch, S. J. (2000). *Equity and science education reform.* Mahwah, NJ: L. Erlbaum Associates.

Lynch, S. J., Burton, E. P., Behrend, T., House, A., Ford, M., Spillane, N., Matray, S., Han, E., & Means, B. (2017b). Understanding inclusive STEM high schools as opportunity structures for underrepresented students: Critical components. *Journal of Research in Science Teaching, 55*(5), 712–748.

Lynch, S. J., Ford, M., & Matray, S. (2017, April). *Comparing student views of opportunity to learn in comprehensive and inclusive STEM high schools.* Paper presented at the American Education Research Education Annual Meeting.

Lynch, S. J., House, A., Peters-Burton, E., Behrend, T., Means, B., Ford, M., Spillane, N., Matray, S., Moore, I., Coyne, C., Williams, C., & Corn, J. (2015). *A logic model that describes and explains eight exemplary STEM-focused high schools with diverse student populations.* Washington, DC: George Washington University OSPrI Project.

Lynch, S. J., Peters-Burton, E., & Ford, M. (2015). Building STEM opportunities for all. *Educational Leadership, 72*(4), 54–60.

Lynch, S. J., & Ross, K. R. (2014). *A Cross-case analysis of exemplar inclusive STEM high schools: Mission and supports for students under-represented in STEM Education.* Paper presented at Critical components of inclusive STEM-focused high schools: A cross-case analysis (E. Peters-Burton, Chair). Symposium conducted at the 2014 Annual Meeting of the American Education Research Association, Philadelphia, PA.

Lynch, S. J., Spillane, N., House, A., Peters-Burton, E., Behrend, T., Ross, K. M., & Han, E. M. (2017a). A policy-relevant instrumental case study of an inclusive STEM-focused high school: Manor new tech high. *International Journal of Education in Mathematics, Science and Technology, 5*(1), 1–20.

Means, B., Confrey, J., House, A., & Bhanot, R. (2008). *STEM high schools: Specialized science technology engineering and mathematics secondary schools in the U.S. Report prepared for the Bill & Melinda Gates Foundation.* Menlo Park, CA: SRI International. Retrieved from http://ctl.sri.com/publications/ displayPublicationResults.jsp

Means, B., Young, V., & Wang, H. W. (2014). *Effects of attending an inclusive STEM high school: A longitudinal study of North Carolina schools.* Paper presented at Comparing studies of inclusive STEM high schools: Three approaches with different findings and policy implications (S. J. Lynch, Chair). Symposium conducted at the 2014 Annual Meeting of the American Educational Research Association, Philadelphia, PA.

Means, B., Wang, H., Wei, X., Lynch, S., Peters, V., Young, V., & Allen, C. (2017). Expanding STEM opportunities through inclusive STEM-focused high schools. *Science Education, 101*(5), 681–715.

Means, B., Wang, H., Young, V., Peters, V., & Lynch, S. J. (2016). STEM-focused high schools as a strategy for enhancing readiness for postsecondary STEM programs. *Journal of Research in Science Teaching.* doi:10.1002/tea.21313

National Academy of Sciences, National Academy of Engineering, & Institute of Medicine. (2011). *Expanding underrepresented minority participation: America's science and technology talent at the crossroads.* Washington, DC: National Academies Press.

National Research Council (NRC). (2010). *Preparing teachers: Building evidence for sound policy.* Committee on the Study of Teacher Preparation Programs in the United States. Washington, DC: National Academies Press.

National Research Council (NRC). (2011). *Successful K-12 STEM education: Identifying effective approaches in science, technology, engineering, and mathematics.* Committee on Highly Successful Science Programs for K-12 Science Education. Board on Science Education and Board on Testing and Assessment, Division of Behavioral and Social Sciences and Education. Washington, DC: The National Academies Press.

Oakes, J. (2005). *Keeping track: How schools structure inequality.* New Haven, CT: Yale University Press.

Obama, B. (2010, September 16). *Remarks by the President at the announcement of the "change the equation" initiative.* Retrieved from http://www.whitehouse.gov/the-press-office/2010/09/16

Peters-Burton, E. E., Kaminsky, S. E., Lynch, S., Behrend, T., Han, E., Ross, K., & House, A. (2014). Wayne school of engineering: Case study of a rural inclusive STEM-focused high school. *School Science and Mathematics, 114*(6), 280–290.

Peters-Burton, E. E., Lynch, S. J., Behrend, T. S., & Means, B. B. (2014). Inclusive STEM high school design: 10 critical components. *Theory into Practice, 53*(1), 64–71.

President's Council of Advisors on Science and Technology (PCAST). (2010). *Prepare and inspire: K-12 education in Science, Technology, Engineering, and Mathematics (STEM) for America's future.* Washington, DC: Author.

President's Council of Advisors on Science and Technology (PCAST). (2012). *Engage to excel: Producing one million additional college graduates with degrees in science, technology, engineering, and mathematics.* Washington, DC: Author.

Robelen, E. W. (2013, April 9). Common science standards make formal debut. *Education Week.* Retrieved from http://www.edweek.org

Roberts, K. (1968). The entry into employment: An approach towards a general theory. *Sociological Review, 16,* 165–184.

Roberts, K. (1984). *School leavers and their prospects.* Buckingham: OU Press.

Rothwell, J. (2013). *The hidden STEM economy. Metropolitan policy program at Brookings.* Retrieved from http://www.brookings.edu

Scott, C. E. (2009). *A comparative case study of characteristics of Science, Technology, Engineering, and Mathematics (STEM) focused high schools.* Retrieved from Proquest

Spillane, N. K. (2014). *Teachers.* Paper presented at Critical components of inclusive STEM-focused high schools: A cross-case analysis (E. Peters-Burton, Chair). Symposium conducted at the 2014 Annual Meeting of the American Education Research Association, Philadelphia, PA.

Spillane, N. K., Lynch, S. J., & Ford, M. R. (2016). Inclusive STEM high schools increase opportunities for underrepresented students. *Phi Delta Kappan, 97*(8), 54–59.

Stake, R. E. (2006). *Multiple case study analysis.* New York, NY: Guilford Press.

Subotnik, R. F., Rayhack, K., & Edmiston, A. (2006). *Current models of identifying and developing STEM talent: Implications for research, policy and practice.* Washington, DC: American Psychological Association.

Tsupros, N., Kohler, R., & Hallinen, J. (2009). STEM education: A project to identify the missing components.

Turn Around Schools. (2018, July 12). Retrieved from http://www.edweek.org/ew/collections/turnaround-schools

U.S. Department of Commerce and Economics and Statistics Administration. (2011, July 14). *STEM jobs now and in the future.* Retrieved from http://www.esa.doc.gov/print/Reports/stem-good-jobs-now-and-future

U.S. Department of Education Magnet Schools Assistance Program. (n.d.) Retrieved from https://innovation.ed.gov/what-we-do/parental-options/magnet-school-assistance-program-msap/

Weis, L., Eisenhart, M., Cipollone, K., Stich, A. E., Nikischer, A. B., Hanson, J., & Dominguez, R. (2015). In the guise of STEM education reform: Opportunity structures and outcomes in inclusive STEM-focused high schools. *American Educational Research Journal, 52*(6), 1024–1059.

Yin, R. K. (2008). *Case study research design and methods.* Thousand Oaks, CA: Sage Publications.

Young, V., Lynch, S. J., Means, B., House, A., Peters, V., & Allen, C. (2017, May). *Bringing inclusive STEM high schools to scale: Policy lessons from three states.* Menlo Park, CA: SRI International. Retrieved from https://inclusivesteminsights.sri.com/downloads/inclusive-stem-high-schools-to-scale-policy-lessons-brief.pdf

CHAPTER 8

Inclusive STEM School Models: A Review of Characteristics & Impact

Justin M. Bathon

Abstract

This chapter summarizes the literature on successful STEM school models and highlights major achievements and challenges of the STEAM Academy in Lexington, KY. The chapter first examines research on inclusive STEM high school characteristics before examining existing, early research on the impact of these models. A particular example of an inclusive STEM high school, co-developed by the author, is then provided as additional context and insight before remaining research questions are shared. Inclusive STEM high school models are showing early promise within limited implementations while questions remain as to the scaling and sustainability of such models.

1   Introduction

Schools are constantly challenged to offer robust learning opportunities for students while achieving one of the critical missions of public education to prepare students for the demands of the broader economy and workforce. Further, schools are challenged to provide these opportunities in an equitable way within the existing policies and norms of public school systems. To meet these simultaneous challenges but still provide a novel approach to the learner experience, a relatively new generation of Inclusive STEM High Schools have developed across the United States. One such school, STEAM Academy, a high school in Lexington, KY, was co-developed by the author and is now in the sixth year of operations.

   This chapter first reviews this emerging framework of high school providing readers a sense of the common elements of these Inclusive STEM High Schools. Next, the early and emerging literature is reviewed for insights into the impact of these approaches on student learning. Throughout the chapter, insights gained while developing STEAM Academy are provided, including some examination of the student learning outcomes that have emerged.

2 Inclusive STEM High Schools

Inclusive STEM High Schools share a given set of characteristics. First, there is an explicit focus on STEM education within the curriculum but access to this curriculum is not limited to high achievers. Frequently, inclusive STEM high schools share a goal of preparing students "to be successful in a STEM college major by providing a program of studies with greater depth and breadth than their states require for high school graduation" (Spillane, Lynch, & Ford, 2016, p. 54).

Inclusive STEM high schools are different from the historically dominant STEM programs which tended to prioritize gifted students (Peters-Burton, Lynch, Behrend, & Means, 2014a). In an examination of the value-added of STEM programs more holistically (not just inclusive STEM high schools) in North Carolina and Florida, Hansen (2014) found a lower rate of participation of underrepresented minority students in successful STEM programs.

This more inclusive approach to STEM emerged in part of a national effort to launch more STEM schools while promoting STEM education to historically disadvantaged students. This priority was captured through the formation and report of a President's Council of Advisors on Science and Technology (2010) that called for "The Federal Government [to] promote the creation of at least 200 new highly-STEM-focused high schools and 800 STEM-focused elementary and middle schools over the next decade, including many serving minority and high-poverty communities."

3 Characteristics of Inclusive STEM High Schools

Within the last ten years, several studies have emerged that examined the characteristics of inclusive STEM high schools. Several of these recent studies have been funded investigations by the National Science Foundation as they seek to promote the literature base around this potentially powerful model. This section examines those emerging common characteristics.

In 2014, Peters-Burton and colleagues found an initial list of ten characteristics of inclusive STEM high schools using a literature, website and document review of identified inclusive STEM high schools. As the authors note, there are no causal links between any particular component of an inclusive STEM high school and direct impacts on student learning. The identified characteristics include: (1) a stem focused curriculum, (2) use of project-based learning, (3) integrated, innovative technology use, (4) blending of informal and external learning, (5) real-world STEM partnerships, (6) early college level coursework,

(7) well prepared STEM teaching staff, (8) inclusive STEM mission, (9) strength amongst school leaders, and (10) supports for underrepresented students (Peters-Burton, Lynch, Behrend, & Means, 2014). The authors suggested future research strategies and data sources to confirm these shared characteristics.

Later, upon a deep examination of eight inclusive STEM high schools funded by the National Science Foundation called Opportunity Structures for Preparation and Inspiration (OSPrI), the ten characteristics were expanded to fourteen as four emergent critical components were added to the initial ten. These four critical components were: (1) dynamic assessment systems for continuous improvement, (2) innovative and responsive leadership, (3) positive school community and culture of high expectations for all, and (4) student agency and choice (Lynch et al., 2015). From this articulation useful tools were developed such as a critical components rating system that schools could use to self-assess their own progress toward meeting the definition of an inclusive STEM high school (OSPRI, 2018).

After further discussion, the research team at OSPrI refined the characteristics of inclusive STEM high schools into a three part logic model. The 14 characteristics were organized and grouped under the labels of (1) structure, (2) what students learn, (3) how students learn, and (4) social dimension and purpose (Lynch et al., 2015). In addition to these characteristics, the updated logic model provides clarity on the expected impacts for high school students during their education and expected outcomes for students beyond high school.

At the same time as the OSPrI team of researchers were developing and refining their characteristics and logic model, a different team, also funded by the National Science Foundation, was also seeking to identify the characteristics of inclusive STEM high schools using a different method. Whereas the OSPrI team did a case study of 8 established STEM schools, the team at the University of Chicago Outlier Research and Evaluation examined 25 inclusive STEM high schools from a broader cross-section of the United States. The characteristics overlap heavily but do have some differences. The 8 identified characteristics of inclusive STEM high schools are: (1) rigorous learning, (2) problem-based learning, (3) personalization of learning, (4) career, technology and life skills, (5) school community and belonging, (6) external community, (7) staff foundations, and (8) essential factors (LaForce et al., 2014). The essential factors were identified as a flexible staff, enrollment of a community representative population, and the provision of professional development resources. Interestingly, there is less of a focus on STEM explicitly in the schools they studied. The researchers found that when school leaders of inclusive STEM high schools referred to STEM they referred to concepts beyond disciplinary subjects such as the impact on school pedagogy and culture. The schools shared a common constructivist vision of learning and utilized inquiry as a primary tool to implement these well known, long established best practices (LaForce et al., 2014).

4   Outcomes of Inclusive STEM High Schools

Barbara Means and colleagues have conducted the most research on the impact of inclusive STEM high schools. Using inclusive STEM schools across North Carolina (n=12) and Texas (n=27) matched against similar non-STEM focused high schools (North Carolina = 10, Texas = 10) the researchers examined the impact across 5 scales: (1) STEM coursework and activities, (2), attitudes toward STEM subjects, (3) students' plans and aspirations, (4) qualities of high school STEM experiences, and (5) high school achievement measures (Means et al., 2017).

Beginning with the achievement metrics, in North Carolina higher grade point averages were found for all students, African American students, and female students. Further, all students performed higher on the science, but not mathematics, component of the ACT. In Texas, the state achievement test was used for comparison and on both science and mathematics there was a small, positive statistical difference for students in the inclusive STEM high schools. This small difference also manifested in particular for African American students on the science exam but no statistically significant difference emerged for African American students on the mathematics exam nor on either exam for Latino/a or female students (Means et al., 2017).

Beyond test score performance, the study found statistically significant differences between inclusive STEM high schools and traditional comprehensive high schools on a number of other variables. Students were more likely to take some STEM focused coursework in both states as well as participate in STEM-based extra-curricular activities. These effects were also present in the African American, Latino/a, and female subgroups. Further, all students responded to the survey with higher perseverance scores as well as higher schools in STEM career interest. These significant effects were also present for female students in both states. Finally, in both states for all students, African American students, and female students, there was significant positive differences for inclusive STEM high schools in that schools had high expectations for students and reported teacher respect for students (Means et al., 2017).

While this study lends some support to the notion that inclusive STEM high schools promote higher outcomes for underrepresented minority students, the calls to expand the model are increasing. Spillane, Lynch, and Ford (2016) found, in a study of 8 inclusive STEM high schools, that cultures of opportunity and open approaches to innovation and leadership provide greater STEM access and opportunities to traditionally marginalized populations. This study has been used, in part, as evidence for promoting more development of inclusive STEM high schools as a way to promote equity, in particularly leading toward STEM careers.

A study of Texas T-STEM schools (n=53 schools, 9,004 students) did not find significant differences between T-STEM and traditional comparison schools when examining all student performance on math, science, and reading state assessments. However, when examined by subgroup some small statistically significant gains were found. While males performed better in T-STEM schools in reading, mathematics, and science, female students only performed better in reading and mathematics but not science. Latino and White students performed slightly better than traditional school counterparts but no significant difference was found for African American students. When examining only economically disadvantaged students a small positive effect was found for these students in the T-STEM schools. For special education students, there was a small positive impact for reading and mathematics assessments but not science, where the students in traditional comprehensive schools performed slightly higher (Erdogan & Stuessy, 2016).

Qualitative studies have also been done on particular inclusive STEM high schools and their components. A two city, eight school study in Denver and Buffalo, examined the deployment of STEM models in high poverty, low performing schools (Eisenhart et al., 2015). Using multiple qualitative methods, the researchers concluded, "when the STEM-focused schools we studied were compared to our comprehensive high schools with similar students, they did not look very different in terms of the STEM course-related opportunities they provided." There was little evidence, outside of adopting Academy models, that major efforts were undertaken to implement or sustain the characteristics of inclusive STEM high schools. They suggest that inclusive STEM models may well be useful but that more serious efforts to understand and adapt to local contextual factors within diverse communities.

Case studies of specific schools also emerged. A study of Manor New Tech High school in Texas found three emergent themes around student acquisition of 21st century skills, the robust relationships the lead to a positive school culture, and the development of a family context to support all learners (Lynch et al., 2017). An inclusive STEM high school in a rural context was examined in North Carolina in the Wayne School of Engineering (Peters-Burton et al., 2014b). The scholars found a school willing to overcome traditional barriers of poverty and access in over performing both the local school district and state on end-of-course exams and graduation rates. The school offered early college opportunities, an extended school day, personalized attention and instruction, and a deep sense of community.

5   STEAM Academy, Lexington Kentucky

STEAM Academy, in which the author was a co-developer, is located in Lexington, Kentucky. Kentucky has been relatively slow to implement inclusive STEM model schools and, thus, STEAM Academy represents the first whole-school inclusive STEM high school in Kentucky. STEAM was started with the support of a Next Generation Learning Challenges grant in the Fall of 2013 (NGLC, 2013). It is a whole-school magnet program for the Fayette County Public School District and has no selection criteria to gain admission other than a parent filling out a form during a student's 8th grade year. The high school program has averaged between 350–400 students per year and has averaged around 40% minority student participation and around 35% free/reduced lunch status.

While no examples of a whole school STEM model existed in Kentucky at the time, STEAM was loosely based on the example of Metro Early College High School in Columbus, OH, a well known and researched inclusive STEM high school (Han, Lynch, Ross, & Han, 2014). This particular school was a result of an investment by the Battelle Corporation and the resulting STEM Network in Ohio has gone on to serve as a model statewide network for STEM programs and schools nationally. In the case of STEAM Academy in Lexington however, Metro provided only a general framework in which to emulate and most of the STEAM program was developed over the course of 2–3 years of innovative development through a partnership between Fayette County Public Schools and the University of Kentucky, College of Education. Thus, STEAM Academy largely represents a homegrown Kentucky model for inclusive STEAM education and the model has already been emulated in Kentucky and a new network of STEAM programs is beginning to develop around the school.

The characteristics of the STEAM Academy, though, largely reflect the national research on characteristics of inclusive STEM high schools cited previously. I will briefly provide an overview of some of the core design characteristics of STEAM Academy, noting overlap with the OSPrI fourteen characteristics (indicated with a *) (Peters-Burton, Lynch, Behrend, & Means, 2014; Lynch et al., 2015) and the University of Chicago S3 team's 8 characteristics (indicated with a ^) (LaForce et al., 2014).

1. *Inclusivity*^*: The student population of STEAM Academy is designed to reflect the surrounding community, including admission of special education students into an full inclusion model.
2. *Project Based Learning*^*: All courses at STEAM Academy are expected to include at least one substantial PBL per semester. STEAM Academy generally follows the Gold Standard PBL format.
3. *Quasi-Mastery Model^*: Using a semester model to increase course taking opportunities, students are expected to get a grade of B (83%) to successfully complete a math, science, or English course on the first attempt.
4. *Early College Model*^*: STEAM works with the local community college to offer dual credit courses both on the STEAM Academy Campus as well as at the dedicated space at Bluegrass Community Technical College. Further, the time structures of high school reflect the college experience.
5. *Career Explorations*^*. All students at STEAM engage in multiple career-based, student selected internship experiences. Students spend between 125–175 hours on the internship per semester.
6. *Digital Learning Environment**: All courses are managed through the Canvas Learning Management system and curricular materials are largely custom curated within the LMS. The school is also 1:1 with Chromebooks and most assessments utilize digital tools.
7. *Staff Autonomy*^*: Staff are largely expected and free to custom build the curriculum and pedagogy for all courses. To support each other, staff participate in weekly project tuning sessions to provide professional feedback on course design.
8. *Student Voice/Choice/Agency**: Students are provided a great deal of ownership over their learning within classrooms including self-designing passion projects. Further, the school culture is highly reflective of student work both formally and informally. For instance, students are provided two permanent positions on the school governing council.

While further and more scientific research is needed, after six years, the initial results emerging from the STEAM Academy are largely reflective of the broader research. While hard to specifically quantify, students from some underrepresented minority and economically disadvantaged backgrounds do seem to be performing better on both traditional and non-traditional metrics at STEAM Academy. On traditional metrics, STEAM has consistently produced amongst the highest ACT outcomes for students from these populations, with a particularly high performance on ACT Math. Further, STEAM's graduation and college-going rates are higher than the larger comprehensive schools in the district. Also, STEAM has lower discipline rates and higher attendance rates than the comprehensive schools in the district. Readers are encouraged to examine the data for themselves by reviewing STEAM Academy results at the Kentucky School Report Card. On non-traditional metics, STEAM Academy students complete, on average, over 400 hours of career-based internships. Students also engage in more dual credit college hours on average obtaining 12–15 hours of college credit while at STEAM Academy.

6 Remaining Questions

Having engaged the challenge of building an Inclusive STEM High School from scratch, the remaining questions are overwhelming. For the sake of brevity, I will break these remaining questions as has been done previously in this chapter first to inclusive STEM high school characteristics and then to outcomes.

6.1 *Inclusive STEM School Characteristics*

1. *The role of STEM:* While inclusive STEM high schools share many common characteristics, those characteristics have more in common with the broader, recent movement toward deeper learning or the historic ideals of progressive education based on constructivist philosophy. While STEM ideals are present in these schools, how much intentionality toward STEM is present in the design of the school model and learner experience?
2. *Details on specific characteristics:* The research thus far has examined characteristics of STEM schools in holistic ways leaving details of particularly implementation choices for future research. This is understandable but these details matter. For instance, a full blown on-the-job internship experience is much different than a within-school career exploration program. Further, the subtle differences between inquiry learning models, problem based learning models, project based learning models, and more are important. Yet, presently, the existing research lumps all of these into broad categories of shared characteristics. Further research on specific characteristics would be welcome.
3. *Whole schools v. embedded programs:* Most of the research done on inclusive STEM high schools has involved whole school models. The one exception is the study that produced the most discouraging results (Eisenhart et al., 2015) where existing schools were either transitioned to a STEM model or had a STEM program embedded in a broader comprehensive model. Based on our experience at STEAM Academy, the massive structural and pedagogical transitions we made would not have likely been possible within a broader existing high school or as a turnaround strategy for an existing school.

4. *Personnel questions:* From students to teachers to leaders the role of the humans in these schools has seemingly taken a backseat in the investigations to date. The qualitative inquiries have provided more insight and as was seen in the OSPrI research (Lynch et al., 2015) the role of students and leaders were added after more qualitative inquiry revealed central roles in the development of these these schools. When visitors come to STEAM they are consistently struck by the ownership of our students in their learning and always ask the questions around teacher retention. More attention to questions of student agency, teacher professionalism, and school leaders is needed as successful school models are the shared result of the hard work of many individuals.

6.2 Inclusive STEM School Outcomes

1. *Small schools:* One common factor amongst most inclusive STEM high schools is a small school nature. What impact, if any, does this small school nature have on the overall performance of all children, but particularly children from underrepresented minority and economically disadvantaged backgrounds. This is a common refrain lobbied at the positive results from STEAM Academy by leaders of larger comprehensive schools.
2. *School choice:* Most whole-school inclusive STEM high schools rely on some choice mechanism to determine enrollment. Whether the school is a charter school or a district magnet, typically parents have chosen to send their child to the STEM program and this choice presupposes a level of parental engagement that may be predictive of higher potential results in student achievement.
3. *Unique learners:* A question which has received no treatment in the literature is what is the impact of inclusive STEM high schools on unique learners, particularly students who have been identified for special education services. While inclusive STEM high schools share a common full inclusion characteristic, what impact are these design choices having on learners that may need additional supports beyond the classroom. Further, the impact of these models on English language learners and other unique sub-populations of students remain open questions.
4. *College persistence & career outcomes:* The desired outcome of STEM programs generally are more, and more diverse, STEM college graduates and more STEM career placements. These questions have yet to be investigated. As state longitudinal data systems continue to strengthen answers to these long-term outcome questions may already be possible, if we have the patience to wait for them.

7 Conclusion

Inclusive STEM high schools are providing a hopeful narrative that links together workforce needs with progressive educational ideals. As such, the potential of these models to help shape the future of America's schools is great. However, like all great ideas, the reality of implementation is more complicated than the purity of theory. So it is with the initial research on inclusive STEM high schools. This research shows promise in such models but perhaps only in more laboratory like conditions such as small, start-up schools with initial parent choice influencing the school population. Much hard work remains to bring the models and research beyond this initial phase to do implementation of the shared inclusive STEM school characteristics to scale with large schools and all students. Perhaps, with some wind at our back, inclusive STEM high schools such as STEAM Academy might help our systems of schools take some small step forward.

References

Eisenhart, M., Weis, L., Allen, C. D., Cipollone, K., Stich, A., & Dominguez, R. (2015). High school opportunities for STEM: Comparing inclusive STEM-focused and comprehensive high schools in two US Cities. *Journal of Research in Science Teaching, 52*(6), 763–789.

Erdogan, N., & Stuessy, C. (2016). Examining the role of inclusive STEM schools in the college and career readiness of students in the United States: A multi-group analysis on the outcome of student achievement. *Educational Sciences: Theory and Practice, 15*(6), 1517–1529.

Han, E. M., Lynch, S. J., Ross, K. M., & House, A. (2014). *Metro early college high school: A case study of an inclusive STEM-focused high school in Columbus, Ohio* (OSPrI Report 2014-01). George Washington University, Opportunity Structures for Preparation and Inspiration in STEM. Retrieved from https://ospri.research.gwu.edu/sites/g/files/zaxdzs2456/f/downloads/OSPr...

Hansen, M. (2014). Characteristics of schools successful in STEM: Evidence from two states' longitudinal data. *The Journal of Educational Research, 107,* 374–391.

Laforce, M., Noble, E., King, H., Holt, S., & Century, J. (2014). *The 8 elements of inclusive STEM high schools.* Chicago, IL: Outlier Research and Evaluation, CEMSE, The University of Chicago.

Lynch, S. J., House, A., Peters-Burton, E., Behrend, T., Means, B., Ford, M., Spillane, N., Matray, S., Moore, I., Coyne, C., Williams, C., & Corn, J. (2015). *A logic model that describes and explains eight exemplary STEM-focused high schools with diverse student populations*. Washington, DC: George Washington University OSPrI Project.

Lynch, S. J., Spillane, N., House, A., Peters-Burton, E., Behrend, T., Ross, K. M., & Han, E. M. (2017). A policy-relevant instrumental case study of an inclusive STEM-focused high school: Manor new tech high. *International Journal of Education in Mathematics, Science and Technology, 5*(1), 1–20.

Means, B., Wang, H., Wei, X., Lynch, S., Peters, V., Young, V., & Allen, C. (2017). Expanding STEM opportunities through inclusive STEM-focused high schools. *Science Education, 101*(5), 681–715.

Next Generation Learning Challenges. (2013). *Fayette county public schools: The STEAM Academy*. Retrieved from https://s3.amazonaws.com/nglc/resource-files/NG1207-pdf.pdf

Opportunity Structures for Preparation and Inspiration. (2018). *STEM inventory: Critical components rating*. Retrieved from https://ospri.research.gwu.edu/stem-inventory

Peters-Burton, E. E., Lynch, S. J., Behrend, T. S., & Means, B. B. (2014a). Inclusive STEM high school design: 10 critical components. *Theory into Practice, 53*, 64–71.

Peters Burton, E., Kaminsky, S. E., Lynch, S., Behrend, T., Han, E., Ross, K., & House, A. (2014b). Wayne school of engineering: Case study of a rural inclusive STEM-focused high school. *School Science and Mathematics, 114*(6), 280–290.

President's Council of Advisors on Science and Technology. (2010). *Prepare and inspire: K–12 education in Science, Technology, Engineering, and Mathematics (STEM) for America's future*. Washington, DC.

Spillane, N. K., Lynch, S. J., & Ford, M. R. (2016). Inclusive STEM high schools increase opportunities for underrepresented students. *Phi Delta Kappan, 97*(8), 54–59.

CHAPTER 9

Informal Learning in STEM Education

Soledad Yao and Margaret J. Mohr-Schroeder

Abstract

Research has shown that exposure to a variety of STEM opportunities will have a long-term effect on individuals and the overall STEM education community (Wai, Lubinski, Benbow, & Steiger, 2010). While many opportunities for informal learning have historically existed within the science and history disciplines, such as museums, nature centers, libraries, and national parks (Fenichel & Schweingruber, 2010), there has been a lot of growth in comparable informal learning environments within STEM education (e.g., engineering, integrated STEM). Participation in informal learning environments allows K-12 grade students to delve deeper into STEM concepts, complementing STEM learning in their traditional school setting. In this chapter, we will review the literature on informal STEM learning and examine the effects these authentic learning experiences in the STEM disciplines will have on their interest and motivation towards STEM careers.

∴

Informal learning should no longer be regarded as an inferior form of learning whose main purpose is to act as the precursor of formal learning; it needs to be seen as fundamental, necessary and valuable in its own right. (Coffield, 2000, p. 8)

∴

1   Introduction

In recent years, informal learning has assumed critical roles in STEM education. Informal STEM learning experiences have the potential to support students' learning and engagement in a formal STEM learning environment. One specific aspect of STEM education is utilizing informal learning environments for demonstrating the integrated nature of STEM in addressing challenges or solving problems. In these efforts, we envision a genuinely integrated, interdisciplinary STEM where the four disciplines of science, technology, engineering, and mathematics are used to strengthen and support each other, rather than being applied as unconnected disciplines. While there is a tendency to put science and mathematics to the forefront, as reflected in the way the four disciplines were first put together as an acronym (SMET, instead of STEM), the more recent trend is towards an integrated STEM use.

An informal learning environment is a place where learning takes place outside of the classroom (i.e., formal learning). There are many opportunities for informal that exist, especially in science and history through places such as museums, nature centers, and national parks. In *Learning Science in Informal Environments*, Bell, Lewenstein, Shouse, and Feder (2009) note many students who have been unsuccessful in school, particularly those from "non-dominant cultural or lower socioeconomic groups," show success in out-of-school contexts (p. 40). Central to successful informal learning environments is "how to integrate experiences across settings to develop synergies in learning—in other words, how to maximize the ecological connections among learning experiences toward outcomes and competencies of interest or of consequence" (p. 40). Many students who were considered struggling students had their first moment of success in an out-of-school activity, and later experienced success in STEM professions (Lauer, Akiba, Wilkerson, Apthorp, Snow, & Martin-Glenn, 2006).

One purpose of STEM informal learning opportunities for K12 students is to develop and increase the number knowledgeable members of the STEM workforce through helping to expose and create more meaningful opportunities between what they learn in the classroom and solving real-world problems. For teachers to be able to help students develop this capability, they should be able to make the connections themselves. Informal STEM learning opportunities are important as they can have a positive effect on students' formal schooling (Roberts et al., 2018). Evidences from various fields of research related to learning outside of school shows that learning of STEM is a result of the dynamic interactions among various factors including the characteristics of the learner (interests, attitudes, values), the diverse settings where learning occurs (schools and out-of-school like the home, museums, libraries, organizations) and the community and culture where the diverse learning settings are embedded (NRC, 2015). Recognizing its important contributing role to STEM learning, efforts have been made to identify what makes for a productive out-of-school program. According to a report from the National Research Council (2015) out-of-school programs that produce positive outcome for learners are those that: (a) engage young people intellectually, socially, and emotionally; (b) respond to young people's interests, experiences, and cultural practices; and (c) connect STEM learning in school, out-of-school, home and other settings.

Research has shown that after school informal learning programs can be very helpful in providing STEM intervention to students (Valentine, 2016). In what may be considered as partnership or cooperation, informal learning in this case assumes a more structured role as it tries to align learning to the school's curriculum, thus forming a link between formal and informal learning. Under this cooperation, students experience STEM content from the school, which is then expanded and enhanced through participation in informal learning after school.

More recently, there has been an increasing trend towards integrating formal and informal learning, particularly in response to concerns about school accountability. This has led, in some cases, to formation of partnerships and collaborations between school administrators and afterschool staff. Through this collaboration, formal and informal learning are integrated for the purpose of helping students reach academic goals. A potential benefit of this collaboration is student access to diverse and quality services that the school may not be able to sustain within school hours, like tutoring and academic enrichment activities (Little, Wimer, & Weiss, 2008). With an added advantage of pooled assets, resources, and perspectives, integration of formal and informal learning programs provides a complementary learning environment where there are more opportunities for development and reinforcement of skills (Afterschool Alliance, 2015).

The essential feature of an integrated formal-informal learning program is alignment of learning that occurs in the two environments, guided by the school's curriculum. It has earlier been proposed that alignment between formal and informal learning can be achieved in the form of interpersonal, curricular and systemic links which serve to bridge the two learning environments (see Figure 9.1). Under this framework, the interpersonal bridge is provided through interactions between school teachers and after-school staff, the systemic bridge takes the form of collaboration in decision making related to the school and after-school programs, and the curricular bridge is achieved through alignment of curriculum and standards.

Expanding on Noam et al.'s framework, Bennet (2015) studied alignment practices and examined the association of those practices with student

FIGURE 9.1   Example of bridging between formal school learning and informal learning via an after school program (adapted from Noam et al., 2002)

academic achievement. Principals and afterschool staff of several schools were surveyed in order to determine the level of alignment between school and afterschool in terms of academic resources (sharing of materials), communication (frequency of collaboration and subjects covered), and partnership (feelings of trust and sense of value). Using the survey results, participating school sites were categorized, based on measured level of alignment, as highly aligned or misaligned. Analysis of results demonstrated a positive association between high alignment and academic achievement in both English Language Arts and mathematics, whereas significant negative associations in mathematics were noted when sites were misaligned.

2 Examples of Informal Learning Environments

There are many types of informal learning environments. The most common can be found in museum and library programs, especially related to science education, and have been around for many years. However, the recent push to produce a STEM literate society and provide more access and opportunity to students in the STEM areas have caused organizations, schools, universities, etc. to also tap into the informal learning arena, especially in the areas of STEM education (e.g., engineering, integrated STEM, mathematics).

For example, Girl Scouts have begun providing numerous STEM informal learning opportunities aimed at increasing opportunity and access to female students. In one particular investigation, middle school Girl Scouts participated in an authentic science experience by working in collaboration with various community members (leaders, parents, university faculty, graduate students) to apply engineering design skills to investigate water quality in a local community river (Burrows, Lockwood, Borowczak, Janak, & Barber, 2018). The researchers defined authentic science as one where participants (a) utilize technology and mathematics in working on a problem in the natural world; (b) develop conclusions based on obtained information and evidences; (c) refine questions and methods for future use; and (d) record and make results accessible to others. Results of the water quality project reflected what the Girl Scout participants learned about the eight essential science and engineering practices identified in the Next Generation Science Standards (NGSS): asking question/defining problems, developing and using models, planning and carrying out investigations, analyzing and interpreting data, using mathematics and computational thinking, constructing explanations and designing solutions, and engaging in argument from evidence (NGSS Lead States, 2013).

In another form of informal learning, an integrated out-of-school STEM education program was implemented in order to improve 6th grade students' perceptions towards STEM fields and careers (Baran, Bilici, Mesutoglu, & Ocak, 2016). The program featured hands-on, collaborative engagement in numerous engineering design activities including egg drop, scaled model of the solar system, application inventor (a basic programming activity), designing a vacuum cleaner, enduring buildings, car, wind turbine, and kaleidoscope, STEM commercial videos, using force and motion probes to interrogate and learn, building structures, and cryptology and Egyptian number systems. The participants evaluated the activities in terms of (a) what students learned from the activities; (b) the skills students developed; (c) activities' future use; and (d) suggestions for improvement. Most frequently cited concepts that the participants learned were on solid structure/bridge construction, cryptology as applied to decimal system, and kaleidoscope construction and its mathematics applications and implications. The skills that the student participants acquired included cognitive skills, mathematics and science applied content, design engineering practices, and computer programming knowledge and skills.

An informal learning environment called Mathematics, Engineering, Science Achievement (MESA) program has been shown to be an effective agency for recruiting underrepresented students into STEM careers (Denson, Hailey, Stallworth, & Householder, 2015). MESA provides educationally disadvantaged students opportunities to succeed in the STEM disciplines by offering underrepresented students trainings in SAT/ACT preparation and study skills, experiences in hands-on activities and competitions, and career exploration through field trips and invited guest speakers. Eight themes emerged as participants described the benefits of participation in the informal learning program: informal mentoring, fun learning, efficient time management, application of mathematics and science, feeling of accomplishments, confidence building, camaraderie, and exposure to new opportunities. In addition to the eight themes, the program also had significant impact on student achievement. Student participants performed better in mathematics and physics courses in the formal school setting and had higher achievement on entrance examinations.

Another area of growth in the informal learning community are those related to summer camp opportunities. In the past ten years, the number of opportunities available for summer STEM camps has grown exponentially. One such camp is the See Blue STEM Camp at the University of Kentucky. The camp began in 2010 at a local middle school with just 8 students and has grown to serving over 400 students, 60% of whom are underrepresented, in 2018. The goal of this informal learning opportunity is to expose rising elementary and middle school students (grades 2–8), especially females and students of color, to positive, hands-on, authentic experiences in STEM in order to increase their interest in a STEM career. Through this work, we have found that as students engaged in the informal learning environment of the See Blue STEM Camp, they (a) develop identities as STEM learners and view themselves as scientists, mathematicians, and/or engineers; (b) develop a more positive disposition toward STEM; (c) obtain an accurate perception of STEM; (d) are exposed to STEM professionals who look like them; and (e) develop an interest in STEM and STEM-related fields (Mohr-Schroeder et al., 2014). Our research has revealed that informal learning environments provide opportunities for students to gain a deeper understanding of STEM concepts they may not have experienced in their traditional, formal school setting. Ultimately, middle grades students interested in STEM content and STEM careers increased statistically significantly after participating in the week-long day camp (Mohr-Schroeder et al., 2014). It is important to note the students we target to attend camp do not necessarily like STEM nor are they necessarily "high-achieving" students. Students also reported gaining more exposure to women and people of color in STEM fields and they liked engaging in authentic, hands-on learning experiences taught by knowledgeable STEM faculty.

3   Implications for K12 Education

Applying integrated STEM in solving real-world problems in an informal setting, such as the examples described above, lends practical applicability to, and thus complements, the traditional K-12 STEM education. It is important to note that none of the above projects were meant to replace formal schooling, but rather supplement and complement the work being done in schools. Through involvement in projects that usually have a community connection, students are afforded opportunities to put into action what they learn in the STEM classroom, thus making them appreciate STEM's practical significance (Burrows et al., 2018). At the same time, project participants realize the transdisciplinary nature of STEM (English, 2016) as they learn and apply it in authentic settings (Robert et al., 2018). In addition, the collaborative nature of community projects instills in students the spirit of cooperation and team work, helps improve students' skills set and allows students access to more role models as they work with various professionals in the STEM fields.

One attractive feature of informal STEM programs are the hands-on, collaborative activities. Typically, these types of activities, specifically those that engage students in design and engineering practices (Baran et al., 2016), are not always made available in the traditional classrooms owing to a variety of reasons: (a) time constraints that are basically dictated by the content coverage of the school curriculum; (b) limited classroom resources/materials for those types of activities; and (c) those activities require collaborative team efforts from experts in various fields whose services may not always be available during regular class hours. Hence, informal STEM programs are great opportunities for students to experience well-planned, hands-on activities that are not readily available in the traditional K-12 classrooms. Moreover, collaborative work that is typical in these types of programs promotes collaborative learning, while experiences working with various STEM professionals provide opportunity and access to broaden students' perceptions regarding all the possibilities of a career in a STEM field.

Informal learning sites, such as the science museum, can provide a contextualized environment where mathematical concepts could be used to explain scientific phenomena (Popovic & Lederman, 2015). The science museum, as an informal contextual setting, provides a suitable environment for connecting mathematics to solving real-world problems. By finding the mathematics behind the various science exhibits, students develop empirical knowledge on the interconnection between science and mathematics. This has meaningful implications to K-12 classrooms which do not always have the capability to provide a contextual learning environment as rich as that which the science museum can offer in making mathematical connections to real-world situations come to life.

However, this does not mean that informal learning environments should replace nor de-emphasize the importance of the formal K-12 classroom. Rather, STEM-based, co-curricular programs conducted in informal settings can offer activities and learning experiences that complement those provided by a traditional, formal STEM classroom. From such programs, students can derive benefits like informal mentoring, fun low-stakes learning, efficient time management, application of STEM content, feeling of accomplishments, confidence building, camaraderie, and exposure to new opportunities which they can add to those that they derive from their own classroom (Denson et al., 2015; Roberts et al., 2018). These themes that have emerged from the research (e.g., Denson et al., 2015; Mohr-Schroeder et al., 2014; Roberts et al., 2018) deserve serious considerations by education policy makers when thinking of possible reforms in the K-12 education program.

4 Conclusion and Future Research Directions

Informal learning in the STEM Education community has continued to gain a bigger following in recent years. Owing to unpredictable trends in work patterns of their parents, children have found themselves spending more and more time in supervised out-of-school, informal learning programs before and after school, and even during summers. With the recent trend towards formal-informal learning integration for added benefits to students, the charge of providing children's education is no longer a monopoly of the school.

K-12 curriculum can be designed to bridge formal and informal learning environments (Valentine, 2016) that leads to a more content-rich, more engaging and more motivating K-12 STEM education (Bevan et al., 2010). Formal environments provide academic structure to the bridged program while informal settings lend support in terms of human, material, and technological resources as well as provide complementary, engaging activities that are aligned with curriculum and standards. The concept of bridged formal and informal learning program can raise STEM literacy and can be applied as well to other content areas such as writing and literacy. With this in mind, we offer the following as future research directions for the field:

1. While much work is emerging on the impact of informal learning environments on student achievement, more longitudinal research studies are needed to look at the long-term impact on student achievement.
2. Informal learning environments have the opportunity to provide more equitable access to high quality STEM experiences for underrepresented populations. How are informal learning environments doing this and what impact does it have on our underrepresented populations?
3. Informal learning environments are meant to complement, not replace, the formal classroom. How are students internalizing what they are learning in the informal setting and how is that transferred to the formal classroom? What kinds of connections are students seeing between the informal learning opportunities and their formal classrooms?
4. Informal learning environments are more often than not aimed at groups of students. These groups vary in size and can be smaller or larger than their traditional classroom settings. Informal learning environments have the opportunity to take a deeper dive into STEM fields and provide students with a more hands-on, immersive experience alongside their peers and with access to experts in the field(s). Using social network analysis, what do those relationships look like? What kind of impact do these relationships have on learning outcomes in the informal environment? What kind of impact do these relationships have on learning outcomes in the formal environment?

The learning of STEM is no longer determined by what happens mainly in the formal classroom. Out-of-school STEM-based programs conducted in informal settings have become an essential component creating additional opportunity and access to STEM learning. STEM-based learning programs conducted in informal settings that have been proven to be productive are those that are engaging intellectually, socially, and emotionally; responsive to learner's interests, experiences, and cultural practices; and connect STEM learning in both school and out-of-school settings (NRC, 2015).

Acknowledgements

This work was supported by the National Science Foundation under Grant Numbers 1348281, 1758447, and 1239968. Any opinions, findings, and conclusions or recommendations expressed in this material are those of the author(s) and do not necessarily reflect the views of the National Science Foundation.

References

Afterschool Alliance. (2015). *Full STEM ahead: Afterschool programs step up as key partners in STEM education*. Retrieved from http://www.afterschoolalliance.org/aa3pm/STEM.pdf

Baran, E., Bilici, S. C., Mesutoglu, C., & Ocak, C. (2016). Moving STEM beyond schools: Students' perceptions about an out-of-school STEM education program. *International Journal of Education in Mathematics, Science and Technology, 4*(1), 9–19. doi:10.18404/ijemst.7133

Bell, P., Lewenstein, B., Shouse, A. W., & Feder, M. A. (Eds.). (2009). *Learning science in informal environments: People, places, and pursuits*. Retrieved from http://www.nap.edu/catalog.php?record_id=12190

Bennett, T. L. (2015). Examining levels of alignment between school and afterschool and associations with student academic achievement. *Journal of Expanded Learning Opportunities, 1*(2), 4–22.

Bevan, B., Dillon, J., Hein, G. E., Macdonald, M., Michalchik, V., Miller, D., ... Yoon, S. (2010). *Making science matter: Collaborations between informal science education organizations and schools*. Washington, DC: Center for Advancement of Informal Science Education.

Burrows, A., Lockwood, M., Borowczak, M., Janak, E., & Barber, B. (2018). Integrated STEM: Focus on informal education and community collaboration through engineering. *Education Sciences, 8*(4). doi:10.3390/educsci8010004

Coffield, F. (Ed.). (2000). *The necessity of informal learning* (Vol. 4). Bristol: Policy press.

Denson, C. D., Hailey, C., Stallworth, C. A., & Householder, D. L. (2015). Benefits of informal learning environments: A focused examination of STEM-based program environments. *Journal of STEM Education, 16*(1), 11–15.

English, L. D. (2016). STEM education K-12: Perspectives on integration. *International Journal of STEM Education, 3*(1), 3.

Lauer, P. A., Akiba, M., Wilkerson, S. B., Apthorp, H. S., Snow, D., & Martin-Glenn, M. (2006). Out-of-school-time programs: A meta-analysis of effects for at-risk students. *Review of Educational Research, 76*, 275–313.

Little, P., Wimer, C., & Weiss, H. B. (2008). After school programs in the 21st century: Their potential and what it takes to achieve it. *Issues and Opportunities in Out-of-school Time Evaluation, 10*, 1–12.

Mohr-Schroeder, M. J., Jackson, C., Miller, M., Walcott, B., Little, D. L., Speler, L., & Schroeder, D. C. (2014). Developing middle school students' interests in STEM via summer learning experiences: See blue STEM camp. *School Science and Mathematics, 114*(6), 291–301. doi:10.1111/ssm.12079

National Research Council (NRC), Division of Behavioral and Social Sciences and Education, Board on Science Education, & Committee on Successful Out-of-School STEM Learning. (2015). *Identifying and supporting productive STEM programs in out-of-school settings.* Washington, DC: National Academies Press. doi:10.17226/21740

NGSS Lead States. (2013). *Next Generation Science Standards: For states, by states.* Washington, DC: The National Academies Press.

Noam, G. G., Biancarosa, G., & Dechausay, N. (2002). *Afterschool education: Approaches to an emerging field.* Cambridge, MA: Harvard Education Press.

Popovic, G., & Lederman, J. S. (2015). *Implications of informal education experiences for mathematics teachers' ability to make connections beyond formal classroom programs in out-of-school settings.* Washington, DC: National Academies Press.

Valentine, D. L. (2016). *Examining bridges between informal and formal learning environments* (Doctoral dissertation). Retrieved from https://digitalcommons.unl.edu/cgi/viewcontent.cgi?article=1106&context=aglecdiss

CHAPTER 10

Bringing out the "T" in STEM Education

Bulent Dogan and Susie Gronseth

Abstract

Technology contributes to the design and implementation of STEM activities in multiple ways, but the "T" in STEM is sometimes overlooked. This chapter outlines the importance and benefits of technology in STEM education and highlights research and teaching strategies for promoting technological literacy. The chapter presents a summary of the types of technology tools, applications, and products that are presently available for conducting STEM activities and how these technologies can support the development of various 21st century skills in students, such as critical thinking, problem-solving, and collaboration. Emerging technologies for supporting STEM, such as coding through game design and programming, mobile app development, 3D modeling and printing, and digital storytelling are discussed. Finally, how these technologies are implemented in two projects as example is described.

1   Introduction

Technology is an indispensable component of STEM and project-based learning (PBL) activities. It may contribute to the design and implementation of STEM activities in a variety of ways. However, there are two primary patterns of technology use for STEM education: (1) *direct integration and embedding* of technology into STEM activities and (2) using technology as a *tool or facilitator* to enrich STEM PBL (Akgun, 2013).

In *direct integration*, technology is embedded into the science, engineering, and math learning activities. In this approach, learners are prompted to use technology to find solutions to the STEM-related problems in creative and challenging ways. Therefore, the use of hands-on technology-integrated activities can encourage students to apply innovative and creative problem-solving approaches as they design and "make" their projects (Hsu, Baldwin, & Ching, 2017). Examples of direct integration include coding/programming, working with robotics, and hands-on design and creation of 3D printed projects (Trust & Maloy, 2017).

Students as young as four to six years old can use structured play with apps, such as *Daisy the Dinosaur* and *Kodable*, to learn entry-level skills in coding (Pila, Aladé, Sheehan, Lauricella, & Wartella, 2019). Programming activities build on these foundational coding skills and provide opportunities for learners to explore computational processes of abstraction and decomposition, and they are especially effective when cognitive aspects involved in the thinking-doing are emphasized (Lye & Koh, 2014). Through robot programming activities, fifth grade students in one study made significant gains in their computational thinking, particularly in the area of working with algorithms (Chen et al., 2017). In another study, 11–12 year old students in Italy showed increases in self-efficacy and engagement following a workshop in which they built a Lego robot and programmed its movements using the visual programming language *Scratch* (Banzato & Tosato, 2017). Similarly, fifth and sixth grade students in Spain made significant improvements in computational thinking and practices during a two-year study of a curriculum that connected programming using *Scratch* with art and history concepts (Sáez-López, Román-González, & Vázquez-Cano, 2016).

Technology also can be used to *facilitate* STEM projects. In this role, technology serves as an enabler for a STEM learning experience and a "tool to move the learning forward" (Constantine, Różowa, & Szostkowski, 2017, p. 342). For example, learners can use productivity tools to perform tasks such as word processing, spreadsheet building, and database managing during a project. Note-taking programs, such as Evernote and OneNote, and project and task management software tools, such as Trello and Asana, are examples of productivity tools. Virtual whiteboard applications, such as Padlet, AWW App, and Sketchboard.io, can facilitate group brainstorming and idea development. Collaboration tools such as wikis, blogs, forums, and cloud-based storage (e.g. *Google Drive*, *Microsoft OneDrive*, and *Dropbox*) can assist learners with working together. Augmented reality (AR), a technology that allows users to interact with enhanced 3D digital objects through overlays on their visual field, is being used to facilitate scientific discovery. Through AR, learners explore STEM concepts, processes, and environments to gather information, make inferences, and construct deeper understandings (Ibáñez & Delgado-Kloos, 2018).

2   Integral Need for and Benefits of Technology in STEM Education

In many ways, computer science advancements and the technology revolution have impacted economy, daily life, and scientific enterprise (Sharp, 2017). With digital age transformations taking place globally, many in the U.S. contend that its K–12 education has fallen woefully behind in empowering students with the fundamental computer science knowledge and skills they will need for future success (Kasulis, 2017). According to a 2013 Brookings Institution study, for example, it is projected that 20 percent of all jobs will require a high level of knowledge in a related STEM field (Rothwell, 2013). At the same time, recent data from the U.S. Bureau of Labor Statistics shows that STEM jobs will grow "faster than other jobs over the next decade and will pay higher wages overall for qualified employees" (Rhodes, 2015, p. 8).

The U.S. Congress passed the STEM Education Act of 2015, which was an official designation that computer science is a key part of STEM (Guzdial & Morrison, 2016). Guzdial and Morrison (2016) argue that computer science classes should be available as easily as mathematics and science classes. It is concerning that around two-thirds of the U.S. states have minimal computer science education standards for secondary school education, and high school computer science courses, when they are offered as electives and not part of the core curriculum in most states (Wilson, Sudol, Stephenson, & Stehlik, 2010). It is estimated that only a quarter of U.S. high schools are even offering computer science courses, and these tend to be concentrated in affluent areas (Nager & Atkinson, 2016). This is concerning, as students must have deeper understandings of the fundamentals of computer science in order to be well-educated citizens in an increasingly computing-intensive world and to be positioned for the jobs of the 21st century.

Enhanced functionalities in computer and network technologies are fostering more student-centered, constructivist, and cooperative learning environments (Bottino & Robotti, 2007; ChanLin, 2008). Previous studies have reported numerous benefits of technology-focused, project-based learning activities in regards to student achievement in science (Barab, Thomas, Dodge, Carteaux, & Tuzun, 2005; Barak & Dori, 2005; Bottino & Robotti, 2007). Technology-rich PBL projects can promote development of technological literacy, defined as one's "ability to use, manage, evaluate, and understand technology" (ITEA, 2006, p. 4). It can also contribute to mastery of 21st century skills, such as critical thinking, problem-solving, communication, creativity, productivity, and adaptability (Partnership for 21st Century Learning, 2016; Trust & Maloy, 2017).

Technology can foster student-directed scientific inquiry of problems in real-world settings (Barak & Dori, 2005). When students participate in PBL projects, they often learn from each other cooperatively through group interactions, and they have opportunities to develop scientific knowledge through various collaborative investigations and explorations (ChanLin, 2008). Students with reading difficulties, though, may need additional supports to be successful in

PBL, and technology can be used to address this. For example, *Alien Rescue*, a technology-enhanced astronomy curriculum that contained embedded virtual probe building tools, instruments, and multimedia resources, enabled middle school students with reading difficulties to perform similarly in the activities to their reading proficient peers (Marino et al., 2010). Study results describe how the technology facilitated greater curricular accessibility by providing multiple means of engagement, representation of content, and action and expression, the core principles of the Universal Design for Learning framework (Meyer, Rose, & Gordon, 2014). Overall, research indicates that the use of technology in PBL may increase student achievement, interest, and motivation, as well as improve students' attitudes (ChanLin, 2008; Cobbs & Cranor-Buck, 2011; Harada, Kirio, & Yamamoto, 2008; Hayden, Ouyang, Scinski, Olszewski, & Bielefeldt, 2011).

STEM-focused summer camps can offer enrichment PBL-structured opportunities that investigate and solve real-world problems with hands-on tools. For instance, a recent summer STEM camp designed for high school students at New York University focused on fundamentals of bioengineering and incorporated technology with science curricula in engaging ways (Chacko et al., 2015). Technology integration included exploration of videos, articles, and websites on student laptops and iPads, virtual playgrounds, hands-on experiments, and touch screen-enabled activities in a paperless classroom setting. Some of the reported benefits included increases in student interest in STEM fields, cognitive gains in the topics taught in the camp, and development of self-regulation and inquiry skills.

3 A Deeper Look into Direct Integration of Technology in STEM Activities

Direct integration involves the use of hands-on, technology-integrated activities. Such activities provide opportunities for problem-solving, "making," and working directly with technology components, processes, and applications. Coding and mobile app development, digital storytelling, and 3D printing are three current areas of direct integration of technology as part of STEM activities. Each of these will be described in turn in the sections that follow.

3.1 *Coding and Mobile App Development for STEM*

Coding and mobile app development skills are increasingly needed in the digital age. The employment outlook for applications developer jobs is bright, projected to grow 24% in the U.S. (over three times the national employment growth average) during the next decade (Bureau of Labor Statistics, 2018). Organizations like Code.org, a non-profit with a mission to expand participation in computer science in schools, urge schools to teach young people about computer programing and coding (Lindegren, 2013). Though computer science bachelor's degrees were on the decline in the U.S. from about 2004–2009, there has been a steady upward trend since then, averaging about a 7% increase in degrees conferred each year (U.S. Department of Education, 2016).

Coding and use of mobile "apps" (i.e., small programs on mobile devices) for learning are growing areas. While worldwide increases in mobile phone usage are also fueling the emergence of mobile learning in education (Koszalka & Ntloedibe-Kuswani, 2010), benefits such as the ability for powerful and authentic "anywhere/anytime" learning make the development and implementation of mobile apps ideal for individualized informal learning (Godwin-Jones, 2011). Mobile apps are being used in K-12 education in various ways, such as for reflection, discussion, and sharing of learning (Leinonen, Keune, Veermans, & Toikkanen, 2014).

Students in K-12 have the capacity to not only be users of mobile apps but creators as well. Teaching mobile app design and development skills to the younger generation will enable them to create apps that can facilitate their own self-learning. App design can provide students with opportunities to translate content knowledge into gameplay elements, such as providing choice, scaffolding levels of complexity, incorporating vocabulary supports, and embedding student-authored feedback (Israel, Marino, Basham, & Spivak, 2013). An example of how the teaching of app design and development is being fostered in schools involves a partnership between Dell and the British initiative Apps for Good (Apps for Good, 2018). Volunteers from these organizations have worked with students in the school setting, guiding their development of smartphone apps to solve common problems in the UK.

Few studies, though, have examined educational uses of mobile app development or the impact of mobile app design and development on student STEM motivation. As coding and programming skills involved in app development are relevant to many emerging careers (such as software application developer, web developer, data scientist, and mobile developer), it is important that scholarly work address these areas to support the preparation of learners for such jobs. Programming expert and *Facebook* founder, Mark Zuckerberg, is an advocate for teaching coding. During the promotion of an "Hour of Code" initiative, he said, "In 15 years, we'll be teaching programming just like reading and writing ... and wondering why we didn't do it sooner" (Computer Science Education Week, 2014).

3.2   *Digital Storytelling*

Digital storytelling is another STEM-oriented activity that is facilitated through technology. Digital storytelling refers to a process of creating a short, purposeful movie containing a variety of multimedia components that work together to tell a story. An original script (often in the author's own voice) combined with multimedia elements are key components of the process (Dogan & Almus, 2018). Digital storytelling can be used as a PBL approach, and it has been found to positively impact learning motivation, attitude, problem-solving capability, and learning achievement (Chun-Ming, Hwang, & Huang, 2012).

Creating digital stories facilitates the development of 21st century skills (Jakes, 2006). In fact, 18 out of the 20 identified 21st century skills (including media literacy, technical skills, information literacy, visual literacy, creativity, risk taking, and others) can be achieved when students actively participate in the process of digital storytelling (Dogan & Almus, 2018; Jakes & Brennan, 2005; Robin, 2008). In addition, student motivation and engagement levels have been reported to increase with the use of digital storytelling (Banaszewski, 2005; Dogan 2010; Paull, 2002; Salpeter, 2005). It has also been reported that students generally enjoy learning through digital stories, and they tend to learn a great deal about the subject on which they create their digital stories (Dogan, 2011, 2012, 2014). Further, teachers report that one of the best features of digital storytelling is that it motivates users to learn more about the subject (Dogan, 2014).

3.3   *3D Printing*

Though 3D printing technology was developed in 1981, it has just been within the last decade that it has become sufficiently accessible and affordable for schools to adopt (Gross, Erkal, Lockwood, Chen, & Spence, 2014; MakerBot, 2015). 3D printing can be used to produce teaching aids, such as 3D models of unit cells for the study of crystal structure in materials science and engineering courses (Rodenbough, Vanti, & Chan, 2015). 3D printing provides students with authentic learning experiences that involve invention, design, and production, and it has shown to have a significant impact on student motivation and mathematical and technical skills (Kwon, 2017). 3D printing can facilitate the integration of creative arts by enabling students to model structures and objects (McGahern, Bosch, & Poli, 2015).

Advocates of 3D printing as part of learning experiences go so far as to claim that "students who learn about 3D printing in school will have a significant competitive advantage when they start looking for jobs after they graduate" (MakerBot, 2015). Empirical support for such a claim is beginning to emerge. The demand for engineers with 3D printing skills continues to rise as 35% of all engineering jobs require 3D printing skills (Rowe, 2014). Universities are increasingly incorporating 3D printing into engineering and other coursework. A 2013 New Media Consortium Horizon Report forecasted that 3D printing would reach mainstream use in the next several years, describing how it "allows for more authentic exploration of objects that may not be readily available to education universities" (Johnson, Adams Becker, Estrada, & Martín, 2013, p. 9). Indeed, 3D printing has since become more commonplace on campuses and is a component of the presently trending "makerspace" movement (Adams Becker et al., 2017).

With all the potential benefits of this disruptive technology in education, research is needed concerning how 3D printing can help motivate students in STEM areas and how it can be used instructionally. Some initial studies appear promising, illustrating viable educational applications of 3D modeling and printing in areas such as pre-service science teacher education (Novak & Wisdom, 2018) and secondary and college bioscience classes (Jittivadhna, Ruenwongsa, & Panijpan, 2010). There is also a need for the development of curricular materials prepared for teachers and students in K-12 STEM settings to maximize their experience with this emerging tool (Brown, 2015). This chapter will later describe one such project called "iTECH-STEM" that aims to investigate the potentials of this tool in STEM education.

4 Examples of Projects That Use Technology in STEM Activities

4.1 Students on the Stage (S.O.S.) Model

The Students on the Stage (S.O.S.) model is a PBL STEM initiative that addresses the increased need for students to acquire 21st century skills (Sahin, 2015). Through PBL activities, students are exposed to deep learning experiences that are inquiry-based, student-centered, and curriculum-integrated. Students experience PBL projects at various levels through S.O.S., identified as *Level I*, *Level II*, and *Level III*. Projects at Level I tend to be short-term and primarily target 21st century skills within the context of the curriculum. Level I activities are curriculum-based learning experiences that are essential activities in a course (Dogan & Robin, 2015). This is contrasted with Level II and Level III projects, which are designed to be interdisciplinary and structured to provide opportunities for the application and development of critical skills at greater depths through engagement in meaningful inquiry of personal interest. They tend to be semester-long projects that students begin during the first quarter of a course and complete by the end of the semester (Dogan & Robin, 2015).

The S.O.S. model structures activities so that students complete a Level II project in addition to curriculum-based Level I investigations. While all students are required to complete Level I projects, students may complete either a project from a set of Level II projects whose handouts are prepared by a PBL team or they can come with their project ideas and driven question as a Level III project (see Figure 10.1). As described in Dogan and Robin (2015), a program framed by the S.O.S. model may include technology-integrated activities such as
- Creating projects and achieving productivity through Chromebooks and Google tools
- Creating digital stories as final video presentations and collaboration
- Building websites to showcase completed work and collaboration
- Designing and publishing project brochures to share information
- Providing technology training for students

The sections that follow will describe components and recommendations for implementing these activities as part of an S.O.S. program.

FIGURE 10.1 Students working on PBL projects in S.O.S model

4.1.1 Using Chromebooks

Chromebooks can be quite useful tools for PBL projects in the S.O.S. model. Once set up with Google accounts, students can access a variety of Google services, including Gmail for communication, YouTube for storage of digital stories, Google Docs for productivity applications, Google Drive for data and file storage, and Google Sites for website development. School-provided Google accounts are recommended, as they can facilitate more seamless integration of the Chromebooks within PBL projects and digital presentations due to the Google accounts already being incorporated into the Chromebook operating system (OS).

4.1.2   Digital Storytelling Final Presentations

Creating digital stories is the recommended approach for the final video presentation component of the S.O.S. model (Dogan & Robin, 2015). Digital stories allow students not only to acquire the technical video editing skills but also to be actively engaged in the process of PBL. They also serve as visible final artifacts and end-products in which students showcase their work through shared means on varied digital platforms (see Figure 10.2). Upon completion of the projects, students can be directed to upload their digital stories to their YouTube channels and Google Drive cloud storage spaces for sharing and archiving purposes.

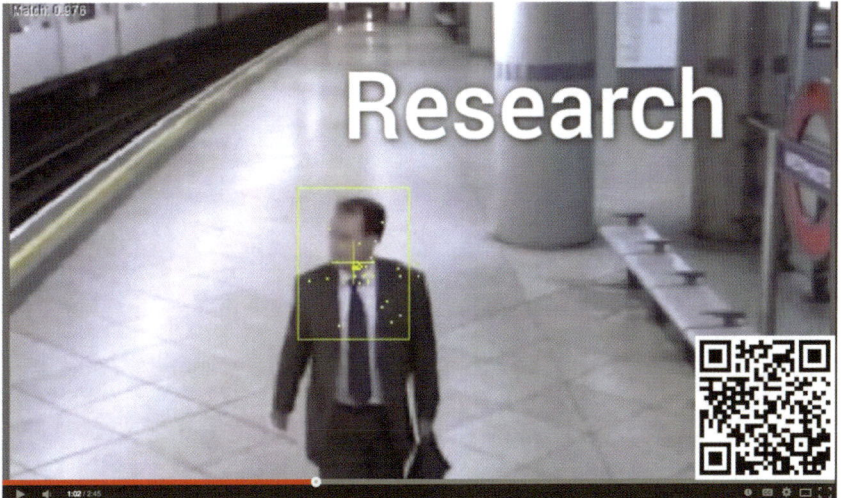

FIGURE 10.2   A digital story created for a PBL project

4.1.3   Developing Websites with Google Sites

In Level II and III projects, students can create websites using Google Sites to compile all of their PBL work in one place (see Figure 10.3). This helps with communication between the students and their instructors, and it also provides a means to showcase their work to a broader audience (Dogan & Robin, 2015). Google Sites is a free and easy-to-use website development tool that is available as part of the Google suite. It is a natural choice for website development in the

S.O.S. model, as it can be used seamlessly on Chromebooks. Using Google Sites, students can build websites individually or collaborate and edit site content together, similar to the editing of a document. Google Sites enables controls regarding who has viewing and editing access, and pages can be made private to specific users or publicly viewable for wider sharing through the Internet.

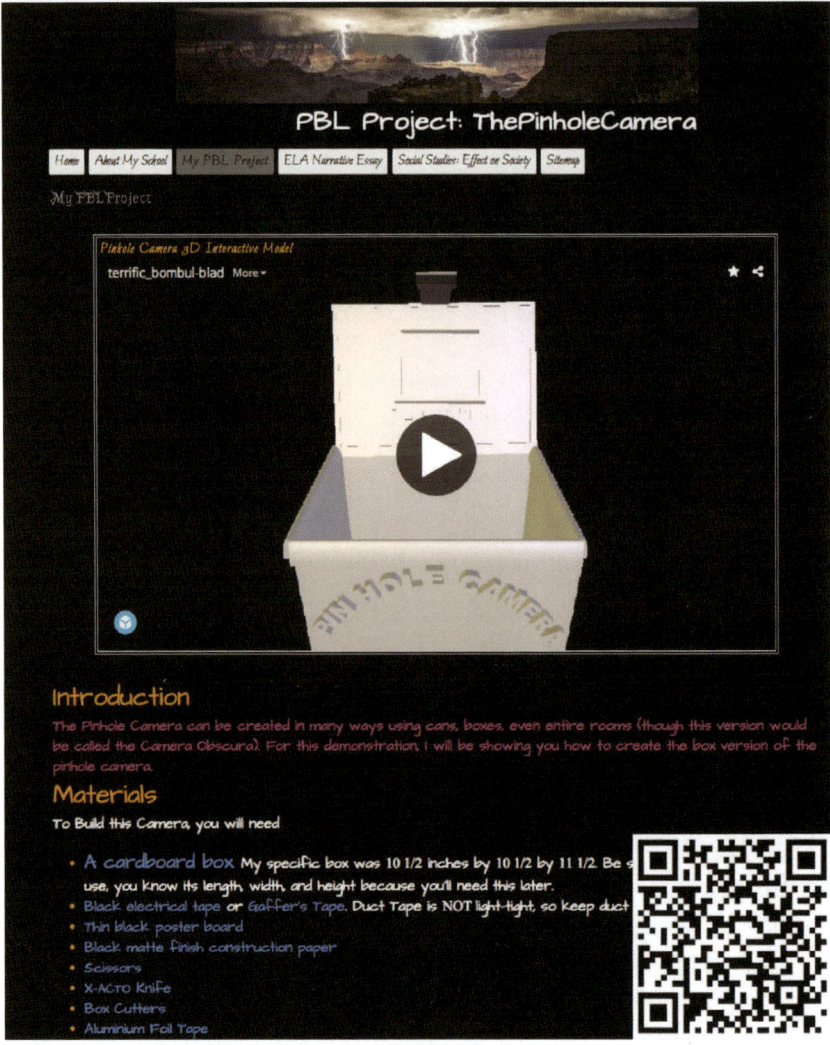

FIGURE 10.3 A website created for a PBL project

4.1.4   Designing and Publishing Project Brochures

Another activity in the S.O.S. model involves student design of project brochures that detail information on local science fairs, academic competitions, exhibitions, and presentations (Dogan & Robin, 2015). The brochures include information on the PBL projects, QR codes for sharing of digital story videos, and links to Google Sites websites (see Figure 10.4). Students can use cloud-based productivity tools, such as Google Docs or Microsoft Word Online, to create their brochures. Once their designs are completed, students can showcase their brochure work on their websites.

FIGURE 10.4   A brochure design for a PBL project in the S.O.S model

4.2   iTECH-STEM Project

The "Innovative Technology Challenges for Science, Technology, Engineering, and Mathematics" (iTECH-STEM) project is a STEM-focused learning initiative sponsored by the University of Houston (UH) STEM Center. It is hosted at the UH – Sugar Land campus and directed by faculty from the UH College of Education's Department of Curriculum and Instruction. The main purpose of this project is to provide an innovative program that promotes and teaches computer science and technology topics to students in third through sixth grades (see Figure 10.5). The selected STEM-focused topics are taught through the following "Domains:"
- Digital Storytelling
- 3D Modeling and Printing
- Programming and Coding
- STEM Art

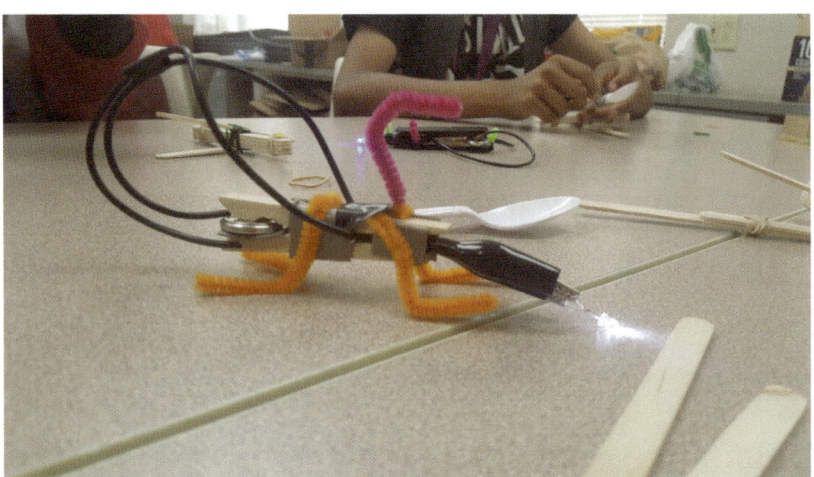

FIGURE 10.5   A circuit bug created in STEM Art class as part of the iTECH-STEM program

The iTECH-STEM project includes several components – a summer academy, an ambassador program, and challenges throughout the school year. The summer academy involves teaching the iTECH domains to elementary age students with hands-on technology projects. Further information and current project updates are available on the iTECH-STEM website (http://www.itech-stem.org).

4.2.1   Digital Storytelling in iTECH-STEM

Students in the iTECH-STEM summer camp create digital stories on topics that are meaningful to them, and the digital storytelling projects are interconnected with their projects that they complete in other iTECH-STEM domains. For instance, students learn how to create a stop motion video as part of a STEM Art class (see Figure 10.6). The activity guides students to work with clay and draw pictures on paper in "frames" for their stop motion projects. Students also take pictures of their work to use in a Digital Storytelling class, where they incorporate the pictures, story, and narration into larger video projects.

FIGURE 10.6 Students working on stop-motion Digital Storytelling projects during iTECH-STEM

As part of Digital Storytelling classes, students also research about the history of stop motion and the process of the stop motion technique. Students complete their video projects by writing unique stories of their own and presenting their stop motion work within their digital stories.

4.2.2   Programming in iTECH-STEM

In iTECH-STEM, students are taught fundamentals of computer programming through graphical and visual programming tools, such as *Scratch* (see Figure 10.7). *Scratch* is a visual programming language developed by the Lifelong Kindergarten Group at the MIT Media Lab (Monroy-Hernández & Resnick, 2008). It aims to simplify the process of creating and programming animations, games, music, and interactive stories (Resnick et al., 2009).

The *Scratch* programming language is designed to teach computational thinking using a simple building-block approach to software development that focuses more on problem-solving than on specific syntax (Resnick et al., 2009). As part of iTECH-STEM, students create games and short programs using

FIGURE 10.7   Students working on programming projects in iTECH-STEM

Scratch, providing opportunities to apply problem-solving skills and further develop computational thinking.

4.2.3   3D Printing in iTECH-STEM

Within the 3D Modeling and Printing domain of iTECH-STEM, students are taught 3D printing fundamentals using programs such as TinkerCad (see Figure 10.8). This facet of iTECH-STEM addresses spatial thinking and mental rotation skills through hands-on projects.

Students work on individual projects (such as designing key rings and sculptures) and interdisciplinary collaborative projects (such as creating models of Leonardo da Vinci's catapult design). Students simultaneously work on creating "maquettes" (preliminary models) of the catapult in STEM Art class. They

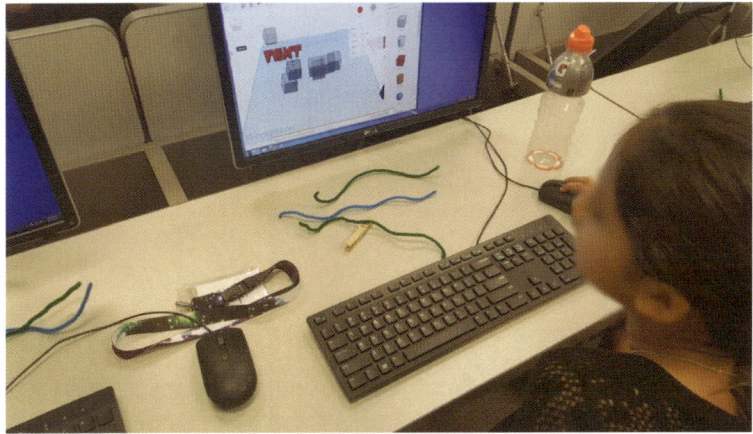

FIGURE 10.8   3D Modeling in iTECH-STEM

FIGURE 10.9   Students learn how to use 3D-printers in iTECH-STEM

also learn about how 3D printing works when they go through the process of printing their models (see Figure 10.9).

5   **Limitations and Future Research Questions**

Scholarly work on uses of technology for STEM learning has demonstrated some learning benefits such as increases in learner motivation, engagement, and attitudes toward STEM (Ibáñez & Delgado-Kloos, 2018). Because the technology tends to support instruction that is constructivist and collaborative, researchers in studies thus far have often designed their own instruments for their specific contexts. Robust evaluation measures that could verify learner cognitive gains across contexts would help to strengthen the knowledge base for using specific technologies in STEM activities. Oftentimes, innovative, technology-rich STEM projects and programs are conducted outside of regular school hours, such as in after-school workshops and summer camps, which can add a "novelty" factor that may bias results (Banzato & Tosato, 2017). Some schools and teachers, though, are finding ways to incorporate computer science concepts and specific STEM technologies into the curriculum (Trust & Maloy, 2017), and research that incorporates more longitudinal work over a school semester or longer is emerging (e.g., Chen et al., 2017). Schools are faced with the challenge of adequately equipping their teachers to effectively integrate these technologies with their pedagogy and content (Constantine et al., 2017). This could somewhat be addressed during a teacher preparation program (e.g., Novak & Wisdom, 2018), but professional development that updates and expands their understanding and skill in using these and future emerging technologies will continue to be needed.

There are, therefore, numerous opportunities for future lines of research to investigate the use of technology in STEM learning. While many prior studies have been conducted using small samples and case study methodology, these could be replicated in additional contexts to identify cross-case themes and contribute to greater understanding regarding implementation of these activities. There are other facets of inquiry that could be explored, such as –

- In what ways do learning experiences that involve using specific technologies in STEM-related activities impact learner cognitive gains?
- To what extent do STEM-related activities that incorporate specific technologies contribute to students' development of 21st century skills?
- How can specific technologies in STEM-related activities address curricular differentiation that supports diverse learner needs?
- What instructional strategies facilitate effective integration of specific technologies to support STEM learning?
- What potential do technology-rich STEM curricular materials have for enhancing academic achievement, self-efficacy, critical thinking and problem solving?
- How do K-12 STEM experiences contribute to students with disabilities' interest in pursuing STEM-related careers?
- What curricular design approaches can target female learner needs in developing STEM-career ready skills?

6 Conclusion

Strategic incorporation of technology into STEM instruction provides opportunities for students to develop technological literacy and 21st century skills that can help to position them for future careers. Projects like the S.O.S. model and iTECH-STEM demonstrate how the teaching of coding, digital storytelling, and 3D printing can be integrated with content for inquiry-oriented, project-based learning experiences for students. Though initial studies on applications of these technologies in education are promising, scholarly inquiry is needed to further develop frameworks for how to approach the "T" in STEM education.

References

Adams Becker, S., Cummins, M., Davis, A., Freeman, A., Hall Giesinger, C., & Ananthanarayanan, V. (2017). *NMC Horizon report: 2017 Higher education edition*. Austin, TX: The New Media Consortium.

Akgun, O. (2013). Technology in STEM project-based learning. In R. M. Capraro, M. M. Capraro, & J. R. Morgan (Eds.), *STEM project-based learning: An integrated Science, Technology, Engineering, and Mathematics (STEM) approach* (2nd ed.). Rotterdam, Netherlands: Sense Publishers.

Apps for Good. (2018). *What we do*. Retrieved from https://www.appsforgood.org/public/about-us

Banaszewski, T. M. (2005). *Digital storytelling: Supporting digital literacy in grades 4–12* (Unpublished master's thesis). Georgia Institute of Technology, Atlanta, GA. Retrieved from http://techszewski.blogs.com/techszewski/files/TBanaszewski_DS_thesis.pdf

Banzato, M., & Tosato, P. (2017). An exploratory study of the impact of self-efficacy and learning engagement in coding learning activities in Italian middle school. *International Journal on E-Learning, 16*(4), 349–369.

Barab, S., Thomas, M., Dodge, T., Carteaux, R., & Tuzun, H. (2005). Making learning fun: Quest Atlantis, a game without guns. *Educational Technology Research and Development, 53*(1), 86–107.

Barak, M., & Dori, Y. J. (2005). Enhancing undergraduate students' chemistry understanding through project-based learning in an IT environment. *Science Education, 89*(1), 117–139.

Bureau of Labor Statistics, U.S. Department of Labor. (2018, April). *Occupational outlook handbook. Software developers*. Retrieved from https://www.bls.gov/ooh/computer-and-information-technology/software-developers.htm

Bottino, R. M., & Robotti, E. (2007). Transforming classroom teaching & learning through technology: Analysis of a case study. *Educational Technology & Society, 10*, 174–186.

Brown, A. (2015). 3D printing in instructional settings: Identifying a curricular hierarchy of activities. *TechTrends, 59*(5), 16–24.

Chacko, P., Appelbaum, S., Kim, H., Zhao, J., & Montclare, J. K. (2015). Integrating technology in STEM education. *Journal of Technology and Science Education, 5*(1), 10.

ChanLin, L.-J. (2008). Technology integration applied to project-based learning in science. *Innovations in Education and Teaching International, 45*(1), 55–65.

Chen, G., Shen, J., Barth-Cohen, L., Jiang, S., Huang, X., & Eltoukhy, M. (2017). Assessing elementary students' computational thinking in everyday reasoning and robotics programming. *Computers & Education, 109*, 162–175. https://doi.org/10.1016/j.compedu.2017.03.001

Chun-Ming, H., Hwang, G.-J., & Huang, I. (2012). A project-based digital storytelling approach for improving students' learning motivation, problem-solving competence and learning achievement. *Journal of Educational Technology & Society, 15*(4), 368–379.

Cobbs, G. A., & Cranor-Buck, E. (2011). Getting into GEAR. *Mathematics Teaching in the Middle School, 17*(3), 160–165.

Computer Science Education Week. (2014). *Hour of code*. Retrieved from https://csedweek.org/files/Zuckerberg_poster.pdf

Constantine, A., Różowa, P., & Szostkowski, A. (2017). The "T" in STEM: How elementary science teachers' beliefs of technology integration translate to practice during a co-developed STEM unit. *Journal of Computers in Mathematics and Science Teaching, 36*(4), 339–349.

Dogan, B. (2010). Educational use of digital storytelling: Research results of an online Digital Storytelling Contest. In D. Gibson & B. Dodge (Eds.), *Proceedings of Society for Information Technology & Teacher Education International Conference 2010* (pp. 1061–1066). San Diego, CA: AACE. Retrieved from http://www.editlib.org/p/33494

Dogan, B. (2011). Educational uses of digital storytelling: Results of DISTCO 2010, an online Digital Storytelling Contest. In M. Koehler & P. Mishra (Eds.), *Proceedings of Society for Information Technology & Teacher Education International Conference 2011* (pp. 1104–1111). Nashville, TN: AACE. Retrieved from http://www.editlib.org/p/36434

Dogan, B. (2012). Educational uses of digital storytelling in K-12: Research results of Digital Storytelling Contest (DISTCO) 2012. In P. Resta (Ed.), *Proceedings of Society for Information Technology & Teacher Education International Conference 2012* (pp. 1353–1362). Austin, TX: AACE. Retrieved from http://www.editlib.org/p/39770

Dogan, B. (2014). Educational uses of digital storytelling in K-12: Research results of a Digital Storytelling Contest (DISTCO) 2013. In M. Searson & M. N. Ochoa (Eds.), *Society for Information Technology & Teacher Education International Conference*

2014 (pp. 520–529). Jacksonville, FL: AACE. Retrieved from http://www.editlib.org/p/130802

Dogan, B., & Almus, K. (2018). Developing and assessing media literacy through digital storytelling. In M. Yildiz, S. Funk, & B. De Abreu (Eds.), *Promoting global competencies through media literacy* (pp. 65–78). Hershey, PA: IGI Global.

Dogan, B., & Robin, B. (2015). Technology's role in STEM education and the STEM SOS model. In A. Sahin (Ed.), *A practice-based model of STEM teaching: STEM Students on the Stage (SOS)* (pp. 77–94). Rotterdam, The Netherlands: Sense Publishers.

Godwin-Jones, R. (2011). Emerging technologies: Mobile apps for language learning. *Language Learning & Technology, 15*(2), 2–11.

Gross, B. C., Erkal, J. L., Lockwood, S. Y., Chen, C., & Spence, D. M. (2014). Evaluation of 3D printing and its potential impact on biotechnology and the chemical sciences. *Analytical Chemistry, 86*(7), 3240–3253.

Guzdial, M., & Morrison, B. (2016). Growing computer science education into a STEM education discipline. *Communications of the ACM, 59*(11), 31–33.

Harada, V. H., Kirio, C. H., & Yamamoto, S. H. (2008). *Collaborating for project-based learning in grades 9–12*. Columbus, OH: Linworth Publishing.

Hayden, K., Ouyang, Y., Scinski, L., Olszewski, B., & Bielefeldt, T. (2011). Increasing student interest and attitudes in STEM: Professional development and activities to engage and inspire learners. *Contemporary Issues in Technology and Teacher Education, 11*(1), 47–69.

Hsu, Y.-C., Baldwin, S., & Ching, Y.-H. (2017). Learning through making and maker education. *TechTrends, 61*(6), 589–594.

Ibáñez, M.-B., & Delgado-Kloos, C. (2018). Augmented reality for STEM learning: A systematic review. *Computers & Education, 123*, 109–123.

International Technology Education Association (ITEA). (2006). *Technological literacy for all: A rationale and structure for the study of technology*. Retrieved from http://www.iteaconnect.org/TAA/TAA_Literacy.html

Israel, M., Marino, M. T., Basham, J. D., & Spivak, W. (2013). Fifth graders as app designers: How diverse learners conceptualize educational apps. *Journal of Research on Technology in Education, 46*(1), 53–80. https://doi.org/10.1080/15391523.2013.10782613

Jakes, D. (2006, March 1). *Standards-proof your digital storytelling efforts*. Retrieved from http://www.techlearning.com/tech/media-coordinators/0018/standards-proof-your-digital-storytelling-efforts/43347

Jakes, D., & Brennan, J. (2005). *Capturing stories, capturing lives: An introduction to digital storytelling*. Retrieved from http://www.jakesonline.org/dstory_ice.pdf

Jittivadhna, K., Ruenwongsa, P., & Panijpan, B. (2010). Beyond textbook illustrations: Hand-held models of ordered DNA and protein structures as 3D supplements to enhance student learning of helical biopolymers. *Biochemistry and Molecular Biology Education, 38*(6), 359–364.

Johnson, L., Adams Becker, S., Estrada, V., & Martín, S. (2013). *Technology outlook for STEM+ education 2013–2018: An NMC Horizon project sector analysis.* Austin, TX: The New Media Consortium.

Kasulis, K. (2017). *The US desperately need computer science majors, so keep coding.* Retrieved from https://mic.com/articles/182644/the-us-desperately-needs-computer-science-majors-so-keep-coding#.GhSgdCQxF

Koszalka, T. A., & Ntloedibe-Kuswani, G. S. (2010). Literature on the safe and disruptive learning potential of mobile technologies. *Distance Education, 31*(2), 139–157.

Kwon, H. (2017). Effects of 3D printing and design software on students' overall performance. *Journal of STEM Education, 18*(4), 37–42.

Leinonen, T., Keune, A., Veermans, M., & Toikkanen, T. (2014). Mobile apps for reflection in learning: A design research in K-12 education. *British Journal of Educational Technology, 47*(1), 184–202.

Lindegren, R. (2013). *Teaching children how to code.* Retrieved from http://opensource.com/education/13/4/teaching-kids-code

Lye, S. Y., & Koh, J. H. L. (2014). Review on teaching and learning of computational thinking through programming: What is next for K-12? *Computers in Human Behavior, 41*, 51–61. https://doi.org/10.1016/j.chb.2014.09.012

MakerBot. (2015). Changing the way things are designed, made, and distributed. *Campus Technology.* Retrieved from https://campustechnology.com/whitepapers/2015/08/makerbot-changing-the-way-things-are-made-082115.aspx?tc=page0

Marino, M. T., Black, A. C., Hayes, M. T., & Beecher, C. C. (2010). An analysis of factors that affect struggling readers' achievement during a technology-enhanced STEM astronomy curriculum. *Journal of Special Education Technology, 25*(3), 35–47.

McGahern, P., Bosch, F., & Poli, D. (2015). Enhancing learning using 3D printing: An alternative to traditional student project methods. *The American Biology Teacher, 77*(5), 376–377.

Meyer, A., Rose, D. H., & Gordon, D. (2014). *Universal design for learning: Theory and practice.* Wakefield, MA: CAST Professional Publishing.

Monroy-Hernández, A., & Resnick, M. (2008). Empowering kids to create and share programmable media. *Interactions, 15*(2), 50–53.

Nager, A., & Atkinson, R. (2016). *The case for improving US computer science education.* Washington, DC: Information Technology & Innovation Foundation. Retrieved from https://itif.org/publications/2016/05/31/case-improving-us-computer-science-education

Novak, E., & Wisdom, S. (2018). Effects of 3D printing project-based learning on preservice elementary teachers' science attitudes, science content knowledge, and anxiety about teaching science. *Journal of Science Education and Technology, 27*(5), 1–21. https://doi.org/10.1007/s10956-018-9733-5

Partnership for 21st Century Learning. (2016). *Framework for 21st century learning*. Retrieved from http://www.p21.org/storage/documents/docs/P21_framework_0816.pdf

Paull, C. N. (2002). *Self-perceptions and social connections: Empowerment through digital storytelling in adult education*. Retrieved from ProQuest Dissertations & Theses Global. (251754612)

Pila, S., Aladé, F., Sheehan, K. J., Lauricella, A. R., & Wartella, E. A. (2019). Learning to code via tablet applications: An evaluation of *Daisy the Dinosaur* and *Kodable* as learning tools for young children. *Computers & Education, 128*, 52–62. https://doi.org/10.1016/j.compedu.2018.09.006

Resnick, M., Maloney, J., Monroy-Hernández, A., Rusk, N., Eastmond, E., Brennan, K., ... Kafai, Y. (2009). Scratch: Programming for all. *Communications of the ACM, 52*(11), 60–67.

Rhodes, R. (2015). Reshaping the educational environment for tomorrow's workforce. *Educause Review, 50*(4). Retrieved from http://er.educause.edu/articles/2015/6/reshaping-the-educational-environment-for-tomorrows-workforce

Robin, B. (2008). The effective uses of digital storytelling as a teaching and learning tool. In J. Flood, S. B. Heath, & D. Lapp (Eds.), *Handbook of research on teaching literacy through the communicative and visual arts* (Vol. 2, pp. 429–440). New York, NY: Lawrence Erlbaum Associates.

Rodenbough, P. P., Vanti, W. B., & Chan, S.-W. (2015). 3D-printing crystallographic unit cells for learning materials science and engineering. *Journal of Chemical Education, 92*(11), 1960–1962.

Rothwell, J. (2013). *The hidden STEM economy*. Washington, DC: Brookings Institute. Retrieved from http://www.brookings.edu/~/media/Research/Files/Reports/2013/06/10%20stem%20economy%20rothwell/SrvyHiddenSTEMJune3b.pdf

Rowe, A. Z. (2014, September 4). Demand for 3D printing skills soars. *Wanted Analytics*. Retrieved from https://www.wantedanalytics.com/analysis/posts/demand-for-3d-printing-skills-soars

Sáez-López, J.-M., Román-González, M., & Vázquez-Cano, E. (2016). Visual programming languages integrated across the curriculum in elementary school: A two year case study using "Scratch" in five schools. *Computers & Education, 97*, 129–141. https://doi.org/10.1016/j.compedu.2016.03.003

Sahin, A. (2015). How does the STEM SOS model help students acquire and develop 21st century skills? In A. Sahin (Ed.), *A practice-based model of STEM teaching: STEM Students on the Stage (SOS)* (pp. 173–188). Rotterdam, The Netherlands: Sense Publishers.

Salpeter, J. (2005). Telling tales with technology: Digital storytelling is a new twist on the ancient art of the oral narrative. *Technology & Learning, 25*(7), 18–24.

Sharp, T. (2017). *9 Ways computer science has had a positive impact on society*. Retrieved from https://www.eucossa.org/9-ways-computer-science-positive-impact-society/

Trust, T., & Maloy, R. W. (2017). Why 3D print? The 21st-century skills students develop while engaging in 3D printed projects. *Computers in the Schools, 34*(4), 253–266.

U.S. Department of Education, National Center for Education Statistics. (2016). *Digest of education statistics, Integrated Postsecondary Education Data System (IPEDS)*. Retrieved from https://nces.ed.gov/programs/digest/d16/tables/dt16_322.10.asp

Wilson, C., Sudol, L. A., Stephenson, C., & Stehlik, M. (2010). Running on empty: The failure to teach K-12 computer science in the digital age. *Association for Computing Machinery*. Retrieved from https://runningonempty.acm.org/fullreport2.pdf

PART 3

Bringing out the "E" in STEM in K-12 Settings

CHAPTER 11

Engineering Education in K-12: A Look Back and Forth

Christine Guy Schnittka

Abstract

Engineering is a way of thinking. It is a process of problem solving, a process of designing solutions, and a process of using what we know to make the world a better place. Over the past 20 years, engineering has made its way from solely a career for adults to a subject taught in K-12 schools across the country and embedded in most state standards. While early research on teaching engineering in K-12 schools focused on only the E in Science, Technology, Engineering, and Mathematics (STEM), today there is a focus on using engineering as a conduit through which to teach science, mathematics, technological skills, to truly integrate the four silos in the acronym, and teach 21st century skills like communication and collaboration. Guidelines have been established to help teachers choose or create engineering-infused lessons, and future research will focus on how to best prepare teachers, how to use engineering across all fields of science and mathematics, and how to use engineering to level the playing field for diverse learners and broaden participation in STEM careers.

1   Introduction

Decades ago, engineering education was something only for college students. In high school, prospective engineers took science and math courses to prepare them for their intended futures. Most future engineers knew there were different fields of engineering, and each field focused on a different topic or field. Electrical engineers dealt with electricity, civil engineers built roads and bridges, and mechanical engineers designed devices one could see and touch. Teaching K-12 science through the lens of engineering was not common in those days. If engineering was taught in a school science classroom twenty years ago, the task was often delegated to the occasional guest speaker (Pols, Rogers, & Miaoulis, 1994; Carroll, 1997).

Some youth did learn about engineering outside of school through Saturday enrichment programs, summer camps, or after-school clubs, and some visited museums like the Museum of Science and Technology in Chicago, or perhaps attended an engineering open-house at a local university and participated in a university-sponsored engineering design competition. When organized by universities or engineering organizations, these experiences for youth were deemed "outreach" from the engineering professionals delivering the experience and were often aimed at increasing the number and/or diversity of people pursuing engineering careers. Many universities sponsored "E-day" in the month of February during National Engineers Week so that high school students could come and learn about careers, providing tours for teens and their parents. National Engineers Week was established in 1951 by the National Society of Professional Engineers and occurs every year during the week of George Washington's birthday since he is considered to be the nation's first engineer due to his surveying work (DiscoverE, 2018).

In 1993, in the AAAS publication, *Benchmarks for Science Literacy*, a chapter was devoted to the designed world, and recommendations were made for science students to have tinkering experiences to help them understand the laws of nature (AAAS, 1993). Suggestions included having students make solar collectors, design wind turbine blades, understand how microphones and speakers work, and manipulate tabletop robots. While the word, "engineering," was not used in this chapter, the concept was definitely there. This marked the beginning of engineering integration into K-12 classrooms outside of Career Tech or shop classes. The Journal of Engineering Education began publishing articles on K-12 engineering in 1994 (Pols, Rogers, & Miaoulis, 1994; Crawford, Wood, Fowler, & Norrell, 1994). The National Science Teachers Association (NSTA) began publishing a few articles for teachers, with ideas for engineering integration in 1995, but it was not until 2004 that engineering curriculum ideas became more accessible to K-12 teachers.

2   Literature Review

2.1   *The Early Years*

In 1996, the National Research Council published the National Science Education Standards, which included a section on science and technology, and teachers were encouraged to complement science activities with straightforward activities that solve a human need or problem through design such as making electric circuits that act as alarm (NRC, 1996). The Egg Drop challenge was described in depth. Subsequently, the first research studies on engineering design-based interventions published in journals on K-12 education, were from 1995–1998 by Wolff-Michael Roth, who studied 4th and 5th graders as they interacted during a civil engineering design activity. Roth (1996) was working with a pair of elementary school teachers and their one class of 29 fourth and fifth grade pupils in British Columbia, Canada, as they became immersed in a 13-week unit on structures. The curriculum, called "Engineering for Children: Structures," was designed by one of the teachers for the Canadian non-profit group, *Association for the Promotion and Advancement of Science* (Roth, 1995). Through an open-ended process, students worked with partners of their choosing to design, build, and test towers, bridges, and domes. The children in this study were learning science through the design process. They were working with loosely defined problems in authentic environments, which allowed them to experience a community of practice with socially constructed knowledge mediated by expert peers, teachers, and others. Roth directly observed and videotaped lessons, took field notes, interviewed students and teachers, and examined students' engineering logbooks which consisted of design ideas, drawings, photographs, reflections, and definitions. He gave a pre- and post-assessment which consisted of the open-ended question, "Engineering is ...," to which students wrote text or drew pictures. Additionally, the results of a pre-assessment challenge to build a bridge from a single sheet of paper were contrasted with the final bridge design challenge. Discourse, conversational, and interaction analyses further enriched the data corpus.

Roth worked within the theoretical perspective of situated cognition, which seeks to explain knowledge as a product of the context, activity, and culture of the learner (Brown, Collins, & Duguid, 1989). He found that while children claimed sole authorship of their designs, the entire learning environment contributed in fundamental ways. The tools and materials available, teacher and peer influences, even current and previous states-of-design contributed to the final products. In keeping with the philosophy of situated cognition, the design processes did not emerge from solitary ideas and skills, but were situated within the interactions of the community, the materials and tools at hand, and the current and past states-of-design. He found that the process of design gave students a way to integrate thinking in the abstract with acting in the concrete. The students learned how to cope with the complex nature of open-ended design, their self-confidence grew, and they learned to think flexibly about tools and materials, putting them to new and creative uses. They learned about the process of design, how to negotiate with one another, and how to write and communicate ideas about engineering design. They did not learn merely a collection of facts and definitions related to civil engineering constructions, they learned a broad variety of concepts through a set of meaningful experiences, set in context. "The collective design of artifacts is a powerful context of learning because it allows children to unload cognition into the physical and social environment" (p. 161).

Roth did not directly assess students' science content knowledge related to the physics of structures, although he expected that the design activities would help the students learn about properties of materials, forces, simple machines, energy, and work. He implied that traditional evaluation of student performance and learning is difficult in design environments, and that constructing artifacts does more for students than merely motivate them to learn; it gives students valuable problem-solving experiences in ill-structured settings. This study is considered to be one of the seminal examples of research on design-based science, however an evaluation of science knowledge was neither planned for nor executed.

Australian educators McRobbie, Stein, and Ginns (2001) found that students engaged in design activities did not implicitly learn science principles. Their study involved students in two sixth grade classes, challenged to build self-propelled model boats. A pre-cut hull and a variety of building materials were provided to each group of student teams. Primarily, students learned about the design process, with "little explicit knowledge ... developed about the natural world, that is, knowledge about the properties of materials, and scientific and engineering knowledge related to the task at hand" (p. 11). The researchers concluded that teachers who incorporate design into the classroom with the goal of teaching science must find a balance between creating authentic design challenges and assisting students in their understandings of scientific knowledge. They proposed that teachers intervene at appropriate times to help students acquire richer understandings of the natural world.

Seiler, Tobin, and Sokolic (2001) conducted a study on design-based science in an urban high school designed for at-risk students. The researchers were specifically seeking to learn whether an active-learning science curriculum centered on students' own interests would motivate them to learn. Students in a physical science class were challenged to design, build and test small model cars that were to be powered by the exhaust from an inflated balloon. This design activity was part of a larger unit on Newton's laws of motion. Researchers noted some initial concern because the design task was imposed on students, not chosen by them. Qualitative data were collected through audiotaped discussions, videotaped classes, observation notes, collected student artifacts, and a follow-up interview with one student. The researchers used discourse analysis to examine their data and develop categories for discourse types. Students were given a set kit of materials from which to build their cars. Students measured distance and time, calculated velocity and discussed acceleration, friction, force, and other relevant concepts with teachers and researchers. The students were deeply involved in the project, but through a close analysis of student discourse, researchers realized that even though a great deal of design-related discourse occurred, scientific vocabulary was not part of typical discussions. Even display posters that students created did not contain references to Newton's laws. Much of the discourse that went on in the classroom had to do with perceived respect from peers. It was not "cool" to play with cars, to be perceived as an achiever, or to step out of stereotypical gender roles. Males were especially reluctant to participate, and often ridiculed those that did choose to participate. Students became possessive over shared materials such as tape, scissors, straws, and bottle-cap wheels. While discourse was not teacher-centered and students were constantly talking with one another, attempts to direct the whole class through teacher-centered discussions were fruitless, and teachers were commonly treated with disrespect. The researchers concluded from this study that one-size does not fit-all when it comes to science teaching and learning, and that educators need to find ways to reach students with methods that match the circumstances. "The activity failed because it neglected to take into account the historical and social environments in which the students live and attend school" (p. 761). Respect was the currency in this classroom, and since respect was not shared between all classroom members, including teachers, science as inquiry – active learning – could not occur. This study resulted in some interesting findings about how teachers need to establish respect and rapport within the classroom and work with the cultural norms, not against them.

At the University of Michigan, Mamlok, Dershimer, Fortus, Krajcik, and Marx (2001) developed a design-based science curriculum called Learning Science by Designing Artifacts (LSDA), a curriculum based on the theoretical framework of social constructivism. They described social constructivism as involving active engagement, contextualized learning, community building, and disciplined discourse. They described design-based science as a form of problem-based science in which students design physical artifacts that solve real-world problems. The artifact design was not the culminating activity but was the actual structure around which all learning was organized. Additionally, the LSDA curriculum incorporated the use of primary source documents and computer technologies. Ironically, in this particular study, the only artifacts students created were posters. When the LSDA curriculum was initially introduced into high school science classrooms, the researchers found that while the students enjoyed the activities and lessons, not enough time was dedicated to science concepts. They began to re-design the curriculum with the goal of improving science content understanding.

In 2002, the Virginia Middle School Engineering Education Initiative (VMSEEI) was established by Dr. Larry Richards at the University of Virginia with help from grants from the National Science Foundation and the Payne Family Foundation (Richards, Hallock, & Schnittka, 2007). The goal was to have university engineering students work in teams to create engineering design-based curricula for middle school teachers to use, provide professional development for teachers, and disseminate kits to local schools. To put the year 2002 in perspective, "STEM" was not an acronym in use until 2001 with American biologist Judith Ramaley coined it while working for the National Science Foundation (Hallinen, 2018). The VMSEEI developed curricula that are still in use today.

In 2006, the National Academy of Engineering (NAE) and the National Research Council (NRC) established the Committee on K-12 Engineering Education which worked to guide the creation and implementation of engineering design-based curricula and teaching practices. The committee looked at how engineering might complement science and math instruction, what appropriate learning outcomes might be, and what policies might need to be in place. The results were published in the book, *Engineering in K-12 Education: Understanding the Status and Improving the Prospects* (NAE & NRC, 2009). The committee recommended that engineering education in K-12 classrooms incorporate important math, science, and technology concepts and skills, emphasize design and the iterative design process, and be set in meaningful contexts. It recommended that students should understand that problems can have multiple solutions, and that they should develop engineering habits of mind such as systems-thinking, creativity, collaboration, and ethical considerations. The committee also recommended that research on K-12 engineering education should be funded to investigate student engagement, content knowledge, career aspirations, design thinking in classroom settings, STEM literacy, and how to create an intersection between scientific inquiry and engineering design. The committee recommended that engineering design-based curricula should be developed that is inclusive of all people, with special attention to students from underrepresented groups, and that professional development should be made available to teachers.

2.2 The Last Decade

Over the past decade, as engineering integration has become more widely adopted and accepted in K-12 settings, researchers have investigated how to help students develop interest in engineering activities and careers (Hylton & Otoupal, 2009; Ing, Aschbacher, & Tsai, 2014; Montfort, Brown, & Whritenour, 2013) and develop engineering identities (Capobianco, French, & Diefes-Dux, 2012), how to help students learn science, technology, or mathematics better through engineering design activities (Mehalik, Doppelt, & Schuun, 2008; Tran & Nathan, 2010), how to help students learn engineering content and practices and understand the nature of engineering (Mentzer, Becker, & Sutton, M, 2015; Schnittka & Bell, 2011), how to broaden participation of diverse learners in engineering (Blanchard et al., 2015; Martinez Ortiz et al., 2018; Thompson & Lyons, 2008), and how to best prepare and support teachers so they can adopt an engineering mindset and integrate it into their typical routine (Lottero-Perdue & Parry, 2017; Martin, Peacock, Ko, & Rudolph, 2015; Sengupta-Irving & Mercado, 2017).

3 Implications for K-12 Education

When the Next Generation Science Standards (NGSS Lead States, 2013) were released in April of 2013, the landscape changed across the country. The need for standards-aligned curricular resources was heightened. Teachers feared not being prepared for such a new shift in their teaching, but the National Science Foundation was funding grants for curriculum development, and the wide dissemination of resources was well under way. Some resources were simply repackaged science or mathematics curricula, and only tangentially included engineering, or simply had the acronym, STEM, branded on the package or front cover. Moore, Tank, Glancy, and Kersten (2014) published a framework that outlined what quality K-12 engineering education entails, and this framework should be consulted when evaluating an engineering design-based lesson or curriculum. The Framework for Quality K-12 Engineering Education includes 12 key indicators which define the characteristics of engineering. For one, engineering in K-12 settings should include a problem that needs solving, an initial plan to solve the problem, and an initial design that weighs the constraints and trade-offs. Students should create a prototype or model, or other type of physical artifact that can be tested. Data should be collected and analyzed in order to evaluate the strengths and weaknesses of the design. Most importantly, the designed solutions should use scientific and mathematical knowledge and not simply "making" or crafting. Engineering design activities should encourage students to think, reflect, communicate, be creative, develop perseverance, and learn from failure. The design activities should also relate to the profession of engineering, and help students understand the work that engineers do, and the types of engineering careers. Engineers use tools, techniques, and processes to help them with their work, and students should become familiar with these also – tools like using Computer Aided Design software, Excel spreadsheets, and physical "shop" tools like hammers and drills and soldering irons. Engineering design activities should be related to real-world issues, include ethical considerations, promote teamwork, and promote communication skills such as technical writing and oral presentations. Teachers are encouraged to consult this Framework when evaluating or designing curriculum for their K-12 students.

In a study of prior experience with K-12 engineering curricula or activities, 141 teachers across the state of Alabama were surveyed about the activities they had done with their students (Schnittka, Turner, Colvin, & Ewald, 2014). Teachers commonly responded that they used the following activities: building fan cars, conducting an egg drop contest, participating in VEX robotics competitions, constructing bridges from balsa wood, using Lego robotics kits, designing roller coasters with marbles and foam tubing, and designing mousetrap cars. Without examining actual lesson plans, it is not possible to evaluate these activities for their alignment with The Framework for Quality K-12 Engineering Education, but most likely, these activities lack a connection to scientific and mathematical knowledge, lack a connection to the actual work engineers do, and lack a connection to real-world issues.

The book STEM Road Map: A Framework for Integrated STEM Education (Johnson, Peters-Burton, & Moore, 2016) is full of ideas for teachers to use as they build engineering-infused lessons. The ideas in the book are aligned with the Framework described above, and they are mapped to the NGSS, the Common Core Mathematics standards, and the Common Core Language Arts standards. The book provides templates for teachers to use as they build curriculum from ideas, and shows teachers how to create test items, rubrics, and other assessments.

When searching for lessons, curricula, kits, units, or other activities that use engineering design, teachers are urged to check for alignment with the Framework described above, and even the book, STEM Road Map, so that modifications can be made to improve existing lesson ideas. Some popular sources for engineering lesson ideas are: the Link Engineering Educator Exchange (https://www.linkengineering.org/) which is sponsored by the National Academy of Engineering, TeachEngineering (https://www.teachengineering.org/), which is sponsored by the National Science Foundation and the University of Colorado Boulder, and Engineering Go For It (http://www.egfi-k12.org), sponsored by the American Society for Engineering Education.

4 Future Work and Research

There is much work to be done to help teachers overcome natural hurdles in integrating engineering into their classes. One major hurdle is teacher self-efficacy, the measure to which a teacher believes he or she can impact students' educational success. Many teachers lack self-efficacy to teach engineering, which can be overcome with professional development, or perhaps online tutorials or videos, or quality curricula that have embedded scaffolds for the teacher (Yoon, Evans, & Strobel, 2014).

Additionally, there is future work to be done to blend engineering with all the science and math disciplines. While it is easier to find activities to fit in with a physical science class, the field of biomedical engineering teaches us that there many opportunities for curriculum developers to align the life sciences with engineering. Chemical engineering is a very popular field of study for college students, yet there is a lack of curricula designed for chemistry teachers to use with their students. Finally, many of the engineering curricula ideas are geared toward design activities traditionally associated with boys' hobbies and interests. There is a need to develop curricular ideas that interest all students, regardless of gender, and portray the engineering career as a helping field like medicine or teaching, a creative field open to all students who are interested in designing a better world

Questions remain as to how to best prepare classroom teachers and informal educators to teach engineering in the context of their science or math classes, or in stand-alone engineering electives. The informal space has so much potential to help youth develop engineering knowledge and skills and dispositions, but there must be ways to better align what happens out of school with what happens in the school classroom. How can afterschool teachers and classroom teachers work together to support all the ideals presented in the Framework for Quality K-12 Engineering Education?

Finally, can engineering be used to level the playing field in the school science or math classroom? Can it be the conduit through which diverse learners can excel (Schnittka, 2012)? Questions remain as to how to broaden participation of students from under-represented racial and ethnic groups, as a more diverse STEM workforce will better represent the needs of us all and bring diverse perspectives to the solutions we need for the grand challenges we face across the Earth, in the depths of the oceans, and in the space we inhabit in order to promote sustainability, health, security, and the joy of living (NAE, 2018).

5   Conclusion

Engineering has made its way over the past 700 years from a job one does while operating a military machine (Engineer, 2018), to a formal occupation divided into two categories: civil and military (Smith, 2017), to a way of thinking taught in K-12 schools across the country and embedded in many state standards. Engineering is a process of problem solving, a process of designing solutions, a process of using what we know to make the world a better place. As our industrialized, technological, and connected world evolves, we can expect to need millions more people in STEM fields over the next decade (Radu, 2018). These future workers are in grade school now, and it is time to make sure that teachers have the skills and knowledge and lesson ideas to prepare the future generation of problem solvers to be critical thinkers and collaborators, to have 21st century skills like creativity and technological literacy, and to know how to *engineer*.

References

American Association for the Advancement of Science [AAAS]. (1994). *Benchmarks for science literacy*. New York, NY: Oxford University Press.

Blanchard, S., Judy, J., Muller, C., Crawford, R. H., Petrosino, A. J., White, C. K., ... Wood, K. L. (2015). Beyond Blackboards: Engaging Underserved Middle School Students in Engineering. *Journal of Pre-College Engineering Education Research, 5*(1), 1–14.

Brown, J. S., Collins, A., & Duguid, P. (1989). Situated cognition and the culture of learning. *Educational researcher, 18*(1), 32–42.

Capobianco, B. M., French, B. F., & Diefes-Dux, H. A. (2012). Engineering identity development among pre-adolescent learners. *Journal of Engineering Education, 101*(4), 698.

Carroll, D. R. (1997). Bridge engineering for the elementary grades. *Journal of Engineering Education, 86*(3), 221–226.

Crawford, R. H., Wood, K. L., Fowler, M. L., & Norrell, J. L. (1994). An engineering design curriculum for the elementary grades. *Journal of Engineering Education, 83*(2), 172–181.

DiscoverE. (2018). Retrieved from http://www.discovere.org/our-programs/engineers-week

Engineer. (2018). In *OED online*. Retrieved from www.oed.com/view/Entry/62225

Hallinen, J. (2018). *STEM: Education curriculum*. Retrieved from https://www.britannica.com/topic/STEM-education

Hylton, P., & Otoupal, W. (2009). Engaging secondary school students in pre-engineering studies to improve skills and develop interest in engineering careers. *International Journal of Engineering Education, 25*(3), 419–425.

Ing, M., Aschbacher, P. R., & Tsai, S. M. (2014). Gender differences in the consistency of middle school students' interest in engineering and science careers. *Journal of Pre-College Engineering Education Research (J-PEER), 4*(2), 2.

Johnson, C. C., Peters-Burton, E. E., & Moore, T. J. (2016). *STEM road map: A framework for integrated STEM education*. Abingdon-on-Thames: Routledge.

Lottero-Perdue, P. S., & Parry, E. A. (2017). Elementary teachers' reflections on design failures and use of fail words after teaching engineering for two years. *Journal of Pre-College Engineering Education Research (J-PEER), 7*(1), 1.

Mamlok, R., Dershimer, C., Fortus, D., Krajcik, J., & Marx, R. (2001, March). *Learning Science by Designing Artifacts (LSDA) – A case study of the development of a design-based science curriculum*. Presented at the annual meeting of NARST, St. Louis, MO.

Martin, T., Baker Peacock, S., Ko, P., & Rudolph, J. J. (2015). Changes in teachers' adaptive expertise in an engineering professional development course. *Journal of Pre-College Engineering Education Research (J-PEER), 5*(2), 4.

Martinez Ortiz, A., Rodriguiz Amaya, L., Kawaguchi Warshauer, H., Garcia Torres, S., Scanlon, E., & Pruett, M. (2018). They choose to attend academic summer camps? A mixed methods study exploring the impact of a NASA academic summer pre-engineering camp on middle school students in the Latino community. *Journal of Pre-College Engineering Education Research, 8*(2), Article 3.

McRobbie, C. J., Stein, S. J., & Ginns, I. (2001). Exploring designerly thinking of students as novice designers. *Research in Science Education, 31*(1), 91–116.

Mehalik, M. M., Doppelt, Y., & Schuun, C. D. (2008). Middle-school science through design-based learning versus scripted inquiry: Better overall science concept learning and equity gap reduction. *Journal of Engineering Education, 97*(1), 71.

Mentzer, N., Becker, K., & Sutton, M. (2015). Engineering design thinking: High school students' performance and knowledge. *Journal of Engineering Education, 104*(4), 417–432.

Montfort, D. B., Brown, S., & Whritenour, V. (2013). Secondary students' conceptual understanding of engineering as a field. *Journal of Pre-College Engineering Education Research (J-PEER), 3*(2), 2.

Moore, T. J., Glancy, A. W., Tank, K. M., Kersten, J. A., Smith, K. A., & Stohlmann, M. S. (2014). A framework for quality K-12 engineering education: Research and development. *Journal of Pre-College Engineering Education Research (J-PEER), 4*(1), 2.

National Academy of Engineering [NAE]. (2018). *NAE grand challenges for engineering*. Retrieved from http://www.engineeringchallenges.org/

National Academy of Engineering and National Research Council [NAE & NRC]. (2009). *Engineering in K-12 education: Understanding the status and improving the prospects*. Washington, DC: The National Academies Press. https://doi.org/10.17226/12635

National Research Council [NRC]. (1996). *National science education standards*. Washington, DC: National Academies Press.

NGSS Lead States. (2013). *Next Generation Science Standards: For states, by states*. Washington, DC: The National Academies Press.

Pols, Y. D., Rogers, C. B., & Miaoulis, I. N. (1994). Hands-on aeronautics for middle school students. *Journal of Engineering Education, 83*(3), 243–247.

Radu, S. (2018, August 23) STEM worker shortage at a crisis, survey shows. *U.S. News & World Report.* Retrieved from https://www.usnews.com/news/best-countries/articles/2018-08-23/americans-think-they-have-a-shortage-of-stem-workers

Richards, L., Hallock, A., & Schnittka, C. G. (2007). Getting them early: Teaching engineering design in middle schools. *International Journal of Engineering Education, 23*, 874–883.

Roth, W. M. (1995). From "wiggly structures" to "unshaky towers": Problem framing, solution finding, and negotiation of courses of actions during a civil engineering unit for elementary students. *Research in Science Education, 25*(4), 365–381.

Roth, W. M. (1996). Learning to talk engineering design: Results from an interpretive study in a grade 4/5 classroom. *International Journal of Technology and Design Education, 6*(2), 107–135.

Schnittka, C. G. (2012). Engineering education in the science classroom: A case study of one teacher's disparate approach with ability-tracked classrooms. *Journal of Pre-College Engineering Education Research (J-PEER), 2*(1), Article 5.

Schnittka, C. G., & Bell, R. L. (2011). Engineering design and conceptual change in the middle school science classroom. *International Journal of Science Education, 33*, 1861–1887.

Schnittka, C. G., Turner, G., Colvin, R., & Ewald, M. L. (2014, June). *A state-wide professional development program in engineering with science and math teachers in Alabama: Fostering conceptual understandings of STEM.* In Proceedings of the American Society for Engineering Education, Indianapolis, IN.

Seiler, G., Tobin, K., & Sokolic, J. (2001). Design, technology, and science: Sites for learning, resistance, and social reproduction in urban schools. *Journal of Research in Science Teaching, 38*(7), 746–767.

Sengupta-Irving, T., & Mercado, J. (2017). Anticipating change: An exploratory analysis of teachers' conceptions of engineering in an era of science education reform. *Journal of Pre-College Engineering Education Research (J-PEER), 7*(1), 8.

Smith, R. J. (2017). Engineering. In *Encyclopaedia Britannica, Inc.* Retrieved from https://www.britannica.com/technology/engineering

Thompson, S., & Lyons, J. (2008). Engineers in the classroom: Their influence on African-American students' perceptions of engineering. *School Science and Mathematics, 108*(5), 197–211.

Tran, N. A., & Nathan, M. J. (2010). The effects of pre-engineering studies on mathematics and science achievement for high school students. *International Journal of Engineering Education, 26*(5), 1049.

Yoon, S. Y., Evans, M. G., & Strobel, J. (2012, June). Development of the Teaching Engineering Self-Efficacy Scale (TESS) for K-12 teachers. In *Proceedings of 2012 ASEE Annual Conference & Exposition* (pp. 25–466).

CHAPTER 12

Classroom Assessment in the Service of Integrated STEM Education Reform

Carol L. Stuessy and Luke C. Lyons

Abstract

Authors make the claim that carefully designed integrated STEM assessment/learning tasks can be used appropriately to lead teachers away from teacher-directed and towards more student-directed instructional practice. A conceptual framework of activity theory emphasizing "doing is learning" grounds a twelve-step, principled design approach to design performance assessments also serving as models of reformed iSTEM instructional practices. At each step, lesson/assessment designers apply a recursive, revise-and-reflect process to self-assess the appropriateness of teacher-directed strategies in every lesson within the performance task. Appropriateness refers to the activity level at which an individual student can perform a learning task with the help of others. Based on the novelty and complexity of what students will be actively doing with the information centering the learning activity, the designer chooses a teacher's role to accommodate students' abilities to cognitively handle the novelty and complexity of the information. Accordingly, designers decide a need for explanation, clarification, prompting, elaboration, or review before and/or during the lesson. While whole-class, teacher-directed instruction can sometimes be the best and most efficient way for students to begin their work with novel or complex information, reform-based designer/practitioners reserve teacher-directed, non-active learning for only these instances. Otherwise, teachers employ instructional strategies that can meet the needs of individual students by encouraging them to actively work on the information with others. These teachers prefer working with students in small groups to identify and address learning difficulties while maintaining as much student-directed activity as possible.

1 Introduction

The cry to use assessment in the service of reform in mathematics and science education was first heard in the early 1990s. The American Association for the Advancement of Science published *Science Assessment in the Service of Reform* in 1991 (Kulm & Malcom, 1991), recognizing that new thinking about assessment at all levels (classroom, district, state, and nation) would be needed to design assessments of student learning aligned with newly published guidelines and standards for teaching science and mathematics (see Kulm & Stuessy, 1991). Publications since that time allow us to trace the progression of the journey of assessment, which began when traditional notions of testing were left behind (e.g., see Stiggins, 1997; Wiggins & McTighe, 2000; Pellegrino, Chudowsky, & Glaser, 2001; Frankland, 2007; Lewin & Shoemaker, 2011; NSTA, 2014; Pellegrino, Wilson, Koenig, & Beatly, 2014; Beatly & Schweingruber, 2017). At this point in the journey, assessment has traveled to center stage and currently plays an essential role in driving decisions about what and how to teach. Assessment is an essential component in new, integrated models of STEM reform, including the three-dimensional model presented in the *Next Generation Science Standards* (Pellegrino et al., 2014). Models integrating classroom assessments with curriculum and instruction require new applications of instructional strategies, new forms of curricula and other learning tools, and multiple assessment types to support learning in relevant, student-centered, engaging learning environments.

> Existing assessments – whatever their purpose – cannot be used to measure the full range of activities and interactions that are happening in science classrooms that have adapted these [i.e., *integrated STEM*] ideas because they were not designed to do so. (Beatly & Schweingruber, 2017, pp. vii–viii)

Thinking about classroom-based assessment in regard to integrated STEM learning begins with the adoption of a strategic approach to the design, implementation, and assessment of STEM learning tasks. If the desire is to facilitate more stimulating and relevant learning experiences for students, adoption of a conceptual framework for STEM experiences is essential to increase educators' abilities to assess student learning in STEM contexts. Conceptual framework building requires;

> a deep understanding of the complexities surrounding how people learn, specifically teaching and learning STEM content. ... Instead of teaching content and skills and hoping students will see the connections to real-life application, an integrated approach seeks to locate connections between STEM subjects and provide a relevant context for learning the content. Educators should remain true to the nature in which science,

technology, engineering, and mathematics are applied to real-world situations. (Kelley & Knowles, 2016, n.p.)

Kelley and Knowles (2016) present a convincing argument that most content in STEM can be grounded within situated cognition theory (Brown, Collins, & Duguid, 1989; Lave & Wenger, 1991; Putnam & Borko, 2000). These authors recognize that the increased integration of STEM subjects may not be effective if a strategic approach to the design and implementation of content and process does not exist. Furthermore, situated cognition theorists recognize that both physical and social contexts are critical to the learning process (e.g., see Jonassen & Land, 2000; Vygotsky, 1978; Walker, Leary, Hmelo-Silmer, & Ertmer, 2015). Kelley and Knowles (2016) propose the term *situated STEM learning* to emphasize the role of authentic context:

> We define integrated STEM education as the approach to teaching the STEM content of two or more STEM domains, bound by STEM practices within an authentic context for the purpose of connecting these subjects to enhance student learning. (p. 5)

This notion is also embraced by instructional designers such as Jonassen, who use activity theory to develop robust, situated, problem-centered, student-centered learning environments (SCLEs; see Jonassen, 2000; Jonassen & Land, 2000, 2012; Land, Hannafin, & Oliver, 2012).

Our intent in this chapter is to support the claim that carefully designed integrated STEM learning tasks used appropriately as classroom-based STEM assessment tasks can lead teachers to reformed classroom practices. Our conceptual framework for developing integrated STEM classroom assessment tasks is grounded in Jonassen's interpretation of activity theory (2000), that "doing is learning," coupled with Kelley and Knowles' conceptual framework of integrated STEM education, and Vygotsky's (1978) early work on social learning. We have taken into mind recent guidelines, examples, and recommendations (Beatly & Schweingruber, 2017; Pellegrino, Wilson, Koenig, & Beatly, 2014) for designing assessments for the *Next Generation Science Standards,* adapting the ten-step process outlined by Krajcik, Codere, Dahsah, Bayer, and Munn (2014) for designing instruction to build students' understanding of the performance expectations appearing in the NGSS. We modified this the Krajcik group's ten-step process to emphasize that evidence of students' performances is paramount:

> Assessment tasks, in turn, have to be designed to provide evidence of students' ability to use the practices, to apply their understanding of the

cross-cutting concepts, and to draw on their understanding of specific disciplinary ideas, all in the context of addressing specific problems. (Pellegrino et al., 2014, p. 32)

2 A Principled Design Approach

The National Academies defines classroom assessment as, "assessments designed or selected by teachers and given as an integral part of classroom instruction" (Pellegrino et al., 2014, p. 83). This definition supports new thinking about assessment: that formal assessments need to be purposefully designed, while intentionally incorporating the factor of scaffolding assessment into instruction. *Developing Assessments for the Next Generation Science Standards* recommends using a principled design approach (Pellegrino et al., 2014), which is both methodological and systematic. We have found that the template provided by Krajcik and colleagues for designing NGSS-based instruction can be easily adapted to emphasize the role assessment in instructional practice, as these authors propose that authentic STEM lessons are more than a myriad of activities that incorporate student-centered learning. Krajcik's group took advantage of the reasoning behind the development of the three-dimensional NGSS framework for K-12 science education: "students engage in science and engineering practices to develop and use disciplinary core ideas (DCIs) and crosscutting concepts to explain phenomena and solve problems" (p. 158). These authors make the comparison to "strands of a rope" working together to "make learning stronger."

> These three dimensions (i.e., practices, DCIs, crosscutting concepts) work together to help students build an integrated understanding of a rich network of connected ideas. The more connections developed, the greater the ability of students to solve problems, make decisions, explain phenomena, and make sense of new information. (p. 158)

As a result, students consistently, as they do science in their classrooms, demonstrate their knowledge connections. Teachers do not teach science as inert, factual information to be taught in isolation; instead, students use knowledge in the process of doing science, just as scientists do in their own scientific investigations. The architecture of each standard in the NGSS blends together DCIs, science and engineering practices, and crosscutting concepts. Foundation boxes associated with the PE further articulate the practice, the DCI, and the crosscutting concept.

The following example (Figure 12.1) incorporates two PEs under Learning Standard 1 for Grade 1, *Heredity: Inheritance and Variation of Traits*. Standards 1-LS1-1 and 1-LS2 combine the dimensions appearing below them and also show connections to 1-LS1-2, *Scientific Knowledge is Based on Empirical Evidence*, and 1-LS1-1, *Influence of Engineering, Technology, and Science on Society and the Natural World*. Performance Expectation, 1-LS1-1, focuses on increasing a learner's ability to use materials in designing a solution to a human problem dealing with survival, growth, and meeting their needs by mimicking plants and/or animals. This PE intertwines (1) the Scientific and Engineering Practices of Constructing Explanations and Designing Solutions and Connections to Nature of Science, with (2) the Disciplinary Core Ideas of Structure and Function and Information Processing, and with (3) the Crosscutting Concepts of Structure and Function and Connections to Engineering, Technology, and Science on Society and the Natural World. The second Performance Expectation, 1-LS1-2, also focuses on survival, but the survival is limited to the identification of patterns of offspring behavior eliciting responses of the parents.

Krajcik's group used a ten-step, principled approach to designing instruction centered on the NGSS. They began with the biggest idea: *Performance expectations (PEs), as outlined in the NGSS, must work together.* They noted that the selection of coordinating standards and learning outcomes assure a synergy between and among several performance expectations, and that some of the PEs might come from different content domain strands. Through their process of development, the group kept the disciplinary core ideas (DCIs) of the NGSS at the forefront of design. As the NGSS inform teachers what is expected of students, Krajcik's team coordinated and aligned their lesson assessments with the expectation.

In the principled approach to assessment we present here, we adapted Krajcik's ten-step process, re-ordering their steps and adding three new steps to indicate the emphasis on classroom assessment. In the example that follows, we present a K-2 lesson from a learning progression (LP) developed in conjunction with the teaching of an integrated Life-Science/Earth-Science college course designed specifically for preservice elementary teachers. One of us (LCL) chose "Dinosaurs" as the LP topic to demonstrate the use of content of interest and relevance to elementary school children. Preservice teachers experienced the LP as a series of inquiries modeling inquiry learning. In line with the NGSS, preservice teachers experienced lessons on dinosaurs designed for K-12 learners at the levels of K-2, 3–5, middle school, and high school. Aligned with notions of principled design brought forward by the National Academies (Pellegrino et al., 2014), this LP incorporated the NGSS over many topics in

1-LS 1 – From Molecules to Organisms: Structure and Processes	
PERFORMANCE EXPECTATIONS Students who demonstrate understanding can:	
1-LS-1-1. Use materials to design a solution to a human problem by mimicking how plants and/or animals use their external parts to help them survive, grow, and meet their needs.* [*Clarification Statement:* Examples of human problems that can be solved by mimicking plant or animal solutions could include designing clothing or equipment to protect bicyclists by mimicking turtle shells, acorn shells, and animal scales; stabilizing structures by mimicking animal tails and roots on plants, keeping out intruders by mimicking thorns on branches and animal quills; and detecting intruders by mimicking eyes and ears.]	**1-LS1-2. Read texts and use media to determine patterns in behavior of parents and offspring that help offspring survive.** [*Clarification Statement:* Examples of patterns or behaviors could include the signals that offspring make (such as crying, cheeping, and other vocalizations) and the responses of the parents (such as feeding, comforting, and protecting the offspring.)] * *This performance expectation integrates traditional science content with engineering through a practice or disciplinary core idea.*

Science and Engineering Practices	Disciplinary Core Ideas	Crosscutting Concepts
Constructing Explanations and Designing Solutions Constructing explanations and designing solutions in K-2 builds on prior experiences and progresses to the use of evidence and ideas in constructing evidence-based accounts of natural phenomena and designing solutions. (See 1-LS1-1) **Obtaining, Evaluating and Communicating Information** Obtaining, evaluating, and communication in K-2 builds on prior experiences and uses observations and texts to communicate new information. (See 1-LS1-2) **Connections to Nature of Science** Scientists look for patterns and order when making observations about the world. (See 1-LS1-2)	**LS1.A: Structure and Function** All organisms have external parts. Different animals use their body parts in different ways to see, hear, grasp objects, protect themselves, move from place to place, and seek, find, and take in food, water, and air. Plans also have different parts (roots, stems, leaves, flowers, fruits) that help them survive and grow. (1-LS1-1) **LS1.B: Growth and Development of Organisms** Adult plants and animals can have young. In many kinds of animals, parents and the offspring themselves engage in behaviors that help the offspring to survive. (1-LS1-2) **LS1-D: Information Processing** Animals have body parts that capture and convey different kinds of information needed for growth and survival. Animals respond to these inputs with behaviors that help them survive. Plants also respond to some external inputs (1-LS1-1)	**Patterns** Patterns in the natural world can be observed, used to describe phenomena, and used as evidence. (1-LS1-2) **Structure and Function** The shape and stability of structures of nature and designed objects are related to their functions. (1-LS1-1) **Connections to Engineering, Technology, and Science on Society and the Natural World** Every human-made product is designed by applying some knowledge of the natural world and is built using materials derived from the natural world. (1-LS1-1)

FIGURE 12.1   An excerpt from the *Next Generation Science Standards* for two performance expectations and their elaborations from the Life Science Topic 1-LS 1, entitled from molecules to organisms: Structure and processes

an integrated manner. The K-2 grade band focused primarily on life science standards dealing with offspring, parents, traits and survival (i.e., 1-LS1-1, 1-LS1-2, 1-LS3-1) (see Figure 12.2).

INTEGRATED STEM EDUCATION REFORM 195

Guiding Question: HOW DO WE KNOW WHAT WE KNOW ABOUT DINOSAURS?

Sub-question 1-LS1 – From Molecules to Organisms: Structure and Processes: *What do scientists know about dinosaurs' traits and behaviors that helped dinosaurs and their offspring survive, grow, and meet their needs?*

Sub-question 1-LS3 – Heredity: Inheritance and Variation of Traits: *How do scientists know some offspring were like, but not exactly like, their parents?*

PERFORMANCE EXPECTATION: TRAITS, BEHAVIORS, GROWTH AND SURVIVAL OF DINOSAURS AND THEIR OFFSPRING

Student Centered Learning Experience: (1-LS1-1, 1-LS1-2, 1-LS3-1)

Students who demonstrate understanding can:

- Use materials to design a solution to a human problem by mimicking how dinosaurs use their external parts to help them survive grow and meet their needs.
- Read texts and use media to determine patterns in behavior of dinosaur parents and offspring that help offspring survive.
- Make observations to construct an evidence-based account that dinosaurs are like, but not exactly like, their parents.

Summative Assessment

- Explain how scientists work to find out about the dinosaurs' traits and behaviors with regard to their survival, growth, and other needs. Develop a scientific argument for how scientists support claims they make based on fossil evidence.

Science and Engineering Practices	Disciplinary Core Ideas	Crosscutting Concepts
Constructing Explanations and Designing Solutions Constructing explanations in K-2 builds on prior experiences and progresses to the use of evidence and ideas in constructing evidence-based accounts of natural phenomena. **Obtaining, Evaluating and Communicating Information** Obtaining, evaluating, and communication in K-2 builds on prior experiences and uses observations and texts to communicate new information. **Connections to Nature of Science** Scientists look for patterns and order when making observations about the world. (A paleontologist is a scientist who studies fossils to look for patterns in plants and animals	**LS1.A: Structure and Function** All organisms have external parts. Different animals (including dinosaurs) use their body parts in different ways to see, hear, grasp objects, protect themselves, move from place to place, and seek, find, and take in food, water, and air. **LS1.B: Growth and Development of Organisms** Adult animals (dinosaurs) can have young. In many kinds of animals, parents and the offspring themselves engage in behaviors that help the offspring to survive. **LS1-D: Information Processing** Animals (Dinosaurs) have body parts that capture and convey different kinds of information needed for growth and survival. Animals respond to these inputs with behaviors that help them survive.	**Patterns** Patterns in the natural world can be observed, used to describe phenomena, and used as evidence. **Structure and Function** The shape and stability of structures of nature and designed objects are related to their functions. **Connections to Engineering, Technology, and Science on Society and the Natural World** Every human-made product is designed by applying some knowledge of the natural world.

FIGURE 12.2   An adaptation of the *Next Generation Science Standards* to specify dinosaurs as organisms in a bundle of performance expectations for K-2 learners

3 The Twelve-Step Process for Developing iSTEM Performance Tasks

3.1 Some Terminology

The term, *integrated STEM (iSTEM) performance task*, refers to the summative assessment of the PE. We used the term, *iSTEM performance task,* to refer to the assessment and learning activities for the entire, grade-level specific, STEM curriculum unit of instruction. We envisioned the unit of instruction as holding a sequence of iSTEM lessons. The sequence, for which we use Jonassen's term (2000), *student centered learning environment* (SCLE), refers to the sequenced activities by which students are enabled to successfully complete the iSTEM performance task. SCLEs are, as a rule, composed of a sequence of student-centered, activity-based lessons. The iSTEM performance task indicates the endpoint of the sequence, which has enabled students to successfully complete a real-world task. Students develop conceptual understanding of science and engineering content and skills while actively engaging in the lessons. The iSTEM performance task is situated within a scenario or story in which students role-play to act and think like scientists, engineers, STEM professionals, or problem solvers (e.g., see Jonassen & Land's interpretation of activity theory, 2000). The scenario is culturally relevant, problem based, and situated in a dilemma that students must work together to solve. The dilemma presents the need for students as a group to solve a problem, develop an explanation, or answer a scientific question. (General protocols for problem-based learning and inquiry-based learning are totally appropriate here for designing lessons.)

The teacher, at all times during the sequence, scaffolds student-directed, active learning. While the teacher primarily uses student-centered instructional strategies to support students' direction of their own learning, direct instructional approaches can also appropriately be applied at times when the teacher perceives that students need more guidance than they can provide for themselves. Direct instruction can be a part of the teacher's plan, or it can occur spontaneously as the need arises; and it can be applied to the whole class or to a small group, depending on how and when the need arises. The SCLE consists of four to six (or perhaps more at higher grade levels) sequenced lessons, culminating in the final student learning product that demonstrates students' conceptual understanding and performances in collaborative problem solving, learning, writing, speaking, and enacting.

Student work provides the evidence of students' active learning during the SCLE. Student work includes student learning products, formative assessments, and the final student learning product, all of which are used by the teacher and the students to assess progress and final completion of the performance expectations for the SCLE. The iSTEM performance task is the final learning product that supplies the solution, explanation, or scientific finding to the dilemma presented in the initial story.

3.2   The Principled Design

Table 12.1 demonstrates our re-ordered list, arranged to facilitate classroom assessment during instruction, which embeds many of the recommendations made by Pellegrino et al. (2014). These authors present a simple but elegant schematic identifying three elements in conceptualizing assessment as a process of reasoning from evidence. In Figure 12.3, we see a triangle indicating that the process of reasoning from evidence rests on *Cognition*:

> … the design of the assessment should begin with specific understanding not only of which knowledge and skills are to be assessed, but also of how understanding and competence develop in the domain of interests. (p. 49)

The NGSS defines cognition to include the practices, crosscutting concepts, and DCIs as they are integrated into the PEs. A second corner of the triangle is *Observation* of students' capabilities in the context of specific tasks designed to show what they know and can do. These must be tightly linked to the specific inferences about student learning that the assessment is intended to support. "If the intended constructs are clearly specified, the design of a specific task and its scoring rubric can support clear interests about students' capabilities" (p. 49). The third corner of the triangle is *Interpretation*, meaning the methods and tools used to reason from observations that have been collected. They are presented in the form of a triangle because they are interrelated. Every decision should be considered in light of each of these three elements.

3.3   Steps 1 and 2: Selecting and Clustering PEs

Steps 1 and 2 engage the assessment designer in selecting and clustering the PEs and then examining the PEs closely to identify the implications for instruction and assessment. In our Dinosaur example, we combined the two original PEs and rewrote the combination to refer specifically to the topic of Dinosaurs. In regard to the practices and DCIs we made slight changes to reflect the emphasis on Dinosaurs. Refer to Figure 12.2.

3.4   Steps 3–5: Situating the Learning with a Story, Guiding Question, and iSTEM Performance Task

In steps 3, 4, and 5, the designer generates a story and guiding question to situate learners as creators of knowledge, embedding the scenario within a

TABLE 12.1 Adapted 12-step approach to designing classroom task assessments aligned with the Next Generation Science Standards

New	Design task
1	*Select and Cluster PEs.* In regard to the topic or DCI selected, select PEs that work together – a bundle – to promote proficiency in using the ideas expressed. The bundle can include PEs from a single NGSS topic or DCI and it can draw in PEs from other topics or DCIs. (1)
2	*Inspect PEs.* Inspect the PEs, clarification statements, and assessment boundaries to identify implications for instruction and assessment. (2)
3	*Identify Implications for the Scenario and Summative Assessment.* Examine DCIs, science and engineering practices, and crosscutting concepts coded to the PEs to identify implications for the scenario and summative assessment, which should reflexively inform the development of both simultaneously. (3)
4	*Build the Scenario.* With PEs, clarification statements, and assessment boundaries in place, generate a story that situates the learners to act as creators of knowledge. Write the scenario in student language. Creators could include scientists, engineers, museum designers, problem solvers. When possible, embed the scenario within a relevant, personal/social/societal problem. Develop a guiding question that will tie the sequence of lessons together and provide a referent for lessons. (no #)
5	*Design and Embed the iSTEM Performance Task.* With the scenario and guiding question in mind, generate a scenario-appropriate performance task that would provide evidence enabling the teacher to assess scenario-level performance via a rubric. Within the scenario, embed the performance task to make a seamless connection between what students will do and what they will learn. (no #)
6	*Develop the Rubric for the iSTEM Performance Task.* Use the PEs, DCIs, science and engineering practices, and crosscutting concepts and a developmental scale for understanding to develop a scenario-appropriate rubric to assess students' iSTEM performance on the summative assessment. (no #)
7	*Identify and Sequence Practices into Lessons.* Identify science and engineering practices that support instruction of the core ideas as they appear in the scenario and naturally lead to the completion of the iSTEM Performance Task. Develop a coherent sequence of lessons that blend together various science and engineering practices with the core ideas and crosscutting concepts. (5)

(cont.)

TABLE 12.1 Adapted 12-step approach to designing classroom task assessments aligned with the Next Generation Science Standards (cont.)

New	Design task
8	*Consider Novelty of DCIs and PEs in each Lesson.* Look closely at the DCI(s) and PE(s). What understandings need to be developed? What content ideas will students need to know? What must students be able to do? Take into consideration prior PEs that serve as the foundations for cluster of PEs the lessons will address. (4)
9	*Select Related Common Core Connections (Optional).* Select related Common Core Mathematics Standards (CCSS-M) and Common Core Literacy Standards (CCSS-L). Integrate them appropriately into the lesson level PEs. (8)
10	*Develop Lesson PEs.* Develop lesson level PEs. Lesson level expectations guide lesson development to promote student learning; they build to the level of understanding intended in the bundle of PEs. *Apply a lesson-level question that assists students in thinking about the nature of lessons as building blocks leading to the guiding question.* (6)
11	*Design Lesson-level Formative Assessments.* Determine the acceptable evidence for assessing lesson level performances. (7)
12	*Construct the Teacher's Storyline.* Carefully construct a storyline for teachers to help learners build sophisticated ideas from prior knowledge, using evidence that builds to the understanding described in the PEs. Describe how the ideas will unfold. What do students need to be introduced to first? How would the ideas and practices develop over time? (9)
1-12	*Apply a Filter to all Steps of the Process.* Ask: How do the task(s)/lesson(s) help students move towards understanding of the PE(s)? Does the sequence of formative assessments logically lead to the creation of the summative assessment? Are related common core connections embedded in both formative and summative assessments in such a way that they can be assessed? (10)

Note: Numbers in parentheses refer to the original placement of steps that are now reflected in the reordered list in column 1 (adapted from Krajcik et al. 2014, p. 163).

relevant, personal problem. With the scenario and guiding question in mind, appearing in Figure 12.4, we simultaneously designed the iSTEM performance task.

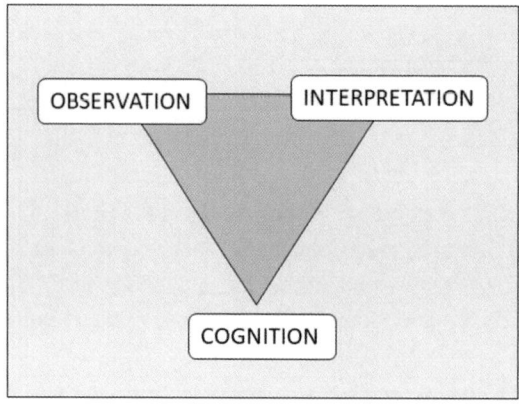

FIGURE 12.3
The three elements involved in conceptualizing assessment as a process of reasoning from evidence (from Pellegrino et al., 2014, p. 49, reproduced with permission)

3.5 Step 6: Development of the Rubric for the iSTEM Performance Task

Step 6 involves use of the PE, DCIs, science and engineering practices, and crosscutting concepts to develop a rubric specific to the requirements of the Performance Expectation. In the rubric for our example (Figure 12.5), we chose a developmental scale to assess first-grade students' understanding of the cognitive elements of the scenario. We have found developmental scales to be particularly useful when you want to help students see how they are moving on a path toward excellent performance while also communicating the highest level of achievement as they move through the science program.

3.6 Steps 7–10: Determining the Sequence of Lessons

Steps 7–10 lead the assessment designer to the final activity of designing lessons to support the completion of the lesson assessments in Step 11. The tasks in Steps 7–10 require the designer once more to look very closely at the PE(s), science and engineering performance(s), DCI(s), and the crosscutting concepts to envision a logical sequence of activities that will lead the learner towards the completion of the PE. In our example, we developed a sequence of four lessons, which we believed would enable students to complete the iSTEM performance task.

In these four lessons, students (1) model the work of scientists, (2) connect external parts and offspring behavior of dinosaurs to their survival, (3) elaborate their understanding of dinosaurs' external parts and offspring behavior to the solutions humans have created to alleviate similar needs, and (4) explain how scientists who study fossils (paleontologists) look for patterns and order when making observations about plants and animals that lived in the past. For each lesson, we crafted a leading question, which refers back to the PE for the SCLE and its iSTEM performance task. See Figure 12.6.

> **A Story about Fossils and Dinosaurs**
>
> **Guiding Question: How Do We Know What We Know about Dinosaurs?**
>
> You attend a school that has a summer science program for students to work with a scientist in their community. For the summer science experience for first graders, you are one of a group working with a paleontologist. In that work, you find out how the work of paleontologists leads us to know some things about dinosaurs, giant reptiles that lived on the Earth a long, long time ago. You will use the tools and procedures that a paleontologist uses to find evidence of dinosaurs at a farm in the country not far from your school. You will find quite a few dinosaur bones, which you have dug out of the ground at the farm. You will also bring these bones to the paleontologist's laboratory. You compare the bones you brought back with specimens and pictures of dinosaurs from the scientist's collection. As a group, you develop some ideas about how the external parts of dinosaurs helped them grow and survive in the environment in which they lived.
>
> When you return to your school in the fall, *you will explain to the new class of first graders how paleontologists' work leads to what we know about the traits, behaviors, growth, survival, and needs of dinosaurs*. You will make this explanation to the new class of first graders who will work on the dinosaur dig the next summer.

FIGURE 12.4   A scenario written in students' language situating the SCLE within a "story" in which students think and act like paleontologists. The guiding question for the SCLE refers to the scientists' work that students will simulate during four sequential lessons. The performance task is indicated (in italics) as part of the story and naturally includes evidence that students have met the student expectation for the SCLE.

3.7   *Steps 10–11: Designing Lessons and Lesson-Level Formative Assessments*

Classroom-based assessment lessons are designed with formative assessment in mind. Lesson-level formative assessments inform both the teacher and students about students' progress with evidence of students' activity within the lesson. While these formative assessments may

> ... involve formal tests and informal activities conducted as a part of the lesson, [the main purpose of formative assessments is] ... to identify students' strengths and weaknesses, assist educators in planning subsequent instruction, assist students in guiding their own learning by evaluating and revising their own work, and foster students' sentence of autonomy and responsibility for their own learning. (Andrade & Cizek, 2010, p. 4, cited in Pellegrino et al., 2014, p. 84)

In that regard, a well-planned lesson reflects not only what the teacher does, but what the students do, consistently, during each task in the lesson. To assure the dual focus on what both teacher and students do in the lesson, we designed a General SCLE Lesson Template (Figure 12.7). For teachers new to

How We Know What We Know about Dinosaurs

Performance	Needs Improvement	Showing Some Improvement	Acceptable Performance	Strong Performance
Articulation of what a paleontologist does in the field and in the laboratory	Provides no or less than adequate explanations of both field and laboratory work	Provides a less than adequate explanation of either field or laboratory work with no or little mention of connections between them	Provides adequate explanation of both field or laboratory work with acceptable explanation of connections between them	Provides a full explanation of field work and of work in the laboratory that leads to hypotheses about how dinosaurs lived
Identification of external parts of dinosaurs	Can accurately identify no more than one external part	Can accurately identify two examples of external parts	Can accurately identify three examples of external parts	Can accurately identify more than three examples of external parts
Identification of offspring behaviors and corresponding parent behaviors enhancing survival	Can identify no more than one offspring behavior with the corresponding parent behaviors	Can accurately identify two offspring behaviors with their corresponding parent behaviors	Can accurately identify three offspring behaviors with their corresponding parent behaviors	Can accurately identify more than three offspring behaviors with their corresponding parent behaviors
Providing evidence of use of text and media in generating connec-tions between external parts and offspring behaviors and survival	Makes no reference to text or media	Makes reference to 1 or 2 sources	Makes reference to 3 or 4 sources	Makes reference to > 4 sources
Explaining and comparing how some dinosaur offspring were similar to their parents, while others were not	Can sufficiently explain no relationships between offspring and adults.	Can sufficiently explain one examples of relationships between offspring and adults	Can sufficiently explain two examples of relationships between offspring and adults.	Can sufficiently explain three or more examples of relationships between offspring and adults
Using comparisons between human solutions and dinosaur traits to elaborate on the explanation	Does not elaborate on the explanation using human and dinosaur solutions	Does elaborate once or twice on the explanation using human and dinosaur solutions	Does elaborate on the explanation 3 or 4 times using human and dinosaur solutions	Does elaborate on the explanation using human and dinosaur solutions > 4 times
Communicating with the audience with minimal reliance on scripted input	Refuses to communicate with the audience or speaks for less than a minute	Speaks to the audience long enough to suggest the speaker has internalized the message	Speaks to the audience giving the impression of both thinking and speaking spontaneously	Gives a consistent impression of communicating *with* the audience
Providing visual representations to appropriately elaborate on the explanation	Does not use visual representations or inappropriately connects them to the explanation	Appropriately connects 1 or 2 visual representations to the explanation	Appropriately connects 3 or 4 visual representations to the explanation	Appropriately connects > 4 visual representations to the explanation

FIGURE 12.5 Developmental rubric for assessing the iSTEM performance task

student-centered teaching, we have found that the dual emphasis allows both designers and teachers to expand their vision of instruction to include specific attention to what students are doing during each of the task in the lesson, thus monitoring when and how much time is spent on both passive and active learning. The General SCLE Lesson Template has been a successful heuristic

How We Know What We Know about Dinosaurs
Student-Centered Learning Experience
Traits, Behaviors, Growth and Survival of Dinosaurs and Offspring (1-LS1-1, 1-LS1-2, 1-LS3-1)
Performance Expectation
Students who demonstrate understanding can: • Use materials to design a solution to a human problem by mimicking how dinosaurs use their external parts to help them survive grow and meet their needs. (1-LS1-1) • Read texts and use media to determine patterns in behavior of dinosaur parents and offspring that help offspring survive. (1-LS1-2) • Make observations to construct an evidence-based account that dinosaurs are like, but not exactly like, their parents. (1-LS3-1) **Summative Assessment** • Explain how scientists work to find out about the dinosaurs' traits and behaviors of dinosaurs with regard to their survival, growth, life cycles and other needs. Develop a scientific argument for how scientists support claims they make based on scientific evidence.
Lesson 1 LQ – *How do scientists discover patterns and make hypotheses about how dinosaurs looked and acted when they were living on the Earth such a long, long time ago?*
• LPE – Students can model working as scientists in discovering patterns in the external parts of dinosaurs and hypothesize behaviors of dinosaurs.
Lesson 2 LQ – *How do external traits and behaviors help dinosaurs and their offspring survive, grow, and meet their needs?*
• LPE – Students can read tests and use media to determine patterns in behavior of dinosaur parents and offspring that help offspring survive.
Lesson 3 LQ – *How are the external traits and behaviors of dinosaurs like the inventions and solutions of humans in terms of their growth, survival, and meeting their needs?*
• LPE – Students use materials to design a solution to a human problem by mimicking how dinosaurs use their external parts to help them survive, grow, and meet their needs.
Lesson 4 LQ – *How do scientists study fossils to make the connections between the external structures and offspring behaviors of dinosaurs with processes of surviving, growing, and meeting their needs?*
• LEP – Students can read texts and use media to determine patterns in behavior of dinosaur parents and offspring that help offspring survive. • Students can make observations to construct an evidence-based account that dinosaurs are like, but not always like, their parents.

FIGURE 12.6   Leading questions (LQ) and lesson-level performance expectations (LPE) for the four lessons in the SCLE, *How We Know What We Know about Dinosaurs*

for teachers to use when reflecting and revising plans for instruction and, therefore, eliminating as much as possible passive learning so characteristic of traditional modes of instruction (Jonassen, 2014).

3.8   Lesson Sequence Example

Five completed templates for the four lessons and the iSTEM performance task for the SCLE entitled, *How We Know What We Know about Dinosaurs*, are given in the Appendix.

3.9   Step 12: Construct the Teacher's Storyline

In this step, the designer constructs a storyline for teachers for each one of the lessons in the sequence. With this background, teachers can help learners build more sophisticated ideas from prior ideas and build understanding

Lesson # in Sequence:	Name of Lesson:	
Leading Question:	Guiding Lesson-Level Leading Question	
Performance Expectation:	Lesson-Level Performance Expectation	
General Description of the Activity:	Overview	
Formative Assessment:	Principled Designed Assessment	
Materials List:	Any materials needed for lesson	
Technology Resources:	A list that can be used by the teacher and/or students to enhance their understanding	
Prep Activities:	What the teacher does to prepare before teaching the lesson	
Total Time for the Lesson:	Amount in minutes	
Lesson		
Time (min)	What the Teacher Does	What the Students Do
	Briefly describe each of the steps in order, using as many steps as needed	Briefly describe what students do to correspond to teacher's ordered list
	1. Describes the lesson for today	1. Listens to the description of the lesson
	2.	2.
	3.	3.
	4.	4.

FIGURE 12.7 General SCLE Lesson Template used to construct the sequence of lessons comprising an iSTEM Performance TASK. Note the dual focus on what both teacher and students do.

required for successful completion of the SCLE-level PE (Krajcik et al., 2014). Figure 12.8 provides a storyline for Lesson 1, demonstrating how the storyline is constructed to describe how ideas and practices will unfold associated with the logical sequence of tasks. We envision placement of the storyline for each lesson occurring directly before the lesson plan to assist teachers in understanding the designers' logic in task completion as well as the connections of the scenario to the lessons that follow.

3.10 Steps 1–12: Applying a Filter to All Steps

3.10.1 Filtering Questions for Designers

For every step in the 12-step process, the designer asks questions about how well all the parts of the SCLE hold together. Overall, the designer of a student-centered learning experience before instruction asks filtering questions such as,

> Does the task in this lesson help students move towards understanding the PE(s) for this lesson? Does the sequence of learning activities and

> **Teacher Storyline: How Do We Know What We Know about Dinosaurs?**
> **Lesson 1**
>
> Lesson 1 begins a lesson sequence entitled, *How Do We Know What We Know about Dinosaurs?* This student-centered learning experience (SCLE) holds four lessons. The approximate total minutes for these four lessons and the preparation of the summative assessment is 420 minutes. (If your science lessons are usually about 40 min. long, the entire sequence would take about 10 days, dividing each lesson up into appropriate "chunks" of about 40 min. long.) In each lesson, students learn as they think and act like paleontologists, scientists who study fossils. At the end of the four lessons, first grade students then complete a performance task to explain what paleontologists do, basing their explanation on what they themselves have experienced while acting as paleontologists.
>
> The teacher introduces the lesson sequence by reading *The Fossil Story* to them. (Note: If students can read, they can read the story to each other in pairs; it they are beginning readers, they can follow as the teacher reads it to them.) After the story, the teacher leads a discussion, showing students pictures of fossil dig sites; distributing tools used by paleontologists, such as chisels, brushes, measuring tapes, and notebooks; and letting students examine physical specimens of all sorts of fossil plants and animals. (See websites below for pictures and information.) After the story and teacher demonstration, students then go to the area of the room designated as the dig site. (It would be helpful if two different spaces in the classroom were set up to be the dig site and the laboratory.)
>
> *How do scientists discover patterns and make hypotheses about how dinosaurs looked and acted when they were living on the Earth such a long, long time ago?* Lesson 1 introduces students to the scientific work of paleontologists. First, students simulate the field work of paleontologists by first "digging" for fossils in the field at a dig site. They "dig" through tubs of sand and rock layers to find parts of dinosaur bones in different layers. They record the layers holding bones by drawing the bones in the layer in which they were found. After students have "found" their dinosaur bones, the teacher then explains that paleontologists go to their laboratories to study the fossils they have found and to make inferences about the way these plants and animals lived long ago. The teacher then directs students to transport the fossils they have found to the classroom laboratory space to learn more about them. In the laboratory, students continue to work as scientists by attempting to match their dinosaur bones with pictures of dinosaurs. As the final part of this lesson, which is used as a formative assessment, pairs of students match a three-dimensional model of a dinosaur skeleton with a picture of a dinosaur in a set of dinosaur cards. The card shows an external feature, or trait, that scientists think enhanced the survival of the dinosaur. Students also hypothesize behaviors of dinosaurs based on the traits that are shown on the cards. Throughout the lesson, the teacher moves from group to group to ask and respond to questions. Direct instruction is limited on purpose to allow students the opportunity to practice independent exploration. By the end of this lesson students show know that
>
> 1. Paleontologists are scientists who study fossils.
> 2. Paleontologists are interested in how ancient plants and animals lived and survived a long time ago.
> 3. Paleontologists use special tools that help them to get fossils out of the ground, transport them to the laboratory, and identify the fossils in the laboratory.
> 4. Paleontologists can learn a lot about dinosaurs and how they lived by studying their skeletons and reading what others have found out about them in their laboratories.
> 5. Although no living dinosaurs exist today, paleontologists can use evidence from fossils and where they are found to make hypotheses about how dinosaurs lived and survived many years ago.
> 6. A dinosaur's skeleton can tell us a lot about what a dinosaur looked like and what features and behaviors the dinosaur used for survival.
>
> Online Resources
> 1. https://www.youtube.com/watch?v=1FjyKmpmQzc [good student-friendly introduction to paleontology]
> 2. https://www.scholastic.com/teachers/articles/teaching-content/career-paleontologist/ [good background information for what paleontologists do]
> 3. https://www.sokanu.com/careers/paleontologist/#what-is-the-workplace-of-a-paleontologist-like [especially good for showing different types of tools and how they are used]
> 4. https://www.gettyimages.com/photos/paleontologist?sort=mostpopular&mediatype=photography&phrase=paleontologist [great photographs of field and laboratory work of paleontologists]

FIGURE 12.8   Storyline for Lesson 1 in *How Do We Know What We Know about Dinosaurs?*

their formative assessments logically lead to the creation of the summative assessment? Are the individual formative assessments for the LPEs embedded in both formative and summative assessments in such a way

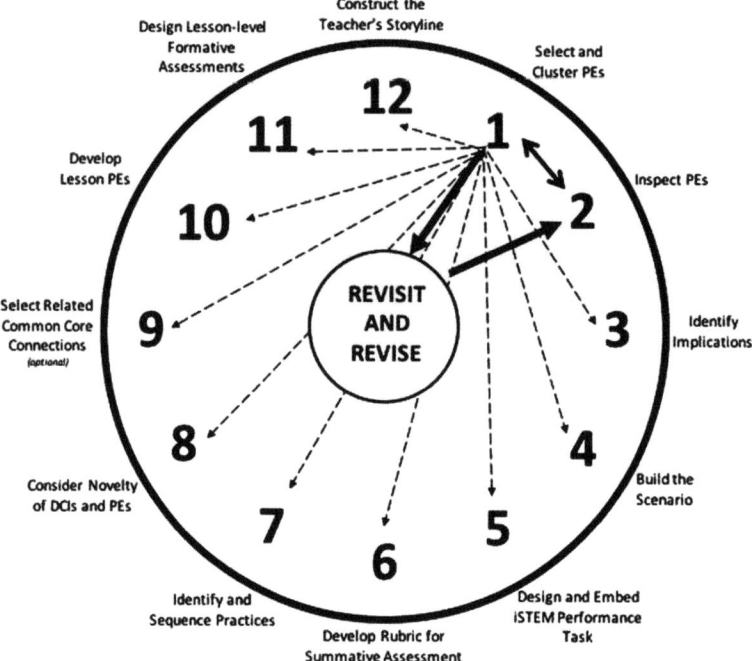

FIGURE 12.9 Schematic of the "revise and reflect clock diagram," indicating the process occurring at all stages of the 12-step process of designing an iSTEM learning performance. In the initial conception of the SCLE (dark arrows), completion of Step 2 after Step 1 requires the designer to revisit Step 1 and to check for coherence between the two steps (indicated by the dual arrow). The process occurs throughout the more or less linear process of moving through the 12 steps. Once all 12 steps are completed, the designer reviews the steps once again and adjusts decision outcomes at each step when necessary to maintain coherence in the system.

that they can be assessed? To the best of my knowledge, what are the levels of familiarity and complexity of the information with which students will be working? What happens to my initial plan for the SCLE if I decide to change this part of the SCLE?

Figure 12.9, the "reflect and revise clock diagram," indicates the recursive nature of the Revise and Review process that occurs throughout all 12 steps of the SCLE design process. This figure demonstrates how a revision at one stage in the process of decision-making may affect all other parts of the SCLE system. A change in Step 2, for example, may require both backward and forward glances at all other decisions that have been, or will be made, at other steps in the process. The designer must carefully consider the effects of a change in the SCLE system on all other parts of the system, once a change is made in one step.

At the risk of reinforcing the notion that the design of classroom iSTEM Performance Tasks is a linear process, we summarize the 12-step process of design in Figure 12.10. The central column provides five types of pages, embedding the 12 design steps (see Table 12.1 for more detail) of the iSTEM Performance Task design process. In the left-hand column, we present a brief summary of the designed page in the design.

Envision the SCLE as a complex system of ideas, processes, and activities that enable students to learn by doing. A basic question regarding every learning activity in the system is whether the design of the activity occurs at an appropriate level of direction by the teacher. The concept, "appropriate level," follows Vygotsky's notions of proximal development (Vygotsky, 1978), which refers to the level at which an individual student can perform a learning task with the help of others. Others could be other students, the teacher, or even accessory materials. We have found Table 12.2 to be helpful in explaining task appropriateness. Table 12.2 provides a heuristic indicating six levels of student-centered interventions by the teacher. This designer's filter allows the designer first (and the teacher later) to ask questions about the levels of novelty and complexity in what students are will be doing with (and therefore learning) the information in the learning activity. Learning experiences in which students are working in groups and the teacher is monitoring provide an excellent opportunity for teachers to listen to students talk about what they are doing and assess their levels of communication and comments about the information they are working with. If students are unfamiliar with the information and/or procedures, or if the information is very complex, the teacher can take the opportunity to scaffold students' learning by offering leading questions, initiating a discussion, or providing resources. What if the information is sort of familiar? Students having difficulties may only need a prompt or a hint from the teacher to allow them to progress. Finally, if students appear to have no difficulty with the information or procedures, students may be ready to elaborate or extend their knowledge. In all instances, the teacher applies a filter to ascertain whether a need exists for explanation, clarification, prompting, or elaboration, in response to the teacher's informal assessment while monitoring students' working in groups. Even proponents of student-directed models of instruction recognize that whole-class, teacher-directed instruction is sometimes the best and most efficient way for students to work with novel or complex information and procedures.

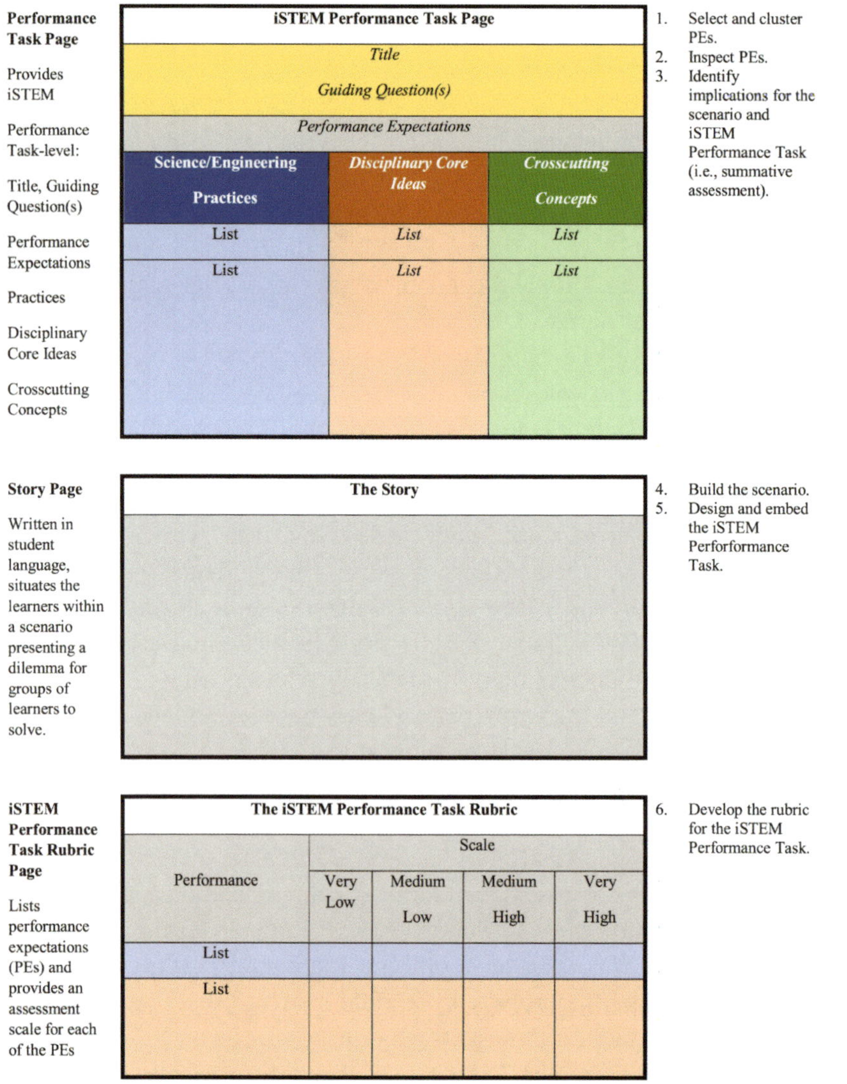

	iSTEM Performance Task Page			
Performance Task Page Provides iSTEM Performance Task-level: Title, Guiding Question(s) Performance Expectations Practices Disciplinary Core Ideas Crosscutting Concepts	*Title*			1. Select and cluster PEs. 2. Inspect PEs. 3. Identify implications for the scenario and iSTEM Performance Task (i.e., summative assessment).
	Guiding Question(s)			
	Performance Expectations			
	Science/Engineering Practices	*Disciplinary Core Ideas*	*Crosscutting Concepts*	
	List	List	List	
	List	List	List	

	The Story	
Story Page Written in student language, situates the learners within a scenario presenting a dilemma for groups of learners to solve.		4. Build the scenario. 5. Design and embed the iSTEM Perforformance Task.

	The iSTEM Performance Task Rubric					
iSTEM Performance Task Rubric Page Lists performance expectations (PEs) and provides an assessment scale for each of the PEs	Performance	Scale				6. Develop the rubric for the iSTEM Performance Task.
		Very Low	Medium Low	Medium High	Very High	
	List					
	List					

FIGURE 12.10   A linear depiction of the 12-step process of iSTEM performance task design for classroom assessments

However, these reformed practitioners still prefer to employ student-directed models and group work to meet the specific needs of students while maintaining student-direction as much as possible. The student-directed teacher works with student groups individually as a result of the filter he/she has applied to identify the learning difficulty and confront it with as little direction as possible.

INTEGRATED STEM EDUCATION REFORM

Lesson List Summary Page Refers to the iSTEM Performance Task while providing lesson-level questions and lesson-level performance expectations	**The Lesson List Summary** Title of iSTEM Performance Task SCLE-Level Performance Expectations Summative iSTEM Performance Task Lesson-Level Question and Performance Expectations List by Question List by Question	1. Identify and sequence practices into lessons. 2. Consider novelty of DCIs and PEs in each lesson. 3. Select related common core connections (opt.).
Lesson Page(s) Provides details for this and every other lesson within the sequence of lessons preparing students for the iSTEM Performance Task	**Lesson Pages** Teacher Storyline Lesson 1 *Lesson 1 Lesson Plan* *Lesson-level Performance Expectation, etc.* | Time | What Teacher Does | What Students Do | | List | List | List | Teacher Storyline Lesson 2 *Lesson 2 Lesson Plan* *Lesson-level Performance Expectation, etc.* | Time | What Teacher Does | What Students Do | | List | List | List | … add more …	4. Develop lesson PEs. 5. Design lesson-level formative assessments. 6. Develop teacher storylines for each lesson. 1-12. Apply a filter to all steps of the process.
Summative Lesson Page Provides details for the final "lesson" in the sequence, while allows time for students to complete the iSTEM	Teacher Storyline Summative Lesson *Summative Lesson Plan* *iSTEM Performance Expectation, etc.* | Time | What Teacher Does | What Students Do | | List | List | List |	1-12. Match this lesson against the performance task rubric to assure that all iSTEM Performance Expectations are met. (Final Filter)

FIGURE 12.10 A linear depiction of the 12-step process of iSTEM performance task design for classroom assessments (*cont.*)

Filtering questions for teachers. Teachers planning to implement the SCLE also apply filtering questions, such as:

- In the planning for implementing for the lesson, what level of direction is most probable for the tasks within this lesson?
- At what points are students likely to need scaffolding and what sort of scaffolding would be most appropriate?

TABLE 12.2 Levels of student-centered instructional strategies in integrated STEM learning environments

Teacher-centeredness	Student-centeredness	Instructional scaffolding	Examples
Very High (5)	Very Low (1)	The teacher directs students to listen as the teacher or another student talks to the entire class; students may be directed to read or do seat work; students' assimilation and/or accommodation occur passively with little or no interaction	Direct instruction models, including those where the teacher asks rhetorical, yes-no or one-word answers; lecture, silent reading, independent practice, seat work; watching a video
High (4)	Low (2)	The teacher directs and encourages students to respond orally or in writing to questions asked by the teacher, in whole group; the teacher facilitates sharing of students' responses	Teacher-led discussion with student input (including recitation or presentations); question and answer; discussion led and directed by the teacher
Medium (3)	Medium (3)	The teacher supervises as students in pairs or small groups work together under the teacher's supervision, with discussion among them; the teacher appropriately responds to individual groups as they indicate a need for assistance; all groups perform the same task	Student discussion in groups; may include task completion, verification laboratories, cooperative learning models

(*cont.*)

- How familiar or complex is this information to the students?
- When is the most appropriate time for me to work with the entire class of students when information is unfamiliar and/or complex?
- What sort of scaffolding is most appropriate for students if they do experience difficulty in understanding?

TABLE 12.2 Levels of student-centered instructional strategies in integrated STEM learning environments (cont.)

Teacher-centeredness	Student-centeredness	Instructional scaffolding	Examples
Low (2)	High (4)	The teacher loosely supervises students individually and in groups, providing informal monitoring and interventions as needed. Students are engaged in varied activities. While the teacher directs students to work in groups on a particular task, students still have opportunities to make decisions about how they will complete the task on their own.	Students work in groups under their own direction; intervals of work can be interrupted either formally (to the whole class) or to groups when students indicate a need for assistance
Very Low (1)	Very High (5)	The teacher facilitates students working in pairs or small groups to discuss, design, and/or formulate their own plans for working in class on their own specified task; supervision is minimal for longer periods of time; little coordination by the teacher except that the teacher allows students to work on their own	Open-ended laboratory or project work in an unrestricted environment; the teacher is available as a consultant when students need help

SOURCE: ADAPTED FROM STUESSY (2015)

– What enhancements and extensions can I suggest for students who appear to be having little difficulty with the tasks they are working on?

Experienced teachers, also, during their planning and implementation of the lessons within the SCLE, constantly undergo a back-and-forth, reflexive process of assessment that assures coherence for all students as they progress through all lessons within the SCLE. It is this process that distinguishes classroom-based assessment in iSTEM learning environments from traditional assessments in traditional classroom instruction.

4 Challenges

Advocates of integrated STEM approaches argue that integration can increase relevance and, over time, can also increase motivation and achievement (Pearson, 2017). Pearson's summary of previous research published by the National Research Council (Honey, Pearson, & Schweingruber, 2014) emphasizes the insights gained by a committee of researchers in regard to the design of STEM education initiatives. Pearson cites the need for research on best ways to integrate STEM disciplines and what factors are favorable for producing positive outcomes. Pearson reports that three factors present both opportunities and challenges for implementing integrated STEM initiatives. Two of the three deal specifically with the issues brought forth in this chapter: (1) the expertise of educators and (2) assessment. While Pearson identifies one challenge as teachers' expertise in content knowledge in the STEM fields, we are just as concerned with teachers' expertise in appropriately using student-centered instructional strategies. While some teachers intuitively understand when students are in need of assistance to remain within reach of the learning goals intended for a particular SCLE, our experience tells us that student-centeredness is a problem space in need of prolonged teacher professional development and research to examine the roles of practice, reflection, revision, and more practice in facilitating student-centered learning. Without the application of student-centered approaches in learning iSTEM classrooms, iSTEM classroom-based assessments based on monitoring and ongoing formative assessment are impossible to enact. The traditional, teacher-centered classroom with its insistence on passive learning and end-of-learning examination is not compatible with iSTEM classroom-based assessment. Based on Jonassen's (2014) concept of "doing is learning," iSTEM performance tasks and the lessons embedded within them, in reality, demonstrate that "doing is learning and assessing." Teachers constantly monitor and scaffold learning as it occurs with learners who work and constantly communicate with their peers as they are "doing science." The first challenge, therefore, we see is that of preparing teachers not only in content knowledge across the STEM domains, but also preparing teachers to teach science *not* as they were taught. Instead, the application of classroom-based iSTEM assessments requires teachers to teach science in a new way, scaffolding learning as the needs arise. Student-centered teachers give up the assumption of the past that learners are blank slates or empty vessel to be "filled up" with information by the more knowledgeable teacher.

The second challenge, assessment, deals with how to measure the outcomes of iSTEM learning. Pellegrino and colleagues (2014) recommend that systems of assessing learning with the NGSS begin from "the bottom up." The comments of these authors show progress in this area:

Assessment tasks that have been designed to be integral to classroom instruction – in which the kinds of activities that are part of high-quality instruction are deployed in particular ways to yield assessment information – are beginning to be developed. They demonstrate that it is possible to design tasks that elicit students' thinking about disciplinary core ideas and crosscutting concepts by engaging them in scientific practices and that students can respond to them successfully. (p. 5)

However, these authors also note that assessments of three-dimensional science learning are challenging to design, implement, and properly interpret. Professional development is identified, once again, as essential in incorporating this type of assessment into teachers' practices. At first glance, our 12-step, principled design approach to the development of NGSS iSTEM performance tasks may appear straightforward and easy to accomplish. At a deeper level, however, the stepped approach to development is really only one part of the complex process of developing iSTEM performance tasks. Extensive professional development before, during, and after the design of these assessments will be required to truly allow the potential for assessment to serve the full extent of integrated STEM reform recommended by the *Next Generation Science Standards*.

References

Beatly, A., & Schweingruber, H. (2017). *Seeing students learn science: Integrating assessment and instruction in the classroom.* Washington, DC: The National Academies Press.

Bransford, J. D., Brown, A. L., & Cocking, R. R. (Eds.). (2000). *How people learn: Brain, mind, experience, and school* (expanded ed.). Washington, DC: National Academy Press.

Brown, J. S., Collins, A., & Duguid, P. (1989). Situated cognition and the culture of learning. *Educational Researcher, 18*(1), 32–42.

Frankland, S. (Ed.). (2007). *Enhancing teaching and learning through assessment.* Dordrecht: Springer.

Honey, M., Pearson, G., & Schweingruber, H. (Eds.). (2014). *STEM integration in K-12 education: Status, prospects, and an agenda for research.* Washington, DC: The National Academies Press.

Jonassen, D. H. (2000). Revising activity theory as a framework for designing student-centered learning environments. In D. H. Jonassen & S. M. Land (Eds.), *Theoretical foundations of learning environments* (pp. 89–121). Mahwah, NJ: Lawrence Erlbaum Associates.

Jonassen, D. H., & Land, S. M. (Eds.). (2000). *Theoretical foundations of learning environments.* Mahwah, NJ: Lawrence Erlbaum Associates.

Kelley, T. R., & Knowles, J. G. (2016). A conceptual framework for integrated STEM education. *International Journal of STEM Education, 3*, 1–11.

Krajcik, J., Codere, S., Dahsah, C., Bayer, R., & Mun, K. (2014). Planning instruction to meet the intent of the new generation science standards. *Journal of Science Teacher Education, 25*, 157–175.

Kulm., G., & Malcom, S. (Eds.). (1991). *Science assessment in the service of reform.* Washington, DC: American Association for the Advancement of Science.

Kulm, G., & Stuessy, C. (1991). Assessment in science and mathematics education reform. In G. Kulm & S. Malcom (Eds.), *Science assessment in the service of reform* (pp. 71–88). Washington, DC: American Association for the Advancement of Science.

Land, S. M., Hannafin, M. J., & Oliver, K. (2012). Student-centered learning environments: Foundations, assumptions and design. In D. H. Jonassen & S. M. Land (Eds.), *Theoretical foundations of learning environments* (2nd ed., pp. 3–25). New York, NY: Routledge.

Lave, J., & Wenger, E. (1991). *Situated learning: Legitimate peripheral participation.* New York, NY: Cambridge University Press.

Lewin, L., & Shoemaker, B. J. (2011). *Great performances: Creating classroom-based assessment tasks* (2nd ed.). Alexandria, VA: ASCD.

NSTA (National Science Teachers Association). (n.d.). Conducting assessments. *NGSS@NSTA STEM Starts Here.* Retrieved from http://ngss.nsta.org/conducting-assessments.aspx

Pellegrino, J. W., Chudowsky, N., & Glaser, R. (Eds.). (2001). *Knowing what students know: The science and design of educational assessment.* Washington, DC: National Academy Press.

Pellegrino, J. W., Wilson, M. R., Koenig, J. A., & Beatly, S. (Eds.). (2014). *Developing assessments for the Next Generation Science Standards.* Washington, DC: The National Academies Press.

Pearson, G. (2017). National academies piece on integrated STEM. *Journal of Educational Research, 110*(3), 224–226.

Putnam, R. T., & Borko, H. (2000). What do new views of knowledge and thinking have to say about research on teacher learning? *Educational Researcher, 29*(1), 4–15.

Stiggins, R. J. (1997). *Student-centered classroom assessment* (2nd ed.). Columbus, OH: Prentice-Hall.

Stuessy, C. L. (2017, October). *The mathematics and science classroom observation protocol system: Understanding student-centered learning environments* (Research Brief #12, Update). Texas A&M University, College Station, TX: Policy Research Initiative in Science Education.

Vygotsky, L. S. (1978). *Mind in society.* Cambridge, MA: MIT Press.

Walker, A., Leary, H., Hmelo-Silver, C. E., & Ertmer, P. A. (Eds.). (2015). *Essential readings in problem-based learning: Exploring and extending the legacy of Howard S. Burrows.* West Lafayette, IN: Purdue University Press.

Wiggins, G., & McTighe, J. (2000). *Understanding by design.* Alexandria, VA: Association for Supervision and Curriculum Development.

Appendix: Five Lesson Plans for Four Lessons (a–d) and the iSTEM Performance Task Completion (e)

Lesson 1 in Sequence:	a. Name of lesson: Fossil remains of dinosaurs
Leading Question:	*How do scientists work to discover patterns in how dinosaurs looked and acted when they were living on the Earth a long, long time ago?*
Performance Expectation:	Students model working as scientists who discover dinosaur bones in the field and use resources in the laboratory to find patterns in the external parts of dinosaurs and hypothesize behaviors of dinosaurs.
General Description of the Activity:	This lesson begins with the *Fossil Story* and an introduction by the teacher of how paleontologists work two places: in the field and in the laboratory. Students then move to the "field" part of the classroom to locate dinosaur bones in tubs holding sand/rocks that have been layered in transparent, plastic tubs. Three colors/textures of sand/rocks represent different geologic layers (Triassic, Jurassic, Cretaceous) of the Mesozoic, although students do not need to know the names of the layers at this time. Students move to the "laboratory" to examine the bones and speculate where the bone is from on the body of the dinosaur by consulting some drawings of complete dinosaur skeletons. They match their skeletons to dinosaur pictures at their tables by comparing traits.
Formative Assessment:	Pairs of students work together. For one student in the pair, a student receives a 3D model of a dinosaur skeleton matched to a picture of the same dinosaur on a dinosaur card that has a description of the external feature. The student receiving the information explains to the individual holding the card what the feature is and how it is used in survival. Student hypothesize behaviors of dinosaurs based on traits by matching cards that coordinate with 3D models of dinosaurs
Materials List:	Dinosaur Skeletons (plastic, maybe cut up into fragments for advanced students); Different colors of sand/rocks (3) to represent different time periods; clear plastic tubs; Matching information cards with dinosaur drawings to coordinate with fossils; 3D models of dinosaurs for assessment; Matching information trait cards for 3D models; Matching information behavior cards for hypothesized 3D models; pencil, paper and typical classroom supplies.

Lesson 1 in Sequence:	**a. Name of lesson: Fossil remains of dinosaurs**
Technology Resources:	1. Dig Into Paleontology – SciShowKids and accessory videos 2. https://www.youtube.com/watch?v=1FjyKmpmQzc [good student-friendly introduction to paleontology] 3. https://www.scholastic.com/teachers/articles/teaching-content/career-paleontologist/ [good background information for what paleontologists do] 4. https://www.sokanu.com/careers/paleontologist/#what-is-the-workplace-of-a-paleontologist-like [especially good for showing different types of tools and how they are used] 5. https://www.gettyimages.com/photos/paleontologist?sort=mostpopular&mediatype=photography&phrase=paleontologist [great photographs of field and laboratory work of paleontologists]
Prep Activities:	Review technology resources to locate videos and pictures to use in the class discussion. Prepare plastic tubs for groups of 3 students to "dig" to find their fossil bones. See Teacher Storyline for more information.
Total Lesson Time:	85–100 minutes; 2–3 class days

Lesson

Time (min)	What the teacher does	What the students do
5	*Reads the Fossil Story. Asks students what they know about what paleontologists do, beginning with responses from students. Writes responses on the board.*	*Students offer conceptions of what paleontologists do, what paleontologists study, how paleontologists work and what paleontologists contribute to science.*

Lesson

Time (min)	What the teacher does	What the students do
10	*Show – Dig Into Paleontology – SciShowKids (Link); Follow up with discussion of new ideas not represented by previous student responses. Use accessory videos.*	*Students watch video and participate in discussion on new ideas from the video that were not suggested before watching.*
5	*Introduce students to the fossil dig inquiry and provide guidelines for exploration.*	*Students listen to directions and get with their lab partners.*
15–25	*Monitor students and offer advice, solutions and guiding questions for student exploration. Be sure to have 3 different colors/textures of sand/rock to represent different layers geologic layers with coordinating representative dinosaur fossils.*	*Students explore different tubs of sand/rocks (different colors/textures represent different layers geologic layers with coordinating representative dinosaur fossils)*
10–15	*Provide information cards and offer guidance during student inquiry. Scaffold assistance in order to informally assess student exploration.*	*Students compare uncovered fossils to provided information cards. Students hypothesize what dinosaurs they found and why base on traits of fossils.*
10	*Allow students to share results.*	*Students share results of their dig with other scientists (students) in the classroom.*
10	*Provide each student group with a 3D dinosaur model (Safari LTD, Schleich, Papo are all acceptable model brands that are anatomically accurate).*	*Students formulate ideas about what role specific features on 3D dinosaur models had in those dinosaurs' lives (adult and offspring). Students also hypothesize what behavior that dinosaur would have based on identified traits. Students write down their ideas for comparison with information cards (next step).*

Lesson

Time (min)	What the teacher does	What the students do
10	*Provide each student group with information cards (more cards than dinosaur models, so students have to sort and match).*	*Students compare hypotheses to information cards provided, which include information about the dinosaurs' name, size, era, and physical traits. Students will match trait information cards to various behavior cards that include diet (carnivore/herbivore), herding, pack hunting, offspring care, and other dinosaur behaviors.*
10	*Allow students to report on findings. (Formative Assessment)*	*Students report their findings to other scientists (students) in the classroom and explain why certain traits on their dinosaur models coordinated with specific behavior cards.*

Lesson 2 in Sequence:	b. Name of Lesson: Exploration of Traits and Behaviors of Dinosaurs
Leading Question:	How do external traits and behaviors help dinosaurs and their offspring survive, grow, and meet their needs?
Performance Expectation:	Read texts and use media to determine patterns in behavior of dinosaur parents and offspring that help offspring survive. (1-LS1-2)
General Description of the Activity:	Students will randomly be assigned a dinosaur and use the field guide provided by PBS, *Dinosaur Train*. Each student will provide a scientific argument with supporting evidence from the field guide. Students will describe how traits (3) depicted on their dinosaur (or ancient marine reptile or pterosaur) lead to particular behaviors for survival based on previously knowledge and evidence present in the dinosaurs' appearance, x-ray, food, facts and size from the field guide. Students will create a poster to tell other expert scientists (students) about their findings. Students may explore other resources provided by the teacher and appropriate media online.

Lesson 2 in Sequence:	b. Name of Lesson: Exploration of Traits and Behaviors of Dinosaurs
Formative Assessment:	Students will draw and create a poster that identifies the traits of their assigned dinosaur for display in the classroom "Dinosaur Gallery." Students will also include hypothesis of what behaviors result from the traits identified (i.e., sharp teeth, carnivore; two-legs, fast; long neck, feed from trees). Finally, students will explain how these traits and behaviors helped dinosaurs and their offspring survive, grow and meet their needs.
Materials List:	Cards with names of dinosaurs written on them, using dinosaurs from the online resource, Dinosaur Train Field Guide. Computers, Internet Access, Paper, Pencils, Colored Markers/Crayons/Pencils, Tape
Technology Resources:	Field guide provided by PBS, *Dinosaur Train* (http://pbskids.org/dinosaurtrain/fieldguide/).
Prep Activities:	Prepare poster materials for students' tables.
Total Time for the Lesson:	90 minutes

Lesson

Time (min.)	What the teacher does	What the students do
5	*Teacher will have each student randomly draw a name of a dinosaur from a basket. Try to NOT select any marine or flying reptiles, as these were not dinosaurs. If class size is large, make this clarification to students clear at the beginning of the lesson.*	*Students select their dinosaur to become an expert scientist.*
5	*Teacher introduces students to PBS Kids Dinosaur Train Field Guide.*	*Students watch introduction of media/text.*
15	*Teacher monitors students exploring field guide.*	*Students explore field guide and supplemental videos supported by Dinosaur Train (age appropriate).*

Lesson

Time (min.)	What the teacher does	What the students do
15	Teacher instructs students to identify 3 traits and behaviors based on identified traits.	Students identify 3 traits for survival of offspring and dinosaurs and hypothesize behaviors of assigned dinosaur based on traits.
30	Teacher assigns students the task of creating a museum display poster for their dinosaur. Teacher provides materials.	Acting as expert scientists, students create a poster for the "Dinosaur Gallery" that includes a drawing of their dinosaur, identified traits and hypotheses about dinosaur behaviors.
20	Teacher will have students present their posters and tell the class why identified traits helped with survival of offspring and dinosaurs. (Formative Assessment)	Students will share their findings by stating 3 traits for survival of offspring and adult dinosaurs, behaviors that resulted from traits and how these traits helped dinosaurs survive when they were young.

Lesson 3 in Sequence:	c. Name of Lesson: Dinosaur and Human Traits
Leading Question:	How are the external traits and behaviors of dinosaurs like the inventions and solutions of humans in terms of their grow, survey, and meeting their needs?
Performance Expectation:	Use materials to design a solution to a human problem by mimicking how dinosaurs use their external parts to help them survive grow and meet their needs. (1-LS1-1)

Lesson 3 in Sequence:	c. Name of Lesson: Dinosaur and Human Traits
General Description of the Activity:	Students will work in pairs with a 3D dinosaur model (2D picture acceptable) to identify traits. Students will examine dinosaur models and list traits. Students will identify a problem that they face on a daily or routine basis (e.g., they are too short, they are a slow runner, they get bruises on the playground). Students will then identify traits they learned about that dinosaurs have that could help them solve their problem. Teacher will provide materials for students to use in their designs of a "solution" (piece of clothing, an accessory of some part, an invention) to the problem using something like what a dinosaur would use. Students will then design (make a drawing) and construct (if possible) their "dinosaur trait invention."
Formative Assessment:	Students will explain to the teacher and group members: How does their invention help them? What dinosaur trait does the invention involve? Why? Why or why not – Would older students or their parents need to use it?
Materials List:	3-D models or 2-D pictures of dinosaurs, Paper, Aluminum foil, Tape (various), Makers/Pencils/Crayons, Recyclables (cans, plastic, cardboard), other various materials teacher can provide
Technology Resources:	Optional
Prep Activities:	Collect dinosaur models, assemble a table holding supplies for constructing the "dinosaur trait invention."
Total Time for the Lesson:	90 minutes

Lesson

Time (min.)	What the teacher does	What the students do
5	*Teacher will provide dinosaur models to students. Please make sure the dinosaurs assigned are different from other activities.*	*Students will examine their assigned dinosaur models.*

Lesson

Time (min.)	What the teacher does	What the students do
5	*Teacher will assist students and challenge them to find all traits on models.*	*Students list traits of assigned dinosaur model.*
10	*Teacher will ask students to share identified traits and how those traits help their assigned dinosaur behave.*	*Students share dinosaur traits and hypothesis about behaviors as a result of identified traits.*
5	*Teacher asks students to think of an everyday problem that an invention that mimics a dinosaur trait(s) could solve.*	*Students identify a problem and think about dinosaur traits that could help solve the problem.*
45	*Teacher informs students to design (draw) and build (if supplies are available) their invention.*	*Students design and engineer their invention inspired by dinosaur traits to solve an everyday issue.*
20	*Teacher asks students to share their inventions with the class and assesses their designed solution.*	*Students share their designs with fellow student scientists and engineers in the classroom. Students identify the problem, dinosaur trait of inspiration and solution.*

Lesson 4 in Sequence:	d. Name of Lesson: Evidence of Changes in Traits: Dinosaurs Skulls, Tracks and Parenting
Leading Question:	How do scientists study fossil evidence to make the connections between the external traits and behaviors of dinosaurs with respect to their life cycle and survival?
Performance Expectation:	Read texts and use media to determine patterns in behavior of dinosaur parents and offspring that help offspring survive. (1-LS1-2) Make observations to construct an evidence-based account that dinosaurs are like, but not exactly like, their parents. (1-LS3-1)

Lesson 4 in Sequence:	d. Name of Lesson: Evidence of Changes in Traits: Dinosaurs Skulls, Tracks and Parenting
General Description of the Activity:	Students will investigate evidence of change in dinosaurs as they age from offspring to juveniles to adults in *Triceratops, Corythosaurus (or other duckbill dinosaur with crest) and Tyrannosaurus*. Students will examine skulls of the aforementioned dinosaur species and arrange them from youngest to oldest of a dinosaur's lifespan. *Dinosaur Train* videos provided by *The Jim Henson Company* channel on YouTube (or other appropriate sources) along with text from reading-ability appropriate texts will accompany this portion of the lesson. Students will also investigate dinosaur tracks using the American Museum of Natural History (AMNH) handout, *Be a Sleuth: How Dinosaurs Behaved*. Students will watch a short clip on dinosaur parenting and nesting from *Dinosaur Train, Hatching Party*. Students will form hypotheses about their favorite dinosaur (or randomly assigned dinosaur) and how traits and behaviors helped it survive from birth to adulthood including if it was taken care by its parents, features it may have had for protection, if it lived in groups and/or if it had developed traits overtime to look more like an adult.
Formative Assessment:	Students will construct an explanation about how dinosaur offspring may or may not resemble their parents and describe traits that baby dinosaurs may have that help them survive. Using the 3 examples, students will organize dinosaurs by age and glue/tape down. As a class, students will share why they organized skulls in a particular order. Using the AMNH, students will answer questions relating to dinosaur tracks
Materials List:	Cut out skulls for different stages of development in dinosaurs (*Triceratops, Corythosaurus, Tyrannosaurus*), paper, glue/tape, AMNH handout, Dinosaur Nest Video, Pencils, Markers/Crayons/Colored Pencils, computer, internet access

Lesson 4 in Sequence:	**d. Name of Lesson: Evidence of Changes in Traits: Dinosaurs Skulls, Tracks and Parenting**
Technology Resources:	http://www.pasttime.org/2014/06/episode-12-field-guide-growing-up-dinosaur/ *Dinosaur Train* videos provided by *The Jim Henson Company* channel on YouTube *Be a Sleuth: How Dinosaurs Behaved* (AMNH) Dinosaur Train, Hatching Party – https://www.youtube.com/watch?v=-Xkg-Kgd8Vw Information about changes in offspring skulls of dinosaurs: T-rex: http://science.sciencemag.org/content/329/5998/1481/F1 Triceratops: https://www.researchgate.net/figure/Five-examples-of-the-four-cranial-ontogenetic-growth-stages-in-Triceratops-skulls-in_fig1_6780399 Corythosaurus: http://eesc.columbia.edu/courses/v1001/twomed.html (near the section on Vocalization in Hadrosaurs)
Prep Activities:	Preview technology resources. Make copies of skulls for (1) different stages of development in dinosaurs and (2) dinosaurs tracks, assemble materials.
Total Time for the Lesson:	90 minutes

Lesson

Time (min)	What the Teacher Does	What the Students Do
10	*Teacher introduces students to Corythosaurus, Triceratops, and Tyrannosaurs with Dinosaur Train, Dr. Scott Videos (Links:* Corythosaurus, Triceratops, Tyrannosaurus 1, Tyrannosaurus 2)	*Students watch short clips for introduction into today's three species of concentration.*

Lesson

Time (min)	What the Teacher Does	What the Students Do
5	*Teacher moderates discussion of dinosaur traits discussed in the videos.*	*Students discuss traits described in the videos that could have been used for survival of dinosaurs and their offspring.*
5	*Teacher passes out a set of 3 different types of dinosaur skull fossils, which have already been cut out by the teacher. Instructs students to arrange skulls from offspring to adult.*	*Each group of students (2–3) gets a set of 3 different species of skulls and groups prepares to arrange them from youngest to oldest based on traits found on dinosaurs' skulls.*
10	*Teacher will provide dinosaur skulls for arrangement, paper and glue. Informally assess students by walking around the room. Lead class share time.*	*Using the 3 examples, students will organize dinosaurs by age and glue/tape down. As a class, students will share why they organized skulls in a particular order.*
10	*Teacher will pass out AMNH handout and lead students through the handout.*	*Students will color different tracks.*
20	*Teacher will lead the class in answering the questions. (If students do not get it, point out the Yellow tracks, this dinosaur traveled with its young based on track evidence.)* *If this is the first time the students have heard the word "hypothesis," the teacher may have to have a short discussion about what a hypothesis is and why scientists find them so useful. Dinosaur Train makes the point that a hypothesis is an "idea you can test." See https://www.youtube.com/watch?v=r0CGhy6cNJE*	*As a class, students will share hypotheses to answer the questions on the AMNH handout.*
5	*Teacher shows short clip of dinosaur nests and care for young from Dinosaur Train.*	*Students watch video clip on dinosaur nests, eggs, and traits of offspring.*

Lesson

Time (min)	What the Teacher Does	What the Students Do
25	*Teacher will instruct students to reflect on findings from the day and develop a scientific argument comparing and contrasting Corythosaurus, Triceratops and Tyrannosaurus.*	*Students construct and explanation of how Corythosaurus, Triceratops and Tyrannosaurus offspring may or may not resemble their parents, how their tracks would have differed, how traits aid in survival and how different dinosaurs care for their young. Students will create a poster for ongoing display in their "Dinosaur Gallery."*

Summative Lesson	8e. Completing the iSTEM Performance Task
Performance Expectation:	Explain how scientists work to find out about the dinosaurs' traits and behaviors of dinosaurs with regard to their survival, growth, life cycles and other needs. Develop a scientific argument for how scientists support claims they make based on scientific evidence.
General Description of the Activity:	Students embed the information and products from Lesson 1–4 to create an explanation about their summer science experience to a new class of first graders. They will develop the explanation as if they were returning to school in the fall after their summer science experience. They will explain to the new class the types of discoveries that paleontologists make that lead to information about the relationship between dinosaurs' growth, survival, and needs and their external features. They will make an oral presentation and use some visual representations in their presentation.

Summative Lesson	8e. Completing the iSTEM Performance Task
Formative Assessment:	For assessment purposes, use the iSTEM Performace Task Rubric, based on the formative assessments of all four lessons. Using their Dinosaur Gallery, students will give one final presentation about what they have learned while being an expert scientist on dinosaurs. Students will provide evidence that they can: Explain how scientists work to find out about the dinosaurs' traits and behaviors of dinosaurs with regard to their survival, growth, life cycles and other needs. Develop a scientific argument for how scientists support claims they make based on fossil evidence.
Materials List:	All available previous materials from prior formative assessments.
Total Time for the Lesson:	50 minutes

Lesson

Time (min)	What the teacher does	What the students do
20	*Teacher asks students to present their findings as an expert scientist.*	*Students prepare final materials (posters, previous assessments) for presentation.*
30	*Teacher summatively assesses students during presentation using iSTEM Performance Task Rubric.*	*Students present findings about dinosaurs using knowledge gained and assessments from previous 4 lessons.*

CHAPTER 13

A Shared Language: Two Worlds Speaking to One Another through Making and Tinkering Activities

Amber Simpson, Jackie Barnes and Adam V. Maltese

Abstract

Making and tinkering is being viewed as an interdisciplinary approach to promote learning of knowledge, practices, and skills across science, technology, engineering, and mathematics (STEM) disciplines in informal (e.g., science and art museums) and formal (e.g., school-based makerspaces) contexts. In this chapter, we present two frameworks for mapping the overlap between making practices developed by Wardrip and Brahms (2015) and standards-based practices developed for PreKindergarten to Grade 12 education. We apply the two frameworks to a making task of one youth who constructed a car using LEGOs and littleBits, electronic building blocks that snap together with magnets. Through the case of Bailey, we highlight how informal and formal learning environments speak to one another to promote STEM learning for students of all ages.

1   Introduction

Making encompasses a variety of activities including building, designing, adapting and/or repurposing objects and material into a physical or digital product of some sort. (Vossoughi & Bevan, 2014). A variety of tools and material can be leveraged including low-tech (e.g., conductive thread), high-tech (e.g., 3D printing), and no-tech (e.g., wood, plastic recyclables, cardboard) tools and material. Making happens in classrooms, basements, garages, museums, libraries, makerspaces, and art studios. As such, it "reaches across the divide between formal and informal learning, pushing us to think more expansively about where and how learning happens" (Halverson & Sheridan, 2014, p. 498). Peppler, Halverson, and Kafai (2017) agree that making provides opportunities for the historical divide between informal and formal learning environments to dissolve and inform one another. With the adoption of content and practice standards across STEM subject areas in Pre-Kindergarten (PK)-12 school settings in the U.S., out-of-school learning sites such as museums, youth-based makerspaces, and afterschool programs have questioned their role in engaging youth as STEM professionals during making-related activities; bridging in-school and out-of-school learning contexts.

As such, making is being considered for it's potential to promote learning across science, technology, engineering, and mathematics (STEM) disciplines (e.g., Simpson, Burris, & Maltese, 2017a; Bevan, 2017; Martin & Dixon, 2017; Quinn & Bell, 2013), as well as the arts (e.g., Clapp & Jimenez, 2016). In terms of learning, the focus is on the process of learning through making as opposed to the final artifact that is made (Gutwill, Hido, & Sindorf, 2015). In this chapter, we too consider learning to be an inherent process of design; and focus on learning practice skills common to STEM professionals (Lee, Quinn, & Valdés, 2013) as enactment of these skills provide youth with authentic opportunities to participate in STEM community (Lave & Wenger, 1991) and to become STEM-literate members of society (Zollman, 2012).

In this chapter, we present two frameworks for mapping the overlap between making practices and STEM practices. The first section describes how eight making practices map onto K-12 standards-based STEM practices (i.e., informal to formal), and the second describing how K-12 standards-based STEM practices map onto making practices (i.e., formal to informal); thus, highlighting how the two learning environments speak to one another to promote STEM learning for students of all ages. We further illustrate this notion by applying the two frameworks to a making task of one youth, Bailey, who decides to build a car. More specifically, youth were engaged in constructing something, anything, using LEGOs and incorporating littleBits[1] (2018) as part of an afterschool making program in an elementary school operated by facilitators from a local science museum.

In presenting the case of Bailey, three things should be considered while evaluating the practices and standards described below. First, our claim is not that specific actions of one child align perfectly or objectively with each standard listed. Our purpose is to show how this activity can potentially support practices valued within STEM standards. Additional standards to those listed might be relevant, but here we present a working example of one short making-interaction. Second, the interaction here is facilitated and scaffolded by facilitator(s); thus, the practices evident below should be considered in relation to the scaffolds present. Third, we consider the case of Bailey to not be representative of other youth his age. Youth provided with the same or similar task may or may not engage in the practices in the same manner as Bailey.

2 Mapping Making Practices to Formal Practice Standards

Makers use mathematical and scientific thinking as part of their craft. Further, it would be extremely difficult to argue that Makers *aren't* designers. Makers solve unstructured problems with available materials, often using innovative methods. Research at the Children's Museum of Pittsburgh resulted in a set of practices that are core to making as a result of multiple iterations in collaboration with museum Teaching Artists who facilitate the MAKESHOP space. This framework (Wardrip & Brahms, 2015) illustrates one way that practices of making are conceptualized and articulated by practitioners. These were developed as emergent values and practices of Teaching Artists, and not driven by academic concerns or comparisons. Nevertheless, the practices have relevance to values and skills embodied within academic and STEM requirements focusing on engineering & design. In the following section, we align Pennsylvania [PA] Common Core academic standards *most* aligned to these eight making practices (see Appendix A). Clearly, the PA standards are only one example for alignment, but productively illustrate the potential for making practices to contribute to academic learning.

2.1 *Learning Practices of Making*

2.2.1 Inquire

Makers inquire through their *openness and curious approach to the possibilities of the context through exploration and questioning of its material properties.* The PA Science, Engineering & Technology standards list "Science as Inquiry" practices that stretch across content and grade levels. The core components of inquiry are reflected in the language of the standards. Students must identify answerable questions, evaluate reasonableness, decide whether and how to conduct an investigation, use inference, make predictions, use appropriate tools to gather data, develop descriptions, explanations, models, analyze alternative explanations and have legitimate skepticism, formulate and revise explanations where investigations may result in new ideas, procedures, or technologies.

2.2.2 Tinker

Learners' show *purposeful play, testing, risk taking, and evaluation of the properties of materials, tools, and processes* during making activities. Tinkering also happens as part of both mathematical and science engagement, as a process similar to troubleshooting. The common core mathematics practices include a standard of students making sense of problems and persevering in solving them—tinkering actually supports persistence in this way. Further, troubleshooting itself is a PA Science & Engineering standard (i.e. "Describe how troubleshooting as a problem-solving method may identify the cause of a malfunction in a technological system").

2.2.3   Seek & Share Resources
Making often happens collaboratively. The process requires *identification, pursuit/recruitment, and sharing of expertise with others as well as recognition of one's own not-knowing and desire to learn.* Common Core Mathematics Practices require the construction of viable arguments and understanding the reasoning of others, including the instructions a collaborator might share. Further, the PA Science & Engineering standards articulate the design process as a collaborative endeavor in which each person in the group presents his or her ideas in an open forum.

2.2.4   Hack & Repurpose
Makers often use obscure materials to create something new. They might *harness and salvage materials, tools and processes to modify, enhance, or create a new product or process; including disassociating object property from familiar use.* Similarly, Common Core Mathematics Practices support learners' use of appropriate tools strategically, including in new or unexpected ways. The PA Science & Engineering standards also support repurposing, stating that learners compare how a product, system, or environment developed for one setting may be applied to another setting and to explain how modifying is used to transform ideas into practical solutions while testing and evaluating the solutions for a design problem. Hacking materials for a new purpose requires exactly these practices.

2.2.5   Express Intention
Through making, learners express their ideas and intentions. They do this through their *discovery, evolution, and refinement of personal identity and interest areas through determination of short- and long-term goals; choice, negotiation, and pursuit of goals alone and with others.* The PA Science & Engineering standards address intention both societally and through the process of design. Learners explore the design process as a collaborative endeavor in which each person in the group presents his or her ideas in an open forum and describe how the design of a message is influenced by such factors as the intended audience, medium, purpose, and nature of the message. They include the recognition that values and interests of individuals, businesses, industries, and societies, as well as cultural priorities and values, drive new technologies.

2.2.6 Develop Fluency

As learners make more and more diverse projects, they *develop comfort and competence with diverse tools, materials, and processes while developing craft*. Math, science, and engineering standards support the understanding and selection of appropriate tools for a variety of tasks. The Common Core Mathematics Practices include the use of appropriate tools strategically. The PA Science & Engineering skills require explaining how different technologies involve different sets of processes, and the selection and safe use of appropriate tools, products and systems for specific tasks.

2.2.7 Simplify to Complexify

Lastly, as learners engage in making practices, they might *connect and combine component elements to make new meaning*. Previous and additional practices can contribute to this understanding, as makers use scientific and technological skills of understanding how invention and innovation, as well as individual processes, lead to changes in society and the creation of new needs and wants within a larger system.

2.3 *Case Study of Bailey Tinkering*

The video of Bailey building a car (see Figure 13.1) shows mathematical thinking, but also strong connections to science, engineering, and design standard.[2] For example, *tinkering* includes testing boundaries and affordances of materials, similar to the scientific process and inquiry through the investigation of properties, materials, and systematically testing hypotheses.

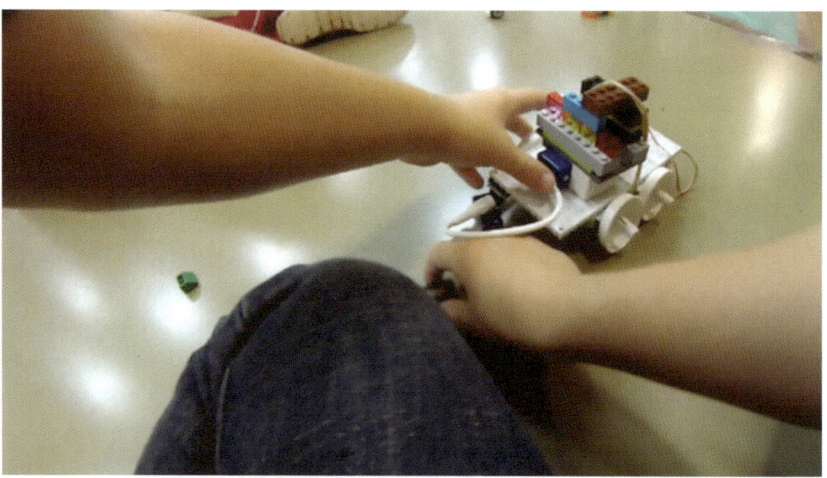

FIGURE 13.1   Image of Bailey's construction of a car using LEGOs and litteBits

In part, Bailey's talk reveals tinkering practices. In the first few seconds of the video, Bailey plays with and manipulates the littleBits, swinging a cord in circles after noticing its affordance of being swingable. In this way, he explores the affordances and constraints of each littleBit piece. Bailey's early questions mirror this exploration of affordances, starting with "Can I ..." and "How do you use ..." or "How do these connect?" (*Identify and collect information about everyday problems that can be solved by technology and generate ideas and requirements for solving a problem*).

When something doesn't fit or work as intended, Bailey changes the littleBits and Legos position to identify the cause of the malfunction by testing and evaluating his current method or strategy. In this way, he shows persistence in making sense of problems (*Sense of problems and persevere in solving them*) and purposeful play. A focused session of tinkering happens between approximately 2:00 and 5:00, beginning with Bailey wondering "How do they connect?" He manipulates the littleBits before saying "How do they go together?" The testing and manipulating of physical pieces show the testing and revision of tinkering practices to accomplish a goal (*Describe how troubleshooting as a problem-solving method may identify the cause of a malfunction in a technological system*). Another instance of Bailey finding a solution can be seen near 6:00. He exclaims "Ah. I'm going to make these so my tires don't come off!" (He puts white covers over the motor bit or tire piece, in an attempt to keep them from sliding off the peg where they are placed.) At 6:45, he says "Oh, I found one way!" The scientific thinking reflected through tinkering is a systematic assessment of a problem and attempt, and revision, of potential solutions.

During the second ten minutes of the video, Bailey does a lot of aesthetic tinkering, but also struggles with dexterity and getting things to snap together. He clearly knows what he wants to make as he picks blocks from the bin, but the intended design doesn't become clear until later. He's showing intention by picking specific pieces from the bin, comparing them against his design, measuring them against each other, and continuing to go back to add to the design.

The third ten minutes depicts the placement of the battery onto the top of his car built during the previous 10 minutes. He troubleshoots how it might fit, and how it might stay secure in the Lego structure, sitting on top of his car. At 20:00, Bailey puts the battery pack in the back of the Lego structure and pushes the structure across the floor as a test. He turns the car on, before (23:00) again seeing if the battery fits in the Lego structure. It falls out. He continues building, and tries to fit it on top of the car (*Test and evaluate the solutions for a design problem*). Around 24:35 Bailey places the battery on top of the car, again, but asks "Where are the rubber bands? Can I use rubber bands?" finding a potential solution (*Explain how modeling, testing, evaluating, and modifying are used to transform ideas into practical solutions*). At 25:25, he quickly places two bands around the top of the car to hold the Lego structure to the littleBit structure, and sticks the battery in back. Saying "Hmmm," he then (25:45) places two Legos on either side of the battery to secure it in its slot. They won't snap in, so he takes them off and turns the car on (*Make sense of problems and persevere in solving them*). Bailey continues to question affordances of the design, asking (at 29:00) "Can it ride upside down like that?" At 29:29, Bailey asks "Where'd those wheels go?" and crawls quickly to another mat, saying "I need this." At 29:34, he says "I know what to do! … Ah ha!" as he rubber bands the battery to its own wheel so that it can roll behind the car independently (*Explain how many inventions and innovations have evolved by using deliberate and methodical processes of tests and refinements*).

In the final (fourth) ten minutes, Bailey turns on the car. When it won't move, he says "I know what to do! If it won't move, then put it on the top!" He continues with "If we put it on the top! It'll work if we put it on the top! 'Cause if you put it with this, it won't move" In this way, Bailey makes predictions about what might happen as he troubleshoots a design problem (*Describe how troubleshooting as a problem-solving method may identify the cause of a malfunction in a technological system*).

3 Mapping Formal Practice Standards to Making Look-Fors

We utilized PK-12 process skills within the U.S. as the foundation of the framework and what to "look-for" in terms of youth's STEM learning in making and tinkering contexts. More specifically, we used the following standards: (1) Science and Engineering Practices in the Next Generation Science Standards ([NGSS], NGSS Lead States, 2013); (2) International Society for Technology in Education [ISTE] standards for students (ISTE, 2015); (3) Standards for Mathematical Practice in the Common Core State Standards (CCSS, 2017); and (4) Computer Science Framework (Association for Computing Machinery [ACM] et al., 2016). We previously utilized science and engineering practices to measure youths' engagement (Simpson et al., 2017a), but found the NGSS Science and Engineering practices to be limited for evaluating informal settings. The framework discussed here is broader in scope and provides educators, evaluators, and/or researchers ways to observe indicators of youth learning through practices common to professionals across STEM fields (NRC, 2012).

Although these standards-based documents frame learning of these practices within a formal setting (e.g., school classroom), we have found evidence of youth enacting these practices within an informal STEM-related making and tinkering setting (Simpson et al., 2017a). Our findings highlight how these practices "looked" different in an informal setting. We also found instances of engagement not explicitly stated within the parameters of these standards-based documents, such as reactions to failure. We built upon our understanding from this study to create an interdisciplinary framework of 11 STEM practices (see Appendix B). While this framework was created with middle grade students in mind (aged 10–14), we contend that these practices are also appropriate for students across grades K-12 (ages 5–18). In the following, we continue with the Case of Bailey from Day 1 to Day 2, highlighting the practice *Analyzing and Interpreting Data* and illustrating its commonality to the making practice *Tinkering*.

3.1 Case Study of Bailey Analyzing and Interpreting Data

On Day 2, Bailey continues re-assembling his car as one of the wheels was not attached to the mounting board, or the base of the car (see Figure 13.2). As Bailey *tests* his design at 3:50, it was turning in a circle as opposed to moving straight. Bailey's expresses his dissatisfaction by picking up the car to make a *change to his design*. In making this change, the wheels continually fell off the base of the car. Yet, Bailey *persists* through his frustration for about 4 minutes before presumably making an *informed decision* at 8:04 to remove the hubs (or small plastic lids with a man-made central hole) from the motor bit as the hubs were hindering Bailey's ability to attach the bits to the mounting board. Bailey's *persistence* continues until 10:15 when he asks for help from a facilitator. In this case, we contend that asking for help does not indicate helplessness or defeat, but a practice common to STEM professionals as Bailey tried multiple times prior to seeking assistance (Simpson & Maltese, 2017b). Bailey again *tests* his design at 11:15. The car moves along a straight path until one of the hubs fall off. In which case, Bailey makes an *informed decision* to add white covers to keep the hubs from falling off; an *informed decision* that dates back to Day 1 (i.e., *uses prior experience*). At 12:15, Bailey again *tests* his design and notes "Mine's leaving us. It's escaping." His *perseverance* led to success.

Around 13:48, Bailey decides to add a littleBit that lights up – "Where's lights? I need a button" (*Planning Statement*) – as he searches for the needed littleBits

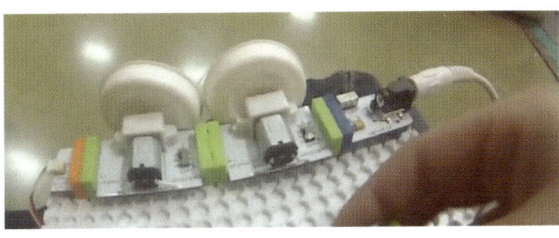

FIGURE 13.2
Bailey continuing construction of his car from Day 1

(*Selecting Appropriate Tools*). In adding these two bits, the car falls apart, yet again. Although Bailey continues expressing his frustrations (e.g., I don't like this. It won't stay together."), he *persists* in re-assembling the car. Moving forward in this case to 26:38, Bailey moves to the middle of the gym floor. He again *tests* his design; yet, the car turns in a circle. As he *examines* his design, Bailey notices he forgot a vital piece of the car, a littleBit that connects the two pairs of wheels on either side of the car together. At 30:47, Bailey seems to experience success in *testing* the design as the car moves in a straight path and he states, "Finally, going!" As the car continues moving straight ahead, Bailey sets a challenge for the car to continue its path and "go to the black circle" located in the middle of the gym floor. As the car begins turning left, Bailey stops the car making an adjustment that leads to the car turning in a circle. Becoming more and more frustrated – "It won't do what I want it to do." – Bailey adjusts the white covers as suggested by a facilitator. As Bailey *tests* this design change at 34:37, the wheels are spinning, but the car is not moving. Considering his design from the previous day, Bailey makes an *informed decision* to place the battery on top of the car (*uses prior experiences*). The car now moves in a circle. At 36:02, he and a facilitator again *test* the design with no changes made to this previous test. Bailey turns the car over (i.e., *examines car*) and notes "This one [is] going right. This one left." He provides an *explanation* that one of the wheels is turning in an opposite direction than the other three wheels. The video ends with Bailey making an *informed design change*, namely flipping the switch on the motor bit to turn in the same direction as the others.

4   Discussion

Here we investigate a case of engineering a car using littleBits and Legos. Through these two case studies we contend that we address the question posed by Peppler and colleagues (2017), "How do we maintain the emergent, messy, whimsical, engaging process of making while adhering to standards and shared practices?" (p. 6). In each case of Bailey, we essentially discuss the same practices and processes from two varying perspectives – one through the making practices and the other through state-mandated practice standards. Hence in many places, we are saying the same thing, but using different language or vocabulary. For example, within the *Tinkering* practice, we observed Bailey *testing and evaluating the solutions for a design problem (3.4.8.D1)*, which is worded in the formal practice standards framework as *Analyzing and Interpreting Data*. Arguably, we are reaching across the divide between formal and informal environments as youth are learning to become members of the STEM community through engaging in making and tinkering activities (e.g., Lave &

Wenger, 1991). The two frameworks presented here are speaking to and with one another as opposed to contradicting one another. Therefore, is the above question by Peppler and colleagues the wrong question to ask? Based on our observations of Bailey, we ask, how can we better articulate the learning inherent in making practices in ways that are translatable to the language of academic knowledge, standards, and paradigms? Additionally, how can educators and policy makers articulate standards to better capture the authentic making practices that we truly care about?

We can also consider our work in terms of understanding how the learning process through the language used by educators, learners, and practitioners supports student learning through making and tinkering. Atman, Kilgore, and McKenna (2008) looked at the importance of language in engineering education to understand the thought and design process of novice engineering students. They found differences in how students responded to choices on a Likert-scale survey versus open response. They interpreted findings as showing the importance of understanding the internal speech that student designers develop through engineering education and working to access and support that language during the learning process. Further, they described students' abilities to express their knowledge of engineering design using appropriate language as limited.

In addition, while the case discussed in this chapter highlights making for fun, the principles of figuring out whether the car is functional, how it works, and whether it adheres to the initial intended vision, compare to professional engineering practices. The project has a functional purpose and completion of the project requires understanding how and why the design works (or doesn't work). Bailey applies scientific and mathematical practices to practical purposes of design. Both making and engineering require the intertwined goals of functionality and design. The two frameworks also align with principals of the engineering design process (e.g., Tayal, 2013); however, the practices in our framework are not framed as a cycle or decision-making process because we contend that engagement in the practices in making activities are "messy" and playful (Simpson et al., 2017a).

5   Conclusion & Implications

We encourage others in the formal and informal worlds of making and tinkering with youths to build upon these frameworks as a means to develop and refine a shared language for making practices that is also applicable to academic standards. Although, there are overlaps between the two frameworks, there are also examples where one framework can inform the other as a practice may not be included or apparent. For example, the formal practice standards framework includes *Communicate/Present Information* to encourage youths to showcase and articulate their process and final object through the making experience. As another instance, the Making Practices framework includes *Simplify to Complexify* to highlight youths' understanding of materials and processes by connecting and combining component elements to make new meaning. Additionally, future research should utilize these frameworks to examine how youths are engaged as STEM professionals and participating in the communities of practice in STEM areas (Lave & Wenger, 1991). Utilizing these as lenses across making contexts will help establish a research-base for highlighting areas of strength and areas of improvement in terms of youths' learning of practice skills inherent in the design process. It will also aid in understanding how the two frameworks speak to one another, as well as areas where they can build upon one another. We further contend that these frameworks will be useful for establishing and maintaining learning communities and video-clubs (e.g., Sherin, 2007) around maker education; as well as other communities that span the informal-formal divide. We also hope that these will be useful to educators who seek to communicate the value of their making, art, and design practices in a STEM-focused academic context.

Notes

1 littleBits are electronic building blocks that snap together with magnets. Each building block has a unique feature (e.g., sound, light, motion, and dimmers).
2 If you would like access to the videos discussed in this chapter, email the first author at asimpson@binghamton.edu

References

Association for Computing Machinery [ACM]. (2016). *K12 computer science framework.* Retrieved from https://k12cs.org/

Bevan, B. (2017). The promise and the promises of making in science education. *Studies in Science Education.* doi:10.1080/03057267.2016.1275380

Clapp, E. P., & Jimenez, R. L. (2016). Implementing STEAM in maker-centered learning. *Psychology of Aesthetics, Creativity, and the Arts, 10*(4), 481–491. doi:10.1037/aca0000066

Common Core State Standards Initiative. (2017). *Standards for mathematical practice.* Retrieved from http://www.corestandards.org/Math/Practice/

Gutwill, J. P., Hido, N., & Sindorf, L. (2015). Research to practice: Observing learning in tinkering activities. *Curator: The Museum Journal, 58*(2), 151–168.

Halverson, E. R., & Sheridan, K. M. (2014). The maker movement in education. *Harvard Educational Review, 84*(4), 495–504.

International Society for Technology in Education [ISTE]. (2007). *ISTE standards for students.* Retrieved from https://www.iste.org/standards/for-students

Lave, J., & Wenger, E. (1991). *Situated learning: Legitimate peripheral participation.* Cambridge: Cambridge University Press.

Lee, O., Quinn, H., & Valdés, G. (2013). Science and language for English language learners in relation to Next Generation Science Standards and with implications for Common Core State Standards for English language arts and mathematics. *Educational Researcher, 42*(4), 223–233.

littleBits Electronics Inc. (2018). *littleBits Education.* Retrieved from http://littlebits.cc/education

Martin, L., & Dixon, C. (2017). Making as a pathway to engineering and design. In K. Peppler, E. R. Halverson, & Y. B. Kafai (Eds.), *Makeology: Makerspaces as learning environments* (Vol. 2, pp. 183–195). New York, NY: Routledge.

Museum of Science. (2018). *The engineering design process.* Retrieved from https://www.eie.org/overview/engineering-design-process

National Research Council [NRC]. (2012). *A framework for K-12 science education: Practices, cross-cutting concepts, and core ideas.* Washington, DC: National Academies Press.

NGSS Lead States. (2013). Appendix F: Science and engineering practices in the next generation science standards. In *Next Generation Science Standards* (pp. 382–412). Washington, DC: The National Academies Press.

Peppler, K., Halverson, E. R., & Kafai, Y. B. (2017). *Makeology: Makerspaces as learning environments.* New York, NY: Routledge.

Quinn, H., & Bell, P. (2013). How designing, making, and playing relate to the learning goals of K-12 education. In M. Honey & D. Kanter (Eds.), *Design, make, play: Growing the next generation of STEM innovators* (pp. 17–33). New York, NY: Routledge.

Sherin, M. G. (2007). The development of teachers' professional vision in video clubs. In R. Goldman, R. Pea, B. Barron, & S. J. Deny (Eds.), *Video research in the learning sciences* (pp. 383–395). Mahwah, NJ: Erlbaum.

Simpson, A., Burris. A., & Maltese, A. V. (2017a). Youth's engagement as scientists and engineers in an after-school tinkering program. *Research in Science Education.* [Advanced Online Publication.] doi: 10.1007/s11165-017-9678-3

Simpson, A., & Maltese, A. V. (2017b). "Failure is a major component of learning anything": The role of failure in the career development of STEM professionals. *Journal of Science and Technology Education, 26*(2), 223–237. doi: 10.1007/s10956-016-9674-9

Tayal, S. P. (2013). Engineering design process. *International Journal of Computer Science and Communication Engineering.* Retrieved from http://static.ijcsce.org/wp-content/uploads/2013/06/IJCSCESI040113.pdf

Vossoughi, S., & Bevan, B. (2014). Making and tinkering: A review of the literature. In *National Research Council Committee on Out of School Time STEM* (pp. 1–55). Washington, DC: National Research Council.

Wardrip, P. S., & Brahms, L. (2015). *Learning practices of making: developing a framework for design.* In Proceedings of the 14th International Conference on Interaction Design and Children (pp. 375–378). ACM.

Zollman, A. (2012). Learning for STEM literacy: STEM literacy for learning. *School Science and Mathematics, 112*(1), 12–19.

Appendix A: Pennsylvania Core and Academic STEM Standards *Most* Aligned to CMP Practices of Making

INQUIRE

Learners' openness and curious approach to the possibilities of the context through exploration and questioning of its material properties

CC Mathematics Practices
- Make sense of problems and persevere in solving them

Science as Inquiry Practices
- Analyze alternative explanations and understanding that science advances through legitimate skepticism
- Understand that scientific investigations may result in new ideas for study, new methods, or procedures for an investigation or new technologies to improve data collection
- Identify questions and concepts that guide scientific investigations
- Formulate and revise explanations and models using logic and evidence

Science, Technology and Engineering (3.4)
- Explain how the type of structure determines the way the parts are put together
- Explain how technology is closely linked to creativity, which has resulted in innovation and invention
- Identify and collect information about everyday problems that can be solved by technology and generate ideas and requirements for solving a problem
- Explore the design process as a collaborative endeavor in which each person in the group presents his or her ideas in an open forum
- Test and evaluate the solutions for a design problem

TINKER
Learners' purposeful play, testing, risk taking, and evaluation of the properties of materials, tools, and processes

CC Mathematics Practices
– Make sense of problems and persevere in solving them

CC Mathematics (2.1)
– Analyze proportional relationships and use them to model and solve real-world and mathematical problems

Science, Technology and Engineering (3.4)
– Explain why some technological problems are best solved through experimentation
– Show how models are used to communicate and test design ideas and processes
– Describe how troubleshooting as a problem-solving method may identify the cause of a malfunction in a technological system
– Explain how many inventions and innovations have evolved by using deliberate and methodical processes of tests and refinements
– Explain how modeling, testing, evaluating, and modifying are used to transform ideas into practical solutions
– Identify and collect information about everyday problems that can be solved by technology and generate ideas and requirements for solving a problem
– Test and evaluate the solutions for a design problem

SEEK AND SHARE RESOURCES
Learners' identification, pursuit/recruitment, and sharing of expertise with others; includes collaboration and recognition of one's own not-knowing and desire to learn

CC Mathematics Practices
– Construct viable arguments and critique the reasoning of others

Science, Technology and Engineering (3.4)
– Identify and collect information about everyday problems that can be solved by technology and generate ideas and requirements for solving a problem
– Explore the design process as a collaborative endeavor in which each person in the group presents his or her ideas in an open forum

HACK & REPURPOSE

Learners harnessing and salvaging of materials, tools and processes to modify, enhance, or create a new product or process; includes disassociating object property from familiar use

CC Mathematics Practices
- Make sense of problems and persevere in solving them
- Use appropriate tools strategically

Science, Technology and Engineering (3.4)
- Apply a design process to solve problems beyond the laboratory classroom
- Identify and collect information about everyday problems that can be solved by technology and generate ideas and requirements for solving a problem
- Select and safely use appropriate tools, products and systems for specific tasks
- Test and evaluate the solutions for a design problem
- Compare how a product, system, or environment developed for one setting may be applied to another setting

EXPRESS INTENTION

Learners' discovery, evolution, and refinement of personal identity and interest areas through determination of short- and long-term goals; includes learners' responsive choice, negotiation, and pursuit of goals alone and with others

Science, Technology and Engineering (3.4)
- Interpret/explain how societal and cultural priorities and values are reflected in technological devices
- Demonstrate how new technologies are developed based on people's needs, wants, values, and/ or interests
- Explain how throughout history, new technologies have resulted from the demands, values, and interests of individuals, businesses, industries, and societies)
- Explore the design process as a collaborative endeavor in which each person in the group presents his or her ideas in an open forum
- Describe how design is influenced by such factors as the intended audience, medium, purpose, and nature of the message

DEVELOP FLUENCY
Learners' development of comfort and competence with diverse tools, materials, and processes; developing craft

CC Mathematics Practices
- Make sense of problems and persevere in solving them
- Use appropriate tools strategically

Science, Technology and Engineering (3.4)
- Recognize that requirements for a design include such factors as the desired elements and features of a product or system or the limits that are placed on the design
- Explain how different technologies involve different sets of processes
- Select and safely use appropriate tools, products and systems for specific tasks
- Analyze the development of technology based on affordability or urgency
- Explain how societal and cultural priorities and values are reflected in technological devices
- Test and evaluate the solutions for a design problem

SIMPLIFY TO COMPLEXIFY
Learners' demonstration of understanding of materials and processes by connecting and combining component elements to make new meaning

Science, Technology and Engineering (3.4)
- Describe how systems thinking involves considering how every part relates to others
- Explain how knowledge from other fields of study (STEM) integrate to create new technologies
- Describe how economic, political, and cultural issues are influenced by the development and use of technology
- Describe how invention and innovation lead to changes in society and the creation of new needs and wants
- Use data collected to analyze and interpret trends in order to identify the positive or negative effects of a technology
- Compare how a product, system, or environment developed for one setting may be applied to another setting
- Explain how throughout history, new technologies have resulted from the demands, values, and interests of individuals, businesses, industries, and societies
- Describe how the design of the message is influenced by such factors as the intended audience, medium, purpose, and nature of the message

Appendix B: Mapping Formal Practice Standards to Making & Tinkering Look-Fors

Practice	Making & tinkering look-fors	Standards alignment
Make Sense of Activity	– Explain the meaning of the activity – Decompose the problem and/or investigation into manageable sub-problems – Analyze and explain constraints, given, unknowns, goals of the activity – Explain how different approaches (including those of peers) may be utilized in carrying out solutions to the activity	Mathematics Practice 1 CS Practice 3
Ask Questions Define Problems	– Pose questions that require further (or new) investigation or research (e.g., How can I make my own solar eclipse glasses?); considered questions of high-cognitive demand – Develop/define a design problem that can be solved through the development of an object, tool, process or system – Identify an interdisciplinary, real-world problem that can be solved computationally – Pose questions that seek clarification of a model/prototype, an engineering problem, an explanation, and/or an argument (e.g., What do you mean? How does adding this part help solve the problem?)	NGSS Practice 1 CS Practice 3
Develop, Use Models, and Select Appropriate Tools	– Sketch/draw a model – Build a model/prototype – Practice technique before final design (e.g., practice a pop-up cut on a scrap sheet of paper) – Select appropriate tools for design solution and/or investigation (e.g., hot glue or tape)	NGSS Practice 2 Mathematical Practice 5

Practice	Making & tinkering look-fors	Standards alignment
Plan Investigation	– Write down or verbally articulate sequence of steps for an investigation or design process – Write down or verbally articulate needed material and tools to carry out an investigation or design – Brainstorm plans and ideas aloud with peer(s) – Select and use digital tools to plan and manage a design process that considers design constraints and calculated risks. – Articulate goals/expectations in relation to the activity	NGSS Practice 3 Mathematics Practice 1 Technology Practice 4 CS Practice 5
Attend to Precision	– Use appropriate vocabulary and definitions in oral and/or written communication – Measure (e.g., length, weight) with precision – Label accurately when measuring, graphing, etc. – Express numerical answers with a degree of precision (e.g., rounding error) – Calculations are accurate	Mathematical Practice 6
Document and Explain Activity	– Synthesize observational notes into an oral and/or written explanation or visual representation. – Utilize prior experiences and/or prior knowledge to construct and/or support explanation (must be explicit) – Explain design solution, including constraints and criteria, and/or decisions made throughout the design or activity. – Document failures and explain how failures led to changes in activity or design – Document process through photographs and/or video files	

Practice	Making & tinkering look-fors	Standards alignment
Analyze and Interpret Data	– Construct a hypothesis or conjecture based on observations (e.g., I think the wheel is the problem because it keeps turning right instead of staying straight.) – Use digital tools to analyze data/information – "Testing" model/object/design (i.e., trials) and make changes to design based on "tests" (and can defend this change – informed decision making as opposed to uninformed decision making) – State and/or write "becauses" in relation to "tests" (e.g, This did not work because …") – Examine object or device (e.g., turning over in hand while "studying" object) – Persevere in solving problem	NGSS Practice 4 CS Practice 6
Use Mathematics and Computational Thinking	– Develop visual representation(s) of observations or investigations (e.g., frequency chart, bar graph) to identify patterns – Apply mathematical concepts and/or processes (prior knowledge) to solve problems and/or investigations (e.g., indirect measurement, estimation, number sense, proportional reasoning, spatial reasoning) – Intuitive precision (e.g., "You've just built this tower with four toilet paper rolls and a flat piece of cardboard. Is the cardboard on the top flat? Why not?"; Are the four columns even?) – Create algorithms that a computer can execute – Recognize patterns and/or repeated sequences in data or code within problem and/or investigation	NGSS Practice 5 Mathematical Practice 4 Technology Practice 5 CS Practice 4

Practice	Making & tinkering look-fors	Standards alignment
Engage in Constructive Feedback and Argumentation from Evidence	– Compare and critique at least two designs and analyze whether they meet the demands of the activity – Provide suggestions to peer(s) in how to improve and/or change design and/or investigation using relevant evidence – Use definitions and appropriate vocabulary to construct arguments and respond to the argument of others – Make conjectures, suggestions and/or use counterexamples to support, improve, refute and/or critique the argument and ideas of others – Defend how the mathematical results is warranted to reach goal(s) of activity	NGSS Practice 7 Mathematical Practice 3 CS Practice 2
Obtain and Evaluate, and Communicate Information in a Responsible Manner	– Engage in positive, safe, legal, and ethical behavior when using technology, including online social interactions – Conduct research (e.g., books, google) to inform design, investigation, interest, curiosities, etc. – Plan and employ effective research strategies to locate information and other resources for their intellectual or creative pursuits. – Evaluate the accuracy, perspective, credibility and relevance of information, media, data or other resources.	NGSS Practice 8 Technology Practice 2 Technology Practice 3
Communicate/ Present Information	– Communicate/showcase final product, design solution, and/or investigation clearly (i.e., appropriate manner that audience can understand) – Explain the mathematical results within the context of the activity – Use technology to demonstrate their learning in a variety of ways (e.g., documentation, social media, portfolio) – Cites the work of online resources (e.g., images) – Creates original digital works – Repurpose or remix digital resources into a new creation	NGSS Practice 8 Mathematical Practice 4 Technology Practice 1 Technology Practice 6 CS Practice 7

Note: NGSS refers to Next Generation Science and Engineering Standards. CS refers to Computer Science Framework.

CHAPTER 14

Educational Robotics as a Tool for Youth Leadership Development and STEM Engagement

Kathleen Morgan, Bradley Barker, Gwen Nugent and Neal Grandgenett

Abstract

When you hear about robotics in education today, you often hear about it in the context of two different purposes: (1) the development of future scientists and engineers capable of leading the science, technology, engineering, and mathematics (STEM) industries, and (2) using robotics as a multi-disciplinary education tool to attract and retain students into the STEM disciplines and the related educational pathways. To support the first purpose, there are a number of youth leadership models that explore ways to develop young capable leaders. To address the second, robotics has been shown to be effective with youth for increasing STEM learning, including youth knowledge, attitudes, and motivation. This chapter examines the intersection of youth leadership development and STEM engagement through educational robotics competitions. The study examines youth perceptions of the importance of leadership and their leadership capacity using a series of leadership survey questions completed by participants in an educational robotics competition. The findings support the use of educational robotics programs toward developing future leaders who promote innovation in STEM.

1 Introduction

The use of robotics in the workplace continues to proliferate in areas where current jobs may be considered dangerous, dull and/or dirty. Globally, robots are found in manufacturing, healthcare, education, transportation, warfare, and even sandwich making (Hatch, 2018; Tsoy, Sabirova, & Magid, 2017).

As a discipline itself, robots integrate many different areas of engineering, computer science, mathematics, artificial intelligence and science. Educational robotics, or learning with or about robotics, is generally divisible into two focus areas; (a) the creation of a workforce in robotics and (b) using robotics as an educational tool to develop and attract a broad range of students interested in the science, technology, engineering, and mathematics (STEM) content areas of primary and secondary education. The burgeoning robotics industry is in search of training and post-secondary educational programs that produce competent and capable roboticists that can support the industry as well as develop scientists and engineers that can advance the field. Due to declining interest at the post-secondary level, in an effort to entice youth to the STEM fields, K-12 education has adopted robotics as a tool to teach broad concepts in the STEM areas in entertaining and hands-on ways (Kennedy, Lyons, & Quinn, 2014).

STEM fields have been economic drivers since the industrial revolution, and our economies today overwhelmingly rely on STEM innovation and advancement to grow the world's economies (Committee on Prospering in the Global Economy of the 21st Century, 2007; Members of the 2005 "Rising Above the Gathering Storm" Committee, 2010) with increasingly important considerations for STEM educational pathways or pipelines. Mead, Thomas and Weinberg (2012) explain that STEM pipeline analogy references the "attempt to get young students into the educational conduit and have them emerge from the other end as professionals with graduate and post-graduate degrees. Much like the trans-Alaskan pipeline that is 800 miles long and has 11 major pumping stations, the educational conduit needs to have its own entrance points and activities that keep the contents flowing" (p. 302). Educational robotics is a foundational tool for STEM career development when supported by formal education entities (primary, secondary, and postsecondary).

In addition, to develop talent the educational robotics pipeline must provide mentors and adequate interconnectedness between learners (Mead, Thomas, & Weinberg, 2012). Robot competitions are well positioned to advance STEM career aspirations because they inherently provide the connectedness for students to work in teams of mixed aged groups working towards a specific outcome under the guidance and tutelage of mentors and other experts. Educational robotics competitions may indeed recruit new youth to pursue STEM careers by increasing self-efficacy of students in many STEM content areas. In addition, these annual competitions ensure the continuous recruitment of students into the STEM pipeline and provide access to mentors and experts.

Another important consideration in the attainment of STEM education and careers for youth is the development of leadership skills. By increasing leadership capacity among youth participants, STEM education programs have the opportunity to impact economic development and prosperity through inspiring more capable people to enter the STEM fields. When people believe in their ability to be successful at a specific task, they are more likely to achieve their goals, leading to an expectation of achievement and additional motivation to continue working toward the goals. In a self-fulfilling model, youth with high science self-efficacy set challenging goals, work hard to achieve their goals, and earn higher grades (Mead et al., 2012). Following the pathway or pipeline model, programs working toward recruiting capable people into the STEM fields begin by engaging students' interest in STEM activities (Metcalf, 2010), expecting that the introduction and engagement in STEM will set them down a pathway of taking science classes in high school, choosing a STEM major in college, and eventually entering a STEM career.

There is a body of knowledge in the research around leadership development in youth (Anderson & Kim, 2009; Hastings, Barrett, Barbuto, & Bell, 2011) and, separately, the impact of educational robotics on STEM knowledge, interests and career development (Grandgenett, Ostler, Topp, & Goeman, 2012; Melchior et al., 2005, 2009; Nugent et al., 2015). A current gap in the literature is the intersection of youth leadership development and STEM engagement. To that end this chapter will focus specifically on the following: (a) youth perceptions of the importance of leadership; (b) perceived levels of leadership capacity; and (c) change in leadership capacity due to participation in a STEM education program associated with educational robotics competitions.

2 Review of the Literature

2.1 *Educational Robotics*

Within primary and secondary education systems, the use of educational robotics in supporting student learning both cognitively and socially has had the interest of educators from a number of perspectives (Afari & Khine, 2017). Robotics as a learning tool has been linked to a wide variety of science inquiry skills like computation, estimation, manipulation and observation (Sullivan, 2008). Studies have also found educational robotics activities improve self-confidence, increase drive to do well in school, provide challenging experiences, allow application of abstract concepts to tangible tasks, and foster learning of 21st century workplace skills (Bers, 2006, 2008; Eguchi, 2012; Melchior et al., 2005, 2009; Miller et al., 2008; Nugent et al., 2012). Studies have also shown that educational robotics increase learning and motivation in multiple areas in problem solving, cooperative learning and interest in STEM (Jojoa, Bravo, Cortes, Tool, 2010; Mauch, 2001; Fagne & Merkle, 2002; Robinson, 2005; Barker & Ansorge, 2007; Williams, Ma, Prejean, Ford, 2007; Nugent, Barker, Grandgenett, Adamchuk, 2010; Beer, Hillel, Chiel, Drushel, 1999; Nourbakhsh et al., 2005).

While the price of educational robotics tools has commonly been a concern of schools and educators, falling costs and rising hardware capabilities have allowed an increase in the number of off-the-shelf educational robotics kits available to youth as STEM learning tools as well as related curriculum materials. The theoretical underpinnings for the use of increasingly affordable robotics in K-12 settings, both formal (in-school) in regular courses and non-formal (outside of the school day in programs and camps), are grounded in the experiential learning model whereby students learn STEM content through authentic problem-solving experiences. The popularity of robotics as engaged learning tools in K-12 STEM learning environments is in large part due to their very interdisciplinary and hands-on nature.

Educators often find robotics appealing because of the robots' ability to catch the attention of youth. Robots are highly engaging and motivating and they encourage learning about STEM concepts (Eguchi, 2012; Hendricks, Ogletree, & Alemdar, 2012; Melchior et al., 2005; Miller, Nourbakhsh, & Siegwart, 2008; Nugent et al., 2012). Further, some educational researchers have identified that robotics activities can promote shared leadership (Scholz & McFall, 2011), collaboration, teamwork, positive youth development, and positive relationships, (Bers, 2006, 2008; Eguchi, 2012; Melchior et al., 2005, 2009; Miller et al., 2008; Nugent et al., 2012).

Theoretically, the use of educational robotics fits well in the philosophy of constructionism, whereby learners use tangible tools to build knowledge (Harel & Papert, 1991; Papert, 1980). Constructionism was developed from Piaget's (1972) constructivism theory which purports learning takes place when knowledge is actively sought through interactions with the environment and the manipulation of some sort of physical object. Constructionism builds on constructivism, and introduces the idea that learners are more successful when constructing knowledge.

Delivery of educational robotics programs typically occurs through a number of different models in K-12 learning environments including formal classes, educational robotics competitions and robotics in informal learning environments. Most recently, robotics competitions have proliferated especially the *FIRST* (For Inspiration and Recognition of Science and Technology) robotics programs. For example, *FIRST* LEGO League (FLL) robotics competition had over 255,000 participants in 2016 (FIRST, n.d.). The popularity of these competitions may be due to the authentic real-world problem base of the competitions, the social aspects of learning as a team, and the excitement of competing with other youth and using the robot as a digital manipulative. In robotics competitions teams work together to build and program robots to accomplish specific tasks, communicate their engineering design processes, and/or complete a related research project (Nugent et al., 2012). *FIRST* organizes four levels of robotics programs, including *FIRST* LEGO League (FLL), which served as the focus of this study. FLL is for youth nine to 14 years old and challenges teams of two to ten members to design, build, and program a LEGO robot to accomplish specific missions, complete a project proposing a solution to a STEM-based problem related to the yearly theme, and work together as a team according to the program's Core Values.

Robotics competitions are highly visible educational robotics contests, usually for teams of middle school or high school youth. Competition programs include those organized by FIRST, BotBall, RoboCup-Junior, BEST, MicroMaze, Sumo, RC Jr. Dance, Trinity College's Firefighting Robot Contest, and the CEENBoT Showcase (Grandgenett et al., 2012; Miller et al., 2008). While differences exist among the competitions, teams generally build and program robots to accomplish specific tasks, communicate their engineering design processes, and/or complete a related research project (Nugent et al., 2012). It has also been found that current educational robotics programs, including *FIRST*, can often disproportionately serve white male students who do not have a disability (Ludi, 2012; Melchior et al., 2005, 2009). However, when structured purposefully to do so, educational robotics activities can be an effective method to engage minority youth, youth in rural areas, and girls in STEM. Rusk, Resnick, Berg, and Pezalla-Granlund (2007) and Ludi (2012) provided recommendations for program designers to increase inclusiveness through purposeful approaches to do so. Strategies for engaging diverse learners included focusing on themes, not only challenges, combining art and engineering, encouraging storytelling, and holding exhibitions rather than competitions.

2.2 *Leadership*

Since the study of leadership began, scholars have been working to understand what influences a person to be a successful leader. Researchers have explored aspects of leadership theory that support the specific and larger goals of this study, including leadership's contribution to innovation, leadership in the STEM fields, and adult and youth leadership development. By increasing leadership capacity among youth participants, STEM education programs have the opportunity to impact economic development and prosperity. According to a committee from the National Academies that studied the competitive status of the United States, leading the world in *innovation* will be the primary source of competitiveness in a global economy (Committee on Prospering in the Global Economy of the 21st Century, 2007). Their work documented that the United States was at a crisis point, at risk of losing its status as the economic and innovative leader and the situation had worsened by their 2010 review.

The committee called for a renewed focus on innovation as a source of prosperity. They identified the ingredients of innovation as "(1) new knowledge; (2) capable people, and (3) an environment that promotes innovation and entrepreneurship" (Members of the 2005 "Rising Above the Gathering Storm" Committee, 2010, p. 44). See Table 14.1 for recommended actions and predicted outcomes.

STEM education programs contribute directly to the second ingredient for building and leading innovation by inspiring more *capable people* to enter the STEM fields. The collective experiences that inspire young people to enter STEM careers (i.e. the STEM pipeline) are often a process consistent with Social Cognitive Career Theory (Fouad & Smith, 1996; Lent, Brown, & Hackett, 1994) and self-efficacy and achievement expectancy models (Mead, Thomas, & Weinberg, 2012). As discussed earlier, STEM education programs strive to inspire student's self-efficacy and to set them onto a path to entering STEM careers (Mead et al., 2012; Metcalf, 2010).

2.3 *Linking Leadership and Innovation*

The link between leadership and innovation is well established in the literature (for a review see Shalley & Gilson, 2004; Mainemelis, Kark, & Epitropaki, 2015). Building off the ingredients for innovation (see Table 14.1), it follows that

TABLE 14.1   Ingredients for leading innovation

Ingredient	Action	Outcome
New knowledge capital	Increase government funding in research and development	Increase in innovative scientific or technological advancements
Capable people	Develop the next generation to enter STEM through the STEM pipeline	More people enter STEM leadership fields
Supportive environment to promote innovation and entrepreneurship	Develop an "innovation ecosystem" through cost of labor, policy changes, capital availability, protection of intellectual property, and enhanced infrastructure	Increased innovation

Note: Summarized from Members of the 2005 "Rising above the Gathering Storm" Committee (2010)

developing a supportive environment for innovation requires leaders in the STEM fields who encourage such an environment.

Research suggests that capable leaders can foster an organizational environment that promotes innovation by providing interpersonal support (Tesluk, Farr, & Klein, 1997). For example, Atwater and Carmeli (2009) reported that perceptions of high quality relationships (leader-member exchange) between leaders and employees was related to higher levels of creative work for employees; the relationship was mediated by individuals' feelings of energy. Furthermore, Reiter-Palmon and Illies (2004) suggest that leaders who facilitate the cognitive processes underlying creative problem-solving enhance innovation. Specifically, leadership efforts that encourage groups to clearly define the problem at hand (problem construction), ensure that enough time is provided to adequately search available information, and appropriately evaluate and select from different ideas are likely to promote creativity. In a study conducted by Carmeli, Gelbard and Reiter-Palmon (2013), leader behaviors that promote knowledge sharing, enhancing the information search element of cognitive processes for creativity, are positively related to enhanced creative problem-solving capacity.

Goal-setting, one of the most widely studied cognitive processes in the field of applied psychology and leadership (Locke & Latham, 2002), is also an important cognitive process related to innovation. Madjar and Shalley (2008) reported that assigning challenging creative goals, for example "generate as many original ideas as possible, at least 14 ideas for each idea-generation task" (p. 793) was related to higher levels of creative task output. These results suggest that leaders should help followers focus on creative goals to promote innovation.

2.4 Leadership in STEM

Some researchers suggest that leadership of people in interdisciplinary STEM organizations may be a unique phenomenon not adequately described by more generalized leadership theories (Mumford, Scott, Gaddis, & Strange, 2002; Robledo, Peterson, & Mumford, 2012). Specifically, scientists are open, conscientious, autonomous, ambitions/achievement oriented, and self-confident/arrogant (Feist, 1999), implying they may be unresponsive to influence attempts from their leaders, more open to professional evaluation than leaders' feedback, and may resist working with others (Robledo et al., 2012). Robledo et al. (2012) proposed a new scientific leadership model that gives special consideration to scientists' and engineers' typical personality traits. The model is tailored to STEM and describes the need for leaders to exert influence across the work itself, the group, and the organization.

In addition to inspiring students to pursue STEM careers, educational robotics programs like FIRST hope to develop innovation. FIRST includes both leadership and innovation development in their mission "to inspire young people to be science and technology leaders and innovators" (FIRST, n.d.). FIRST LEGO League specifically promotes innovation through the Core Value "What we discover is more important than what we win," by evaluating innovative project solutions and robot designs for all teams, and by recognizing the most innovative projects through local and global awards each year.

The relationship between specific models of leadership and innovation have been explored in adults, e.g. empowering leadership (Zhang & Bartol, 2010) and transformational leadership (Eisenbeiss, van Knippenberg, & Boerner, 2008; Gumusluoglu & Ilsev, 2009). However, these models have not been developed for use with youth populations and may not be developmentally appropriate (Murphy & Johnson, 2011). Thus, consistent with social-cognitive theory, an initial step in assessing youth leadership development is to determine leadership self-efficacy and identity (Guillén, Mayo, & Korotov, 2015).

2.5 Leadership Development and Innovation

Rather than only focusing on adult leaders, some researchers advocate viewing leadership development as a life-long activity beginning in childhood and suggest a need for additional studies focused on youth (Avolio & Vogelgesang, 2011; Gottfried et al., 2011; Gottfried & Gottfried, 2011; Lord et al., 2011; Murphy & Johnson, 2011). Several scholars have developed comprehensive models of leadership development that encompass leaders' early experiences (Avolio & Vogelgesang, 2011; Komives, 2011; Lord et al., 2011; Van Velsor, McCauley, & Ruderman, 2010). Like language acquisition, there may be a "sensitive period" in which it is easier to learn how to be a leader earlier in life rather than later (Avolio & Vogelgesang, 2011; Bornstein, 1989; Murphy & Johnson, 2011). Awareness of leadership, identifying one's self as a leader, and building leadership self-efficacy are considered important aspects of leadership development in many leader development models (Avolio & Vogelgesang, 2011; Komives, Longerbeam, Osteen, Owen, & Wagner, 2009; Komives, Longerbeam, Owen, Mainella, & Osteen, 2006; Lord et al., 2011; Murphy & Johnson, 2011; Ricketts & Rudd, 2002; van Linden & Fertman, 1998). Campbell et al. (2003) presented a leadership development model focusing on "the process of acquiring particular personal qualities and skills that create influence independent of the individual's positional influence" (p. 29), a particularly relevant model for youth programs because positional influence is rarely a relevant factor.

FIRST is among the more well known educational robotics competitions that also include leadership development as a goal, in addition to inspiring youth to learn science and eventually enter STEM careers (CREATE Foundation, 2012; FIRST, 2011; KISS Institute for Practical Robotics, 2012). The FLL program as a robotics competition strives to develop participants as leaders who support innovation through inclusion of elements aligned with developing leadership capacity and innovation. The FLL Core Values include the statements "We are a team; we do the work to find solutions with guidance from our coaches and mentors; we know our coaches and mentors don't have all the answers; we learn together; and we share our experiences with others" (FIRST, n.d.) FLL teams are required to demonstrate and answer questions about their team's Core Values at tournaments but teams and coaches are not provided specific activities or curriculum to intentionally develop participants' leadership capacity. FIRST LEGO League develops leadership by using what Parks (2005) describes as learning above and below-the-neck. The process is active, experiential, and includes reflections (Parks, 2005). Using a similar approach as Klau (2006), who analyzed youth programs' alignments with their leadership development intentions, FLL is likely to develop leadership capacity in participants through the use of experiential learning and reflections. As one of the longest-running educational robotics programs, FIRST has been evaluated in several studies focused on attitudes toward the STEM fields, plans to enter STEM careers, and STEM learning (Melchior et al., 2005, 2009; Oppliger, 2001; Skorinko et al., 2010; Tougaw, Will, Weiss, & Polito, 2003; Varnado, 2005). Despite leadership being a central mission of educational robotics, no studies were found investigating youth perceptions of the importance of leadership, their leadership capacity, or their change in leadership capacity as a result of participating in a STEM education program.

Educational robotics programs can contribute to innovation by providing students an outlet for developing both leadership and innovation talents. Past research suggests leadership can create and support innovation (Damanpour & Schneider, 2006; Gumusluoglu & Ilsev, 2009; Jung, Chow, & Wu, 2003; Oke, Munshi, & Walumbwa, 2009; Sarros, Cooper, & Santora, 2008). Building off the ingredients for innovation (see Table 14.1), similar to the workplace, robotics competitions model an "innovation ecosystem" that requires capable leaders in the STEM field and the need for capable leaders in STEM can be addressed by developing and supporting leadership capacity among those entering STEM careers. With a similar theoretical basis as the STEM "pipeline" founded on social cognitive theory (Bandura, 2001) several models of youth leadership development describe a self-reinforcing leadership "pipeline" in which youth learn about leadership, begin to identify themselves as leaders, practice being leaders, prompting them to seek out future leadership opportunities, resulting in increased leadership self-efficacy (Avolio & Vogelgesang,

2011; A. E. Gottfried et al., 2011; A. W. Gottfried & Gottfried, 2011; Lord, Hall, & Halpin, 2011; Murphy & Johnson, 2011; Ricketts & Rudd, 2002).

3   Understanding the Intersection of Leadership and STEM

The purpose of a recent study, consistent with this chapter was to address the research gap between educational robotics and youth leadership development by examining youth leadership development (Campbell, Dardis, & Campbell, 2003; Komives et al., 2006; Murphy & Johnson, 2011) as applied to middle school aged participants in four FLL robotics competitions. For the study, leadership is defined as a process where any individual may influence the group to come together and work toward achieving a common goal (Zula et al., 2010). Although this definition does not include a specific focus on innovation, the elements of effective leadership, as previously discussed, can support creation of an environment for innovation. Further, the chosen definition can be considered as a baseline for youth just entering the leadership "pipeline."

The current study sought to understand how FLL impacted youth perception of leadership importance and leadership capacity and to determine the extent demographic characteristics of FLL participants are related to these two key leadership outcomes. We addressed the following research questions:

1a. To what extent is leadership important to FLL participants?
1b. To what extent do demographic characteristics of FLL participants (age, years of participation, gender, team size) predict their ratings of leadership importance?
2a. At what level do FLL participants rate their own leadership capacity?
2b. To what extent do demographic characteristics of FLL participants (age, years of participation, gender, team size) predict their ratings of leadership capacity?
3a. To what extent do youth report a change in their leadership capacity due to their participation in FLL?
3b. To what extent do demographic characteristics of FLL participants (age, years of participation, gender, team size) predict their perceptions of change in leadership capacity due to FLL?

4   Methodology

This study surveyed participants in four FLL qualifying tournaments in the United States. The survey was cross-sectional, with data collected at the end of four tournaments in December 2012 and January 2013. All 74 teams at the tournaments participated in the study, with 14 to 23 teams at each event. All data was collected through a paper questionnaire.

Rohs (1999) found retrospective assessment, or a "then-post" method, to give a more accurate assessment of leadership skills than traditional pre-post evaluation structures. Rohs (1999) recommends using the then-post model, in which participants give themselves ratings after a program and then provide ratings of their skills before the program, to get a less conservative and more accurate picture of the impacts of leadership development programs. Following this recommendation, a then-post model was used for this study.

A total of 501 surveys were collected from youth that attended the four events. The youth included 340 males (67.9%), 152 females (30.3%), and 9 who did not report a gender (1.8%). The majority of participants identified as White (85.4%), representing a proportion of White (not Hispanic or Latino) youth slightly higher than the population of the state (81.4%; U. S. Census Bureau, 2012). The ages of participants ranged from eight to 15, with a mean of 11.4 years ($SD = 1.419$). Youth had participated in FLL for one to five years, with a mean of 1.5 years in the program. For the majority of participants, it was their first year in FLL (63.1%) and 88.1% had one or two years of FLL experience. FLL teams had a mean of 7.71 members.

4.1 *Instrumentation*

The study was designed to measure leadership capacity and perceived leadership development as a result of experiences in FLL. The instrument used for this study was a compilation of the four items from Shertzer et al. (2005) and a modification of the SPLI (Zula et al., 2010).

Researchers have designed measures of youth leadership development or leadership capacity to help assess the impact of youth leadership programs. Instruments for measuring youth leadership development include researcher-developed or modified instruments (Anderson & Kim, 2009), Roets Rating Scale for Leadership (Chan, 2000), and the Youth Leadership Life Skills Development Scale (YLLSDS) (Seevers, Dormody, & Clason, 1995; Seevers, 1994; Wingenbach & Kahler, 1997). Min and Bin (2010) also listed the Student Leadership Inventory, Leadership Skills Inventory, Multifactor Leadership Questionnaire, and qualitative techniques as methods to assess leadership.

In a study of college students, Shertzer et al. (2005) used four items to measure attitudes about the importance of leadership. Shertzer et al. (2005) define leadership importance as a youth's ratings of the significance of leadership in their lives and serves as an indicator of youth leadership awareness, the first step in leadership development. For this study, the Shertzer et al. (2005) items were chosen for their efficient and direct measurement of perceived importance of leadership. Since the leader development models (Avolio & Vogelgesang, 2011; Komives et al., 2009, 2006; Lord et al., 2011; Murphy & Johnson, 2011; Ricketts & Rudd, 2002; van Linden & Fertman, 1998) begin with awareness and identification of leadership, this measurement suggests whether participants have started down the leadership "pipeline" that will lead to increased future leadership potential. In the Shertzer et al. (2005) study, the leadership importance questions demonstrated internal consistency (Cronbach's alpha = 0.82) and independence from other leadership factors including student government, leadership, community pride, positional leader, civic responsibility, and peer education.

Focusing on college student leadership development, Zula, Yarrish, and Christensen (2010) created a four-factor measure assessing leadership development, the Student Perceptions of Leadership Instrument (SPLI). The measure and factors were based on the work of Campbell et al. (2003) and included the following factors: interpersonal/intrapersonal skills, task-specific skills, cognitive skills, and communication skills. In this study, the SPLI (Zula et al., 2010) was chosen to measure leadership development for its strong psychometric properties (Cronbach's alpha = 0.84 and expert panel review for validity) and connection to research-based factors that contribute to leadership development (Campbell et al., 2003). The SPLI required modifications to 12 of the 18 items to be appropriate for youth ages 9 to 14. For example, "I enjoy relating to others on an interpersonal basis" was changed to "I like getting to know people and making new friends." An example unchanged item is "I am good at planning." Questions were also modified specifically to measure the impact of participation in FLL. For example, "I am comfortable giving directions to others" was changed to "Through FLL, I learned to be more comfortable giving directions to others." All items were placed on a Likert scale, ranging from strongly disagree to strongly agree. For each question, the students reported their current status and had a corresponding question to report how much they changed as a result of FLL. The instrument was compiled and analyzed as three scales, including the importance of leadership, leadership capacity, and self-reported change in leadership capacity.

4.2 Data Analysis

The following demographics and youth participant attributes served as independent variables for the study: age, length of participation in years, gender, and team size. Each scale on the survey was treated as a dependent variable:

importance of leadership, leadership capacity, and perceived change in leadership capacity.

4.2.1 Multilevel Linear Modeling

Research questions 1a, 2a, and 3a were descriptive questions addressed through calculation of means and standard deviations for the three leadership scales using IBM SPSS Statistics software, version 21.

Since participants were clustered within competition teams, a multilevel linear model was created to determine statistical effects of demographics (age, years of participation, gender, and team size) on each of the leadership scales. The multilevel hierarchical linear model provided information about the effect for each demographic variable on the leadership scales between teams and within teams and allowed the analysis to account for all of the variables simultaneously. Multilevel linear models are recommended whenever participants are organized into groups (Tabachnick & Fidell, 2013). Multilevel linear models can also accommodate violation of the assumption of independence in errors that occurs when shared group experiences, such as being on the same team, may affect responses.

The statistical model assessed the effects for each research question at two levels (between teams and within teams) and provided the effects of gender, ethnicity, age, experience, and team size for each of the outcome variables. The individual youth effects were designated as level 1 in the model and the between team effects were considered level 2 in the model. The multilevel modeling was implemented through SAS mixed procedures, version 9.2.

Initially, all demographic variables were included in the multilevel linear models for each of the three outcome variables. In preparation for creating the multilevel linear models, the team mean values were subtracted from each participant's value to center the results and create Level 2 (between team) variables. For example, the team's mean age was subtracted from the individual participant ages to represent difference from team average age.

To determine the reliability of each scale in the study, Cronbach's alphas were calculated: Leadership Importance ($\alpha = .692$), Current Leadership Capacity ($\alpha = .892$), and Change in Leadership Capacity ($\alpha = .938$). Current Leadership Capacity and Change in Leadership Capacity both met the recommended minimum of 0.7 (Field, 2005), Leadership Importance was near 0.7 and was therefore retained. Each scale was visibly negatively skewed, so skewness and kurtosis values were calculated for each variable. Calculations showed skewness and kurtosis would have no impact on analysis, with the exception of Leadership Importance. To accommodate a J-shaped severe negative skew, the Leadership Importance variable was transformed (New $X = 1/(K-X)$; $K = 5$

demographic variables +1). The Current Leadership Capacity and Change in Leadership Capacity variables were not transformed.

5 Results

5.1 Research Question 1a: Importance of Leadership

Four questions composed the Importance of Leadership scale. Results were limited to those who replied to at least three of the four instrument questions, providing 496 valid responses. Participants rated leadership as very important with a mean of 4.45 (SD = 0.59). See Table 14.2 for question means.

5.2 Research Question 1b: Demographic Predictors of Leadership Importance

The multilevel model for Leadership Importance ratings had an intra-class correlation (ICC) of 0.04, demonstrating significant variability between teams. The full model with all demographic variables (gender, ethnicity, age, experience, and team size) accounted for 15.89% of the overall variance in Leadership Importance (x^2 (1, N = 484) = 2.49; p > 0.114). Gender within teams, team age, and experience within teams were significantly related to Leadership Importance. The model is reported in Table 14.3 (x^2 (1, N = 484) = 2.21; p > 0.14). On the transformed scale, on average, girls rated Leadership Importance 0.11 points higher than boys. For each year of increased average team age, ratings increased by 0.03 points. For each year of experience within a team, ratings increased by 0.07. The cumulative significant effects (see Figure 14.1), indicate that a female with more experience compared to others on her team, whose team was older than average, tended to give the highest ratings of Leadership Importance. Correspondingly, a male with less experience than others on his team, who was on a team younger than average, tended to give the lowest ratings of Leadership Importance.

TABLE 14.2　Leadership importance scale question means

Question	N	M	SD
Leadership is important to me.	496	4.55	0.77
I consider myself to be a leader.	493	4.08	1.01
Leadership will be an important part of my life in the future.	495	4.42	0.82
Leaders need to be able to work in teams and groups.	495	4.76	0.56

TABLE 14.3   Multilevel model for leadership importance

Effect	Estimate	Standard error	DF	t value	Pr > |t|
Intercept	0.70	0.02	69.20	41.54	<.0001
Gender within Teams	0.11	0.03	419.00	3.36	0.00*
Gender between Teams (Percent)	0.00	0.00	196.00	-1.52	0.13
Age within Teams	0.00	0.01	419.00	0.21	0.84
Age between Teams	0.03	0.01	80.70	2.56	0.01*
Experience within Teams	0.07	0.02	418.00	3.20	0.00*
Experience between Teams	0.05	0.02	76.40	1.91	0.06

*$p < .05$

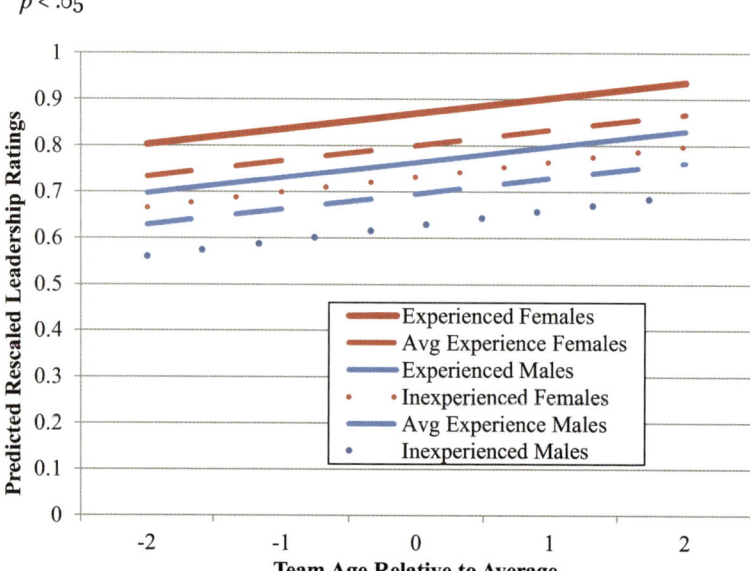

FIGURE 14.1.   Predicted rescaled leadership importance ratings for gender and experience groups

5.3   *Research Question 2a: Leadership Capacity*

Only participants who answered at least half of the 18 Current Leadership Capacity questions were included in the final results, leaving 488 valid responses. Like the importance of leadership, participants rated their leadership capacity high, with a mean of 4.26 (SD = 0.51). Table 14.4 reports the means for each question. In the table, the Current Leadership Capacity questions are paired with the corresponding question about improvement due to FLL participation.

TABLE 14.4 Question means for current leadership capacity and change in leadership capacity scales

Question Theme	Current leadership capacity			Change due to FLL		
	N	M	SD	N	M	SD
Working on teams	497	4.54	0.74	488	4.47	0.78
Getting to know people/making friends	498	4.59	0.67	487	4.50	0.76
Delegating tasks to others	494	4.16	0.87	485	4.23	0.88
Wanting to take charge	495	3.99	1.07	485	4.09	1.04
Giving feedback	497	4.36	0.79	486	4.38	0.79
Desire to be a leader	498	3.70	1.15	486	3.74	1.18
Giving directions	494	4.13	0.99	496	4.13	0.91
Planning	496	4.10	0.90	494	4.35	0.80
Knowing rules and their importance	489	4.59	0.61	493	4.50	0.75
Setting and carrying out goals	487	4.45	0.69	496	4.35	0.80
Solving problems	484	4.35	0.80	497	4.45	0.73
Collecting and interpreting data	486	3.95	0.98	495	4.08	0.98
Doing things in new ways	489	4.27	0.85	496	4.33	0.85
Curiosity	488	4.60	0.69	498	4.27	0.97
Asking for advice	487	4.39	0.80	495	4.30	0.90
Admitting and correcting mistakes	487	4.36	0.77	496	4.23	0.92
Working with diverse people	488	4.51	0.77	497	4.41	0.85
Enjoying change	488	3.55	1.15	498	3.68	1.25

5.4 Research Question 2b: Demographic Predictors of Current Leadership Capacity

The multilevel model for the Current Leader Capacity scale had an ICC of 0.1, demonstrating significant variability between groups. The full model with all demographic variables (gender, ethnicity, age, experience, and team size) accounted for 23.29% of the overall variance in Current Leader Capacity (x^2 (1, N = 475) = 11.95; p >0.0005).

In the first run, only Age between Teams and Years of Participation between Teams had significant contribution to the overall variance. In a second version of the model the non-significant variables (p > .05) at both levels were eliminated. The updated model only included Age and Years variables and is shown in Table 14.5 (x^2 (1, N =487) = 11.48; $p > 0.0007$). On average, for each year of mean team age, the Current Leadership Capacity score increased by 0.08. For each year of experience within a team, Current Leadership Capacity increased by 0.14. See Figure 14.2 for cumulative significant effects. A youth with more experience than others on his or her team, who was on a team older than average tended to give the highest Leadership Capacity ratings. A team member with less experience than others on his or her team, who was on a team younger than average, tended to give the lowest ratings.

TABLE 14.5 Multilevel regression for current leadership capacity – Significant variables

Effect	Estimate	Standard error	DF	t value	Pr > |t|
Intercept	4.26	0.03	71.60	149.94	<.0001
Age within Teams	0.00	0.02	418.00	-0.22	0.82
Age between Teams	0.08	0.03	81.90	2.76	0.01*
Experience within Teams	0.14	0.04	417.00	3.24	0.00*
Experience between Teams	0.05	0.06	78.20	0.83	0.41

*$p < .05$

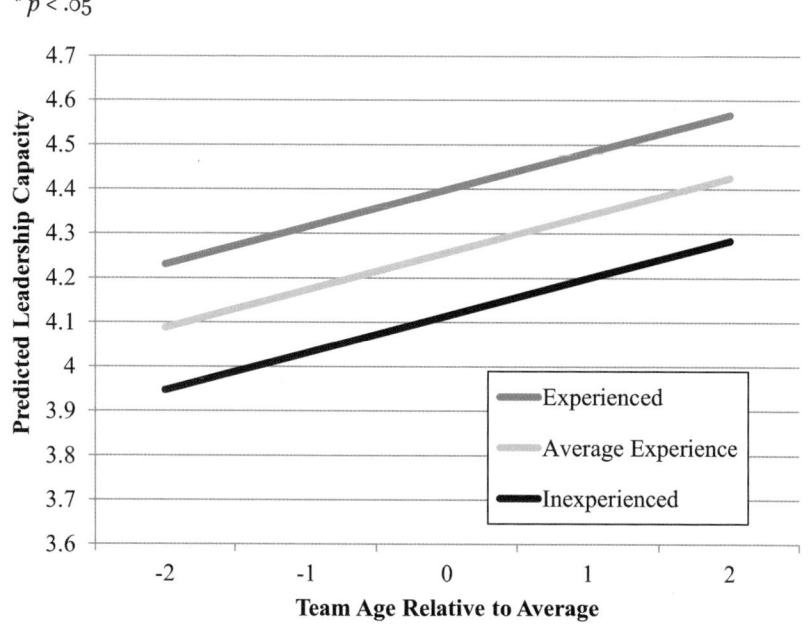

FIGURE 14.2. Predicted current leader capacity for significant effect variables

5.5 Research Question 3a: Perceived Change in Leadership Capacity

The last 18 questions on the survey composed the scale measuring perceived improvement in various elements of leadership capacity due to participation in FLL. Participants were required to answer at least half of the questions, leaving 497 valid responses. Perceived change due to FLL received high ratings, with a mean of 4.25 (SD = 0.623). Only two items had means below four (the "agree" level on the scale) and were related to the desire to be a leader and enjoying change (see Table 14.4).

5.6 Research Question 3b: Demographic Predictors of Leadership Capacity Change

The multilevel model for the Change in Leadership Capacity scale had an ICC of 0.11, demonstrating significant variability between groups. The full model accounted for 24.25% of the overall variance (X^2 (1, N = 484) = 13.25; p > 0.0003). In the first run, none of the demographic variables were significant contributors to the overall variance. Since Experience within Teams was close to the .05 significance level, the model was run again with only the Experience variables used but they remained non-significant effects.

6 Discussion

As mentioned, this research set out to advance understanding of how educational robotics programs develop youth leadership capacity as a method to address the larger goal of developing leaders who create a climate for innovation within the STEM fields. Specifically, the study examined youth leadership identity and capacity development (Campbell et al., 2003; Komives et al., 2006; Murphy & Johnson, 2011) among robotics competition participants by assessing perceptions of the importance of leadership, leadership capacity, and changes in leadership capacity as a result of FLL. Results show youth reported high levels across the three leadership scales.

The first group of research questions focused on youth perceptions of the importance of leadership. Since leader development models include awareness of leadership, recognition of the importance of leadership, and identifying as a leader (Komives et al., 2009, 2006; Komives, 2011; Lord et al., 2011; Ricketts & Rudd, 2002; van Linden & Fertman, 1998), FLL participants' high ratings of leadership importance suggest they are in the initial stages of understanding the need for leadership development. Girls' slightly higher ratings of leadership importance than boys' ratings in this sample suggest FLL may be successful at initially addressing Hoyt and Johnson's (2011) recommendation to develop girls' leadership self-confidence. The significance of experience provides evidence FLL is developing awareness of leadership and encouraging youth to self-identify as leaders. The significance of average team age supports Murphy and Johnson's (2011) model that views age as an important context for youth leadership development.

In the second group of research questions, youth reported high levels of capacity for leadership. Youth on older teams and who had higher levels of experience in FLL rated their leadership capacity higher. Although a number of explanations may be possible, we theorize that youth who had participated in FLL in previous years had developed leadership capacity as a result of participating in the program multiple times. This would align with the self-reinforcing model of leadership development (Avolio & Vogelgesang, 2011; Lord et al., 2011; Murphy & Johnson, 2011). The non-significant effects suggest youth perceive having similar levels of leadership capacity regardless of their gender, ethnicity, average team experience, range of ages on a team, or team size.

In the last group of research questions, youth reported participating in FLL improved their leadership capacity. The lack of significance in the demographic/attribute variables suggests youth perceived similar leadership capacity development while participating in FLL. Additional variables such as coach influence, response bias, prior leader development, and fatigue in testing should be investigated further.

6.1 Practical Implications of the Research

As mentioned previously, the *Gathering Storm* committee set a goal for educators, administrators, and policy makers to develop and lead *innovation* as the primary source for the United States continued economic competitiveness (Members of the 2005 "Rising Above the Gathering Storm" Committee, 2010). FLL has been designed to contribute to the "ingredients" for *innovation* (capable people and supportive environment) as described by the *Gathering Storm* committee through creating future leaders who support and lead innovation within the STEM fields. This study provides initial evidence for FLL's contribution to leadership development. The findings may be used to support the value of participation in FLL and improve youth leadership development within educational robotics programs.

Legislators and policy makers are working to create economic prosperity in the United States. Studies have documented that youth participants in *FIRST* programs show a strong interest in entering the STEM fields (Melchior et al.,

2005, 2009; Oppliger, 2001; Skorinko et al., 2010; Tougaw et al., 2003; Varnado, 2005). Together with this study's findings, the research supports the value of educational robotics programs and their contributions toward developing future leaders who support innovation in STEM.

Educational program administrators, like governmental officials, are tasked with providing ways for youth to become contributors to the economic health of the community. Administrators can use the findings of this study to support educational robotics programs' ability to develop leadership skills among youth and fill the STEM leadership pipeline. The predictors of team age and experience should become factors in educational robotics programs' design and structure. Given that age, experience, and gender were not related to a youth's perceived leadership developing development during FLL, it would be wise to continue offering the program to the current age groups, with the potential of expanding to additional age ranges.

Informal educators often face the same demands for the evidence of impact as do formal educators. Team coaches, parents, afterschool educators, and community program coordinators can use the results in similar ways when recruiting supporters, requesting funding, and planning programs. This study suggests educational robotics programs may contribute to developing youth leadership capacity; youth organizations with leadership development goals can use this study to help support adding educational robotics as a method of leadership capacity development.

The findings provide initial evidence that youth in FLL believe leadership is important, report high levels of leadership capacity, and perceive improvement in leadership capacity as a result of participating. Researchers have reported that youth in educational robotics competitions have high levels of interest in entering careers in the STEM fields (Melchior et al., 2005, 2009; Nugent et al., 2012). With their high levels of engagement and interest in STEM, educational robotics program participants are in the STEM pipeline and are likely to build their self-efficacy in STEM, work harder as they expect achievement, and experience additional success as they continue in STEM (Mead et al., 2012). Based in social cognitive theory (Bandura, 2001), youth leadership development models and the "STEM pipeline" share a common self-reinforcing structure in which youth gain interest, have opportunities for experience and practice, see and expect success, learn from their experiences, and become successful and engaged through additional practice. In accordance with the theoretical models, the combination of leader development and STEM interest suggests youth who participate in FLL are on the path to become leaders in the STEM fields.

6.2 Future Research Directions

Additional research is needed to confirm the findings of this study and further support the reliability and validity of the instruments. As this was the first time these instruments were used in this format within the context of a large robotics competition, further validation should be conducted with youth in a wider geographic area, in other programs, and outside of the studied age range. In addition, the ability of youth in this age group to self-report improvement due to program participation would benefit from further research.

Since the measured variables explained less than 25% of the overall variance, further study is needed to understand the wider set of predictors for each scale. For example, socioeconomic status, academic achievement levels, peer relationships, family focus on youth leadership, prior experience in robotics, self-awareness, prior leader development, and coach training may impact leadership ratings. As the leadership scales were not evaluated for social desirability, future research might determine if responses were elevated due to perceived social pressure to become a leader.

Since all outcomes were measured through a one-time self-report, future research might also investigate youth leadership through coaches' or parents' observations and expand to study the long-term effects of FLL on leadership development. As the definition of leadership and instruments used did not focus explicitly on innovation, further research could more directly measure development that leads to leadership that supports innovation. Beyond tracking if participation in FLL influences youth to enter the STEM fields, future longitudinal studies could examine to what extent former FLL participants are more likely to become leaders who create a climate for innovation in their organizations.

This study provides initial evidence linking educational robotics with youth leadership development. Supported by the parallel structures of the youth leadership development and STEM "pipeline" theoretical models, the findings suggest youth in educational robotics activities are likely to develop into leaders who support innovation in the STEM fields. Educational robotics programs, as documented by this study, appear to be part of creating a future United States STEM workforce that leads the global economy through continuous innovation. This study reinforces the vital contributions of educational robotics programs as stated by *FIRST* Founder Dean Kamen:

It's not about the robots. Robots are just a vehicle. What you are building is way bigger. It's about self-confidence. It's about relationships. It's about making sure the future is better than the past. Don't blow this opportunity by thinking it's about robots. (as quoted in Benedict, 2011, January launch, paras. 2 & 4)

References

Afari, E., & Khine, M. S. (2017). Robotics as an educational tool: Impact of Lego Mindstorms. *International Journal of Information and Education Technology, 7*(6), 437–442. https://doi.org/10.18178/ijiet.2017.7.6.908

Anderson, J., & Kim, E. (2009). Youth leadership development: Perceptions and preferences of urban students enrolled in a comprehensive agriculture program. *Journal of Agricultural Education, 50*(1), 8–20. doi:10.5032/jae.2009.01008

Atwater, L., & Carmeli, A. (2009). Leader-member exchange, feelings of energy, and involvement in creative work. *Leadership Quarterly, 20*(3), 264–275. doi:10.1016/j.leaqua.2007.07.009

Avolio, B. J., & Vogelgesang, G. R. (2011). Beginnings matter in genuine leadership development. In S. E. Murphy & R. J. Reichard (Eds.), *Early development and leadership: Building the next generation of leaders* (pp. 179–204). New York, NY: Routledge Taylor and Francis Group.

Bandura, A. (2001). Social cognitive theory: An agentic perspective. *Annual Review of Psychology, 52*(1), 1–26. doi:10.1146/annurev.psych.52.1.1

Benedict, M. (2011, September). About more than robots. *The Magazine of the Ontario College of Teachers*. Retrieved from http://professionallyspeaking.oct.ca/september_2011/features/robots.aspx

Bers, M. U. (2006). The role of new technologies to foster positive youth development. *Applied Developmental Science, 10*(4), 200–219. doi:10.1207/s1532480xads1004_4

Bers, M. U. (2008). *Blocks to robots: Learning with technology in the early childhood classroom*. New York, NY: Teachers College Press.

Bornstein, M. (1989). Sensitive periods in development: Structural characteristics and causal interpretations. *Psychological Bulletin, 105*(2), 179–97. Retrieved from http://psycnet.apa.org/journals/bul/105/2/179/

Campbell, D. J., Dardis, G., & Campbell, K. M. (2003). Enhancing incremental influence: A focused approach to leadership development. *Journal of Leadership & Organizational Studies, 10*(1), 29–44.

Carmeli, A., Gelbard, R., & Reiter-Palmon, R. (2013). Leadership, creative problem-solving capacity, and creative performance: The importance of knowledge sharing. *Human Resource Management, 52*, 95–121. doi:10.1002/hrm.21514

Chan, D. W. (2000). Assessing leadership among Chinese secondary students in Hong Kong: The use of the Roets Rating Scale for Leadership. *Gifted Child Quarterly, 44*(2), 115–122. doi:10.1177/001698620004400204

Committee on Prospering in the Global Economy of the 21st Century. (2007). *Rising above the gathering storm: Energizing and employing America for a brighter economic future*. Washington, DC: The National Academies Press. Retrieved from http://www.nap.edu/catalog.php?record_id=11463

CREATE Foundation. (2012). *CREATE foundation mission*. Retrieved from http://www.create-found.org/Mission.php

Damanpour, F., & Schneider, M. (2006). Phases of the adoption of innovation in organizations: Effects of environment, organization and top managers. *British Journal of Management, 17*(3), 215–236. doi:10.1111/j.1467-8551.2006.00498.x

Eguchi, A. (2012). Educational robotics theories and practice: Tips for how to do it right. In B. Barker (Ed.), *Robots in K-12 education: A new technology for learning* (pp. 1–30). Hershey, PA: IGI Global.

Eisenbeiss, S. A., van Knippenberg, D., & Boerner, S. (2008). Transformational leadership and team innovation: Integrating team climate principles. *Journal of Applied Psychology, 93*(6), 1438–1446. doi:10.1037/a0012716

Feist, G. J. (1999). The influence of personality on artistic and scientific creativity. In R. J. Sternberg (Ed.), *Handbook of creativity* (pp. 273–296). New York, NY: Cambridge University Press.

Field, A. (2005). *Discovering statistics using SPSS* (2nd ed.). Thousand Oaks, CA: Sage Publications.

FIRST. (2011). *2011 annual report*. Manchester, NH: FIRST. Retrieved from http://www.usfirst.org/sites/default/files/uploadedFiles/Who/Annual_Report-Financials/2011_Annual-Report.pdf

FIRST. (n.d.). *About*. Retrieved from http://www.firstlegoleague.org/about-fll

Fouad, N. A., & Smith, P. L. (1996). A test of a social cognitive model for middle school students: Math and science. *Journal of Counseling Psychology, 43*(3), 338.

Gottfried, A. E., Gottfried, A. W., Reichard, R. J., Guerin, D. W., Oliver, P. H., & Riggio, R. E. (2011). Motivational roots of leadership: A longitudinal study from childhood through adulthood. *The Leadership Quarterly, 22*(3), 510–519. doi:10.1016/j.leaqua.2011.04.008

Gottfried, A. W., & Gottfried, A. E. (2011). Paths from gifted motivation to leadership. In S. E. Murphy & R. J. Reichard (Eds.), *Early development and leadership: Building the next generation of leaders* (pp. 71–91). New York, NY: Routledge Taylor and Francis Group.

Grandgenett, N. F., Ostler, C., Topp, N., & Goeman, R. (2012). Robotics and problem-based learning in STEM formal educational environments. In B. Barker, G. Nugent, N. F. Grandgenett, & S. Adamchuk (Eds.), *Robots in K-12 education: A new technology for learning* (pp. 94–119) Hershey, PA: IGI Global.

Guillén, L., Mayo, M., & Korotov, K. (2015). Is leadership a part of me? A leader identity approach to understanding the motivation to lead. *The Leadership Quarterly*. doi:10.1016/j.leaqua.2015.05.001

Gumusluoglu, L., & Ilsev, A. (2009). Transformational leadership, creativity, and organizational innovation. *Journal of Business Research, 62*(4), 461–473. doi:10.1016/j.jbusres.2007.07.032

Harel, I., & Papert, S. (Eds.). (1991). *Constructionism*. Westport, CT: Ablex Publishing.

Hastings, L., Barrett, L., Barbuto, J., & Bell, L. (2011). Developing a paradigm model of youth leadership development and community engagement: A grounded theory. *Journal of Agricultural Education, 52*(1), 19–29. https://doi.org/10.5032/jae.2011.01019

Hatch, M. (2018). *The maker revolution*. Hoboken, NJ: John Wiley & Sons, Inc.

Hendricks, C., Ogletree, T., & Alemdar, M. (2012). Impact of VEX robotics competition on middle and high school students' problem-solving, collaboration, and communication. In *ISTE Conference* (pp. 1–12). San Diego, CA. Retrieved from http://www.isteconference.org/2012/uploads/KEY_69929389/ISTE_FinalPaper_VEX_Hendricks_Ogletree_Alemdar_Feb10_RP.pdf

Hoyt, C. L., & Johnson, S. K. (2011). Gender and leadership development: A case of female leaders. In S. E. Murphy & R. J. Reichard (Eds.), *Early development and leadership: Building the next generation of leaders* (pp. 205–228). New York, NY: Routledge Taylor and Francis Group.

Jung, D. I., Chow, C., & Wu, A. (2003). The role of transformational leadership in enhancing organizational innovation: Hypotheses and some preliminary findings. *The Leadership Quarterly, 14*(4–5), 525–544. doi:10.1016/S1048-9843(03)00050-X

Kennedy, J., Lyons, T., & Quinn, F. (2014). The continuing decline os science and mathematics enrolments in Australian high schools. *Teaching Science, 60*(2), 34–46.

KISS Institute for Practical Robotics. (2012). *Botball educational robotics*. Retrieved from http://www.botball.org/about

Klau, M. (2006). Exploring youth leadership in theory and practice. *New Directions for Youth Development, 109*, 57–87. doi:10.1002/yd

Kline, R. B. (2005). *Principles and practice of structural equation modeling* (2nd ed.). New York, NY: Guilford Press.

Komives, S. R. (2011). College student leadership identity development. In S. E. Murphy & R. J. Reichard (Eds.), *Early development and leadership: Building the next generation of leaders* (pp. 273–292). New York, NY: Routledge Taylor and Francis Group.

Komives, S. R., Longerbeam, S. D., Osteen, L., Owen, J. E., & Wagner, W. (2009). Leadership identity development: Challenges in applying a developmental model. *Journal of Leadership Education, 8*(1), 11–47.

Komives, S. R., Longerbeam, S. D., Owen, J. E., Mainella, F. C., & Osteen, L. (2006). A leadership identity development model: Applications from a grounded theory. *Journal of College Student Development, 47*(4), 401–418. doi:10.1353/csd.2006.0048

Lent, R. W., Brown, S. D., & Hackett, G. (1994). Toward a unifying social cognitive theory of career and academic interest, choice, and performance. *Journal of Vocational Behavior, 45*(1), 79–122.

Locke, E. A., & Latham, G. P. (2002). Building a practically useful theory of goal setting and task motivation: A 35-year odyssey. *American Psychologist, 57*(9), 705.

Lord, R. G., Hall, R. J., & Halpin, S. M. (2011). Leadership skill development and divergence: A model for the early effects of gender and race on leadership development. In S. E. Murphy & R. J. Reichard (Eds.), *Early development and leadership: Building the next generation of leaders* (pp. 229–252). New York, NY: Routledge Taylor and Francis Group.

Ludi, S. (2012). Educational robotics and broadening participation in STEM for underrepresented student groups.In B. Barker, G. Nugent, N. Grandgenett, & V. Adamchuk (Eds.), *Robots in K-12 education: A new technology for learning* (pp. 343–361). Hershey, PA: IGI Global.

Madjar, N., & Shalley, C. E. (2008). Multiple tasks' and multiple goals' effect on creativity: Forced incubation or just a distraction? *Journal of Management, 34*(4), 786–805.

Mainemelis, C., Kark, R., & Epitropaki, O. (2015). Creative leadership: A multi-context conceptualization. *The Academy of Management Annals, 9*(1), 393–482. doi:10.1080/19416520.2015.1024502

Mead, R. A., Thomas, S. L., & Weinberg, J. B. (2012). From grade school to grad school: An integrated STEM pipeline model through robotics. In B. Barker (Ed.), *Robots in K-12 education: A new technology for learning* (pp. 302–325). Hershey, PA: IGI Global.

Melchior, A., Cohen, F., Cutter, T., & Leavitt, T. (2005). *More than robots: An evaluation of the FIRST robotics competition participant and institutional impacts*. Waltham, MA: Brandeis University.

Melchior, A., Cutter, T., & Cohen, F. (2009). *Executive summary evaluation of the FIRST LEGO® League "Climate Connections" season (2008–09)*.

Members of the 2005 "Rising above the Gathering Storm" Committee. (2010). *Rising above the gathering storm, revisited: Rapidly approaching category 5*. Washington, DC: The National Academies Press.

Metcalf, H. (2010). Stuck in the pipeline: A critical review of STEM workforce literature. *InterActions: UCLA Journal of Education and Information Studies, 6*(2), Article 4. Retrieved from http://escholarship.org/uc/item/6zf09176

Miller, D., Nourbakhsh, I., & Siegwart, R. (2008). Robots for education. In B. Siciliano & O. Khatib (Eds.), *Springer handbook of robotics* (pp. 1283–1301). Berlin: Springer-Verlag. Retrieved from http://scholar.google.com/scholar?hl=en&btnG=Search&q=intitle:Robots+for+Education#5

Min, L., & Bin, W. (2010). Review on researches of youth leadership. In *2010 2nd Conference on Environmental Science and Information Application Technology* (pp. 754–757). IEEE. doi:10.1109/ESIAT.2010.5568430

Mumford, M., Scott, G., Gaddis, B., & Strange, J. (2002). Leading creative people: Orchestrating expertise and relationships. *The Leadership Quarterly, 13*, 705–750.

Murphy, S. E., & Johnson, S. K. (2011). The benefits of a long-lens approach to leader development: Understanding the seeds of leadership. *The Leadership Quarterly, 22*(3), 459–470. doi:10.1016/j.leaqua.2011.04.004

Nugent, G., Barker, B., & Grandgenett, N. (2012). The impact of educational robotics on student STEM learning, attitudes, and workplace skills. In B. Barker, G. Nugent, N. Grandgenett, & V. Adamchuk (Eds.), *Robots in K-12 education: A new technology for learning* (pp. 186–203). Hershey, PA: IGI Global.

Nugent, G., Barker, B., Welch, G., Grandgenett, N., Wu, C., & Nelson, C. (2015). A model of factors contributing to STEM learning and career orientation. *International Journal of Science Education, 37*, 1067–1088.

Oke, A., Munshi, N., & Walumbwa, F. O. (2009). The influence of leadership on innovation processes and activities. *Organizational Dynamics, 38*(1), 64–72. doi:10.1016/j.orgdyn.2008.10.005

Oppliger, D. E. (2001). *University – Pre college interaction through FIRST robotics competition*. In International Conference on Engineering Education, Oslo, Norway.

Papert, S. (1980). *Mindstorms: Computers, children, and powerful ideas*. New York, NY: Basic Books. Retrieved from http://dl.acm.org/citation.cfm?id=1095592

Parks, S. D. (2005). *Leadership can be taught: A bold approach for a complex world*. Boston, MA: Harvard Business School Press.

Piaget, J. (1972). Intellectual evolution from adolescence to adulthood. *Human Development, 15*(1), 1–12. https://doi.org/10.1159/000271225

Reiter-Palmon, R., & Illies, J. J. (2004). Leadership and creativity: Understanding leadership from a creative problem-solving perspective. *The Leadership Quarterly, 15*(1), 55–77.

Ricketts, J. C., & Rudd, R. D. (2002). A comprehensive leadership education model to train, teach, and develop leadership in youth. *Journal of Career and Technical Education, 19*(1), 7–17. Retrieved from http://scholar.lib.vt.edu/ejournals/JCTE/v19n1/ricketts.html

Robledo, I. C., Peterson, D. R., & Mumford, M. D. (2012). Leadership of scientists and engineers: A three-vector model. *Journal of Organizational Behavior, 33*(January 2011), 140–147. doi:10.1002/job

Rohs, F. R. (1999). Response shift bias: A problem in evaluating leadership development with self-report pretest-posttest measures. *Journal of Agricultural Education, 40*(4), 28–37. doi:10.5032/jae.1999.04028

Rusk, N., Resnick, M., Berg, R., & Pezalla-Granlund, M. (2008). New pathways into robotics: Strategies for broadening participation. *Journal of Science Education and Technology, 17*(1), 59–69. https://doi.org/10.1007/s10956-007-9082-2

Sarros, J. C., Cooper, B. K., & Santora, J. C. (2008). Building a climate for innovation through transformational leadership and organizational culture. *Journal of Leadership & Organizational Studies, 15*(2), 145–158.

Scholz, H., & McFall, K. (2011). Comparison of an introductory engineering course with and without LEGO mindstorms robots. *Technology Interface International Journal, 11*(2), 61–64. Retrieved from http://www.tiij.org/issues/summer2011/TIIJ spring-summer 2011-PDW4.pdf#page=63

Seevers, B. S. (1994). Predicting youth life leadership skills development among senior 4-H members: A tri-state study. *Journal of Agricultural Education, 35*(3), 64–69. Retrieved from http://pubs.aged.tamu.edu/jae/pdf/Vol35/35-03-64.pdf

Seevers, B. S., Dormody, T. J., & Clason, D. L. (1995). Devleoping a scale to research and evaluate youth leadership like skills development. *Journal of Agricultural Education, 36*(2), 28–35. doi:10.5032/jae.1995.02028

Shalley, C. E., & Gilson, L. L. (2004). What leaders need to know: A review of social and contextual factors that can foster or hinder creativity. *The Leadership Quarterly, 15*(1), 33–53.

Shertzer, J., Wall, V., Frandsen, A., Guo, Y., Whalen, D. F., & Shelley, M. C. I. (2005). Four dimensions of student leadership: What predicts students' attitudes toward leadership development? *The College Student Affairs Journal, 25*(1), 85–108.

Skorinko, J., Lay, J., McDonald, G., Miller, B., Shaver, C., Randall, C., ... van de Ven, J. (2010). The social outcomes of participating in the FIRST robotics competition community. In *American Society for Engineering Education 2010 Zone I Professional Papers Proceedings*. Retrieved from http://www.asee.org/documents/zones/zone1/2010/professional/The-Social-Outcomes-of-Participating-in-the-FIRST-Robotics-Competition-Community.pdf

Sullivan, F. R. (2008). Robotics and science literacy: Thinking skills, science process skills and systems understanding. *Journal of Research in Science Teaching, 45*(3), 373–394. https://doi.org/10.1002/tea.20238

Tabachnick, B. G., & Fidell, L. S. (2013). *Using multivariate statistics* (6th ed.). Upper Saddle River, NJ: Pearson Education, Inc.

Tesluk, P. E., Farr, J. L., & Klein, S. R. (1997). Influences of organizational culture and climate on individual creativity. *The Journal of Creative Behavior, 31*(1), 27–41.

Tougaw, D., Will, J. D., Weiss, P., & Polito, C. (2003). *Sponsoring a FIRST robotics team*. In American Society for Engineering Education 2003 IL/IN Sectional Conference (pp. 60–62). Valparaiso, IN.

Tsoy, T., Sabirova, L., & Magid, E. (2017). *Towards effective interactive teaching and learning strategies in robotics education*. In Proceedings 10th International Conference on Developments in eSystems Engineering (DeSE) (pp. 267–272), Paris. doi:10.1109/DeSE.2017.38

U.S. Census Bureau. (2012). *State & county quickfacts – Nebraska*. Retrieved from http://quickfacts.census.gov/qfd/states/31000.html

Van Linden, J. A., & Fertman, C. I. (1998). *Youth leadership: A guide to understanding leadership development in adolescents*. San Francisco, CA: Jossey-Bass.

Van Velsor, E., McCauley, C. D., & Ruderman, M. N. (Eds.). (2010). *The center for creative leadership handbook of leadership development* (3rd ed.). San Francisco, CA: Jossey-Bass.

Varnado, T. (2005). *The effects of a technological problem solving activity on FIRST™ LEGO™ League participants' problem solving style and performance*. Retrieved from http://scholar.lib.vt.edu/theses/available/etd-04282005-101527/

Wingenbach, G. J., & Kahler, A. A. (1997). Self-perceived youth leadership and life skills of Iowa FFA members. *Journal of Agricultural Education, 38*(3), 18–27. doi:10.5032/jae.1997.03018

Zhang, X., & Bartol, K. M. (2010). Linking empowering leadership and employee creativity: The influence of psychological empowerment, intrinsic motivation, and creative process engagement. *Academy of Management Journal, 53*(1), 107–128.

Zula, K., Yarrish, K., & Christensen, S. D. (2010). Initial assessment and validation of an instrument to measure student perceptions of leadership skills. *Journal of Leadership Studies, 4*(2), 48–56. doi:10.1002/jls

PART 4

Factors Affecting Students' Choice of STEM Majors

CHAPTER 15

Factors Affecting Students' STEM Choice and Persistence: A Synthesis of Research and Findings from the Second Year of a Longitudinal High School STEM Tracking Study

Adem Ekmekci, Alpaslan Sahin and Hersh Waxman

Abstract

In this chapter, we first synthesize the last two decades of research on students' intentions, choices, and persistence in the STEM pipeline (i.e., earning a postsecondary degree in STEM and/or joining the STEM workforce). In doing so, we frame the literature review by the social cognitive career theory (SCCT; Lent, Brown, & Hackett, 1994) with a special focus on underrepresented minorities (URMs) in STEM (Yu, Corkin, & Martin, 2017). SCCT centralizes three variables in forming career decisions: personal inputs, motivational factors, and contextual factors. We then introduce a 4-year longitudinal study tracking STEM intentions of high school students. The longitudinal study started with the 9th grade students and will conclude when the same students are in the 12th grade. As the study is in its third year, findings from the first and second year are summarized here. From 9th grade to 10th grade, there are positive changers (i.e., changing intention from non-STEM to STEM college major), negative changers (i.e., changing intention from STEM to non-STEM college), and non-changers. Focusing on negative and positive changers, findings from data collected through interviews of students delving into reasons for changing intentions from 9th to 10th grade are presented, describing the most critical factors influencing students' decision making/changing. We discuss the results, implications, and limitations of the study informing both the research field (e.g., study design, future research areas) and the practice (e.g., focus on school level factors and steps to improve individual motivation).

1   Introduction

The last decade has seen a serious concern pertaining to a shortage of workforce in science, engineering, and mathematics (STEM). For instance, not long ago, President's Council of Advisors on Science and Technology (2012), projected a need for approximately one million more STEM professionals than the U.S. will produce at the current rate until 2022. Another important finding comes from U.S. Department of Commerce based on the data produced by the U.S. Census Bureau and the Bureau of Labor Statistics: STEM occupations are projected to grow by 8.9% from 2014 to 2024, compared to 6.4% growth for non-STEM occupations (Noonan, 2017). Recent discussions, however, challenge this concern as to whether a shortage really exists to this extent (e.g., Lohr, 2017; Rincon, 2017). Given that STEM encompasses a wide range of areas, on average STEM degree holders may even surplus the need for the demands of the labor market. However in general, whether there is still a shortage or a surplus in STEM, it is agreed that the demand of workforce is still higher than the available qualified workers in specific areas such as computer science (Xue & Larson, 2015). On the other hand, a critical concern that persists for the STEM workforce is the underrepresentation of diverse populations – in particular, women and historically underrepresented racial and ethnic groups. For instance, according to the National Science Board (2018), African American and Hispanic persons, remain underrepresented in the STEM workforce.

To help address these concerns regarding the shortages and underrepresentation in STEM workforce, it is vital to study factors affecting K-12 students' future college and career plans in STEM fields. The vast majority of existing evidence in the extant literature on student persistence on STEM pipeline is based on college-level experiences (Sass, 2015). This narrow scope may fail to consider important impacts of pre-college experiences, preparation, and resources on pursuing STEM areas during post-secondary years (Archer et al., 2012; Crosnoe & Muller, 2014; Maltese & Tai, 2011; Tai, Liu, Maltese, & Fan, 2006). Therefore, the goal of this study is to investigate the factors during the high school years that contribute to student motivation and interest in continuing in a STEM field. More specifically, this chapter focuses on the first two years of a longitudinal study and reports on: (1) how 9th and 10th grade students' high school STEM experiences, teacher and parental expectations, and students' motivational beliefs affected students' interest in selecting STEM majors; and (2) whether those who indicated interest in majoring in STEM or lack of interest in majoring in STEM changed their decisions in the second year (10th grade) and what might have affected these changes. The study utilizes a social cognitive career theoretical (SCCT; Lent & Brown, 1996; 2017) approach to frame the research.

2 Theoretical Framework

This study is grounded in a social cognitive career theoretical (SCCT) framework. As an extension and application of social cognitive theory (Bandura, 1986) to career choice, Lent, Brown, and Hackett's (1994) SCCT posits that one's career choice is influenced by the beliefs the individual develops and refines through the complex interaction between himself/herself as the individual, environment, and his/her behavior (Lent et al., 1994; Yu, Corkin, & Martin, 2016). According to SCCT, the most important factors influencing career decisions relate to student motivation (e.g., task value, self-efficacy, interest, outcome expectations; see Wigfield & Eccles, 2002). These psychological variables are considered as the mediators that connect other personal and contextual factors to future career choice and decisions (Lent & Brown, 2006; Yu et al., 2016). Empirical research has shown that students with higher mathematics and/or science self-efficacy and outcome expectations for engaging in mathematics and science are more likely to persist and be successful in these areas (e.g., Andersen & Ward, 2014; Lee, Min, & Mamerow, 2015).

In addition to personal motivation, the SCCT framework also recognizes several contextual influences (e.g., supports and barriers at school and at home) that mold individuals' career aspirations and choices (Maltese & Tai, 2011; Lent & Brown, 1996; Yu et al., 2016). Specifically, several groups, including parents, peers, and teachers have socializing influences on students' academic and career-related outcomes. This study integrates SCCT and previous research on factors closely connected with academic choices of 9th and 10th grade students from a STEM-focused public school district. This study addresses the interconnectedness among personal and environmental factors and STEM choice under the SCCT framework.

3 Research on Factors Affecting Students' STEM Outcomes: A Brief Look into the Last Two Decades

3.1 *Gender and Racial Gaps in STEM Motivation and Persistence*

The reason behind the myriad of calls and efforts to increase diversity and broaden participation in the STEM workforce is undoubtedly the persistent gap among certain subpopulations in their educational attainment and employment in STEM areas. Although this gap seemed to be somewhat closing since the turn of the millennium, statistics around the world show that it stills remains significant (e.g., BusinessEurope, 2011; National Science Foundation, 2017). Existing gaps relate primarily to sex/gender, social class/status, and race/ethnicity (Eccles & Wang, 2016; Mau, Perkins, & Mau, 2016; Riegle-Crumb, Farkas, & Muller, 2006; Yu et al., 2017). Research related to students' career plans revealed factors associated with student STEM persistence can vary substantially by students' gender and ethnically underrepresented minority status (Ing, 2014; Frank et al., 2008; Riegle-Crumb et al., 2006; Uitto, 2014). Moreover, these gaps can also be magnified by the moderation effect of other factors. For example, research indicates that associations between the expectations of parents and the learning motivation and achievement of students were stronger for boys than for girls (Taskinen, Dietrich, & Kracke, 2016).

In a STEM motivation study at the college-level, Hardin and Longhurst (2016) measured SCCT-relevant variables in a gateway introductory science course and found that female students had lower STEM self-efficacy, coping self-efficacy, and STEM interest than did the male students. In a secondary level longitudinal study, Sahin, Ekmekci, and Waxman, (2017a) examined school and out-of-school, and motivational factors that may have affected 9th grade students' intention of choosing a STEM major in college. Results indicated significant gender and racial differences in students' contemplation of choosing STEM fields. In another longitudinal study of science aspirations and careers of young adolescents (age 10–14), DeWitt et al. (2011) and Archer et al. (2012) found that Asian students demonstrated a highly positive set of attitudes towards science and aspirations in science when compared with students from other ethnic/racial backgrounds. In brief, gender, racial, socioeconomic, and other demographic characteristics of students are significantly influential in their development of STEM interest and career outcomes (Eccles & Wang, 2016; Sahin et al., 2017a; Yu et al., 2017).

3.2　*Motivational Factors*

According to SCCT, the most important factors influencing career decisions are those relating to student motivation (i.e., task value, self-efficacy, interest, outcome expectations; see Eccles & Wigfield, 2002). These motivation variables mediate the effect of personal and contextual factors on future career choice and decisions (Lent & Brown, 2006; 2013; Yu et al., 2016). Individuals' behavior and actions are influenced primarily by their sense of personal capability (self-efficacy), their beliefs about the likely consequences of performing particular actions, and the extent they find certain academic domains useful and/or interesting (Bandura, 1986; Eccles & Wigfield, 2002; Lent & Brown, 1996). The main SCCT-related motivational factors include self-efficacy, expectancy, and interest (Yu et al., 2016). Empirical research around the world has shown that students with higher self-expectations and self-efficacy in STEM subjects are more likely to persist and be successful in these areas (Andersen & Ward, 2014; Lee et al., 2015; Mujtaba & Reiss, 2014; Sahin et al., 2017a, 2017b; Uitto, 2014; Wang, 2013).

Within the SCCT research context, it has been most common to assess content- or task-specific self-efficacy (e.g., beliefs about one's ability to complete the requirements of a particular academic major; Lent & Brown, 2013). In a 9-year longitudinal study of middle school students, Le and Robbins (2016) found that quantitative ability and STEM interest fit were reciprocally related and meaningfully predicted the probability that students obtained a college degree in STEM within 6 years from their first enrollment in college. Lent et al. (2016) found that freshman engineering students' intentions to persist in their degree, satisfaction with the major, and self-efficacy strongly and positively related to their persistence in the major by the end of their junior year. There is evidence indicating that the pursuit of STEM degrees is strongly related to the choice made during the high school years and that choice may relate more strongly to a growing interest in mathematics and science rather than enrollment or achievement in STEM-focused courses (Matlese & Tai, 2011). For low-income students, in particular, motivation and interest in math and science could be even stronger precursors to develop STEM interest, resulting with pursuit of a career in STEM area (e.g., Maltese & Tai, 2011; Shumow & Schimdth, 2013). Moreover, Eccles and Wang (2016) found that math ability self-concepts more strongly predicted both individual and gender differences in the likelihood of entering STEM careers than math scores on aptitude tests. Therefore, motivational factors are central to persistence in STEM and can even account for gender and socio-economic differences.

3.3 *External Expectations*

Others' expectations can be considered as part of contextual factors, as they may serve as supports and barriers during the choice-making process regarding future career paths (Lent & Brown, 1996, 2006; Mujtaba & Reiss, 2014; Sahin et al., 2017a). Others' expectations concerning how an individual would do in a particular situation or how he or she would be influenced by a particular experience may actually influence that individual's actions (DeWitt et al., 2011; Sahin et al., 2017a; Schunk, Meece, & Pintrich, 2014). Decades of research has focused on parents' and teachers' expectations about their children's educational attainment. In a study of sixth-grade students from low-income schools about their academic beliefs, Jackson, Suizzo, and Harvey (2017) asked 13 math and science teachers to report their perceptions of the students achievement and compared these to student grades provided by the school district.

According to logistic regression analysis, students who were positively viewed by their teachers were more likely to have STEM career goals, controlling for grades and math self-efficacy. Support from parents has also been shown to be a significant factor in students' pursuit of STEM degrees (Archer et al., 2012; Garriott, Navarro, & Flores, 2017; Hardin & Longhurst, 2016; Hui & Lent, 2017; Sahin et al., 2017a; Thomas & Strunk, 2017). DeWitt et al. (2011) also found that students' aspirations in science were most strongly predicted by parental attitudes to science and engagement in science-related activities outside of school. The directions for future research on students' STEM aspirations suggested by researchers includes investigation of the links between parents' and teachers' beliefs and children's motivation for STEM (Gunderson, Ramirez, Levine, & Beilock, 2012), from the SCCT perspective, in particular (Garriott et al., 2017).

3.4 Formal and Informal STEM Experiences

Formal and informal STEM activities can also contribute to contextual factors in STEM, since they function as opportunities for students to participate in STEM in their current educational context or environment. Education research has documented several important factors related to learning experiences in STEM: (a) success in mathematics and science classes (e.g., Andersen & Ward, 2014; Sahin, Erdogan, Morgan, Capraro, & Capraro, 2012), (b) access to AP STEM courses (e.g., Ehrenberg, 2010), (c) opportunities to participate in STEM related projects during high school (e.g., Gottfried, & Williams, 2013), (d) interaction with successful peers (e.g., Frank et al., 2008), and (e) high self-efficacy (e.g., Anderson & Ward, 2014). Schools and local organizations offer programs for STEM learning and often include sustained, self-organized activities for STEM enthusiasts. There is a large body of evidence suggesting that structured in-and out-of-school STEM programs can stimulate the science-specific interests of its participants (e.g., Sahin, Gulacar, & Stuessy, 2014; Sevdalis & Skoumios, 2014), positively influence academic achievement for students (e.g., Sahin, 2013), and expand participants' sense of future STEM career options (e.g., Sahin et al., 2017a).

Eccles and Wang (2016) uncovered that high school course-taking more strongly predicted both individual and gender differences in the likelihood of entering STEM careers than math scores. Additionally, a high school STEM intervention study (Rozek, Svoboda, Harackiewicz, Hulleman, & Hyde, 2017) revealed that greater high-school STEM preparation (STEM course-taking and ACT scores) was associated with increased STEM career pursuit (i.e., STEM career interest, the number of college STEM courses, and students' attitudes toward STEM) five years after the intervention. In another study comparing the high school experiences and achievement of students in inclusive STEM high schools and other high schools in North Carolina, researchers found that "attending STEM schools raised the likelihood of completing pre-calculus or calculus and chemistry in high school, led to increased involvement in STEM extracurricular and out-of-class activities, and enhanced interest in science careers and aspirations to earn a master's or higher degree" (Means, Wang, Young, Peters, & Lynch, 2016; p. 709). Long term effects of high school STEM learning experiences have been also documented by empirical research. For example, Steenbergen-Hu and Olszewski-Kubilius (2017) studied high school students who participated in supplemental enriched or accelerated math and science learning activities by following up with them four to six years after high school. This study revealed through online surveys that students' experiences in the supplemental outside-of-school STEM programs helped students intensify their interests in STEM and increased their likelihood of earning a college degree in STEM. All these research findings are parallel with the findings of Maltese and Tai (2011) who found that the majority of students who pursue STEM areas made that choice during high school.

3.5 Research Questions
Our research questions were:
1. What are the impacts of school and out of school-related STEM activities on students' intention to pursue a STEM degree?
2. What are the impacts of both teacher and parental educational expectations on students' intentions to pursue a STEM degree?
3. What are the impacts of students' self-educational expectations and self-efficacy in math and science on their intentions to pursue a STEM degree?
4. For students who changed their intensions from year 1 (9th grade) to year 2 (10th grade), which one(s) of the all aforementioned factors relate to the changes in their decisions?

4 Methods

4.1 Participants
In this study, we worked with a charter schools system – Harmony Public Schools (HPS) – with a STEM focus. We invited 23 Harmony high schools to be part of this project but 20 of them agreed to participate in the study. We sent an online survey to 2,157 9th grade students from the class of 2019, representing 20 HPS campuses in year 1. A total of 1,520 (~ 70%) students completed the survey. For the year 2, the instrument was sent to 2,100 10th grade students from 20 HPSs. Total of 1,595 (76%) 10th graders completed the survey. Of these, 995 (~65%) continued to participate in the study from year 1. A total of 232 students indicated a change in their intention of selecting their college major

from 9th to 10th grade. Of these, 168 switched from STEM majors to non-STEM majors (negative changers); and 64 changed from non-STEM to STEM fields (positive changers). We randomly chose 8 negative and 4 positive changers for the interviews to investigate why they changed their college major plans from year 1 to year 2. Of these, 7 negative and 3 positive changers accepted to be interviewed and turned their parent consents for the interviews.

4.2   Instruments and Data Collection

We developed an online survey consisting of 43 questions to request information about four categories of variables: (a) student demographics, (b) family context, (c) school and out-of-school related activities, and (d) parent and teacher expectations (see Appendix A). We benefited from previously developed valid and reliable scales (e.g., Faber, Unfried, Wiebe, Corn, & Townsend, 2013; Lee et al., 2015). For the qualitative component, we developed a structured interview protocol (see Appendix B) to see what might have caused students' switch between majors in addition to what has been found through quantitative analyses. All 10 interviews were audio recorded and transcribed verbatim by a research assistant. First, to develop the codes for this study, we read the interview transcriptions separately. Then, we met to discuss themes, determined a common list of themes, and developed a codebook with all codes discussed along with definitions. The interviews were coded based on the codebook. After individual coding process, we came together one more time to discuss each interview and each researcher's codes. We also resolved any discrepancies we had in coding of all the interviews in order to achieve inter-rater reliability.

We used Naviance to collect student data. Naviance is a comprehensive K-12 college and career readiness platform (Naviance, 2016). Parent consent forms were obtained in year 1. Each year's survey is released separately in Naviance so that students can see it under their account as a task to be completed. We administered the survey between early March and late April each year.

4.3   Data Analyses

We conducted multiple logistic regressions to investigate which group of variables predicted students' probability of contemplating about choosing a STEM major in college. Before we ran multiple logistic regressions, we verified the assumptions of absence of multicollinearity, independence of errors, and linear relationship between the independent variables and the log odds (Meyers, Gamst, & Guarino, 2006). We also used independent-samples t-test between positive and negative changers to investigate whether there were any significant differences between the two groups in terms of predictive variables, which might have played important role in students' change of decisions about their college major from STEM to non-STEM or from non-STEM to STEM.

5 Findings

We categorized our findings under four groups: (1) Individual and in-and-out of school-related factors, (2) Pygmalion factors, (3) Motivational factors, and (4) Changers.

1. *Individual and in-and-out of School-related Factors*

We conducted separate logistic regression analysis for each year. Table 15.1 shows our findings for year 1 and 2 separately. Because the findings from year 1 was published in a peer-reviewed journal (Sahin et al., 2017a), 9th grade findings were pulled and cited from that source.

In both 9th grade (Sahin et al., 2017a) and 10th grade results, gender was a significant factor in STEM major selection where we found that male students were 1.5 to 1.9 times more likely to consider choosing a STEM major in college than were female students (see Table 15.1). Although students whose parents earned a degree from a U.S. college seemed to be in an advantaged situation in

TABLE 15.1 Impacts of school and out of school-related activities on STEM-major intentions

	Year-1 (9th Grade)[a]			Year-2 (10th Grade)		
	B	Sig.	Exp(B)	B	Sig.	Exp(B)
Gender	0.649**	.004	1.915	.430**	.005	1.537
Lunch Status	-0.151	.418	0.860	.026	.684	1.027
College Degree in the US	0.585*	.012	1.795	-.118	.346	.889
Asian	.416*	.045	1.516	1.048***	.000	2.853
African American	-0.348	.389	0.706	.422	.058	1.525
White	-0.369	.323	0.691	.606*	.014	1.833
Count STEM Clubs	N/A	N/A	N/A	.331**	.002	1.392
Total STEM PBL Projects	0.246***	.000	1.279	.045	.099	1.046
DISTCO Contest	0.284	.301	1.329	.136	.273	1.145
STEM Summer Camp	0.856*	.013	2.353	-.409**	.007	.665
Count Science Fair	-0.223	.069	0.800	.121*	.013	1.129
Count STEM Internship	-0.120	.776	0.887	.262*	.028	1.427
Count AP Courses Taken	0.058	.822	1.060	.099	.171	1.104
Current GPA	0.627***	.000	1.872	-.028	.774	.973
Constant	-1.992**	.009	0.136	-1.635	.053	.195

***$p < .001$, **$p < .01$, *$p < .05$.
[a] Source: Sahin et al. (2017a)

year 1, there was no statistical difference in year 2 when students were at 10th grade. Asian students were more willing to major in STEM in both years where they were 1.5 to 2.8 times more likely to consider majoring in STEM-related fields compared to students from other ethnicities. Grade-9 students completing multiple STEM PBL projects were found to be more likely to intend on choosing a STEM major in college. However, this was not statistically significant in 10th grade. As most STEM educators and researchers may expect, participation in a STEM summer camp was also a significant factor in students' STEM major contemplation in both 9th and 10th grade. However, 10th grade results showed that STEM summer camp participation was found to be negatively associated with the likelihood of considering a STEM major in college, while it was a positive predictor in 9th grade results. In 10th grade, students' completion of more science fair projects and internships at universities or medical institutes became important factors in students' plans for selecting a STEM major in college. Interestingly, students' GPA was one of the most influential factors in students' development of STEM major interest in 9th grade, yet the impact of GPA did not continue to sustain significant in 10th grade.

2. *Pygmalion Factors*

To explore the effects of Pygmalion variables on students' intention of selecting a STEM major in college, we conducted additional logistic regression analysis. Looking at the conglomerate data from both years, we found that male students were 1.5 to 1.7 times more likely to consider choosing a STEM major in college than were female students. Again, in year 1 and 2, both Asian and White students were more likely to contemplate about choosing a STEM major than were Hispanic students. African American students were 1.5 times more likely to choose a STEM major than were Hispanic 10th graders. Even though parent and STEM teacher expectations were significant factors in year-1 study, only parent expectation continued to be a significant predictor for students who plan to major in STEM in year 2.

3. *Motivational Factors*

We conducted separate logistic regression analyses to find out which motivational factors significantly influenced Harmony 9th and 10th grade students' intentions for selecting STEM majors. We found that in both 9th and 10th grades male and Asian students were still more likely to contemplate about choosing a STEM-related field in college than were female and Hispanic students, respectively. Grade-10 students with higher self-expectations were 1.2 times more likely to consider majoring in STEM fields than were their peers

TABLE 15.2 Impacts of Pygmalion effect variables on STEM-major intentions

	Year-1 (9th Grade)[a]			Year-2 (10th Grade)		
	B	Sig.	Exp(B)	B	Sig.	Exp(B)
Gender	0.371**	.002	1.449	.541***	.000	1.718
Lunch Status	0.092	.364	1.097	-.010	.871	.990
College Degree in the US	0.102	.435	1.107	-.152	.206	.859
Asian	.724***	.000	2.063	1.007***	.000	2.738
African American	-0.322	.140	0.724	.447*	.046	1.501
White	.421*	.021	1.524	.352*	.039	1.582
Parents Expectation	0.213*	.011	1.238	.264**	.004	1.302
STEM Teachers Expectation	0.389***	.000	1.475	.178	.085	1.146
Constant	-1.482***	.000	0.227	-.829**	.002	.416

***$p < .001$, **$p < .01$, *$p < .05$
[a] Source: Sahin et al. (2017a)

with less self-expectation. Self-expectation was a not significant factor for the 9th grade students. Having high math efficacy was a significant factor in students' development of STEM major interest in both grades while science efficacy was only a significant factor for 10th grade students' contemplation about choosing a STEM major.

4. *Changers*

To investigate why students changed their majors from STEM to non-STEM or non-STEM to STEM, we conducted a mixed method analysis. For the quantitative component, two comparisons were made between two groups of students: (1) *negative* changers – students who expressed they would consider a STEM major in year 1 but said no in year 2 (outcome variable changed from 1–Yes to 0–No); and (2) *positive changers* – students who made the change in the opposite direction (outcome variable changed from 0–No to 1–Yes). A close attention was paid to the predictive factors and to identifying significant changes in these factors for each group of students. To accomplish this, we conducted independent-samples *t*-test to compare the changes in variables of interest from 9th grade to the 10th between the *positive* and *negative* changers. There were only two statistically significant differences in how the two groups compared in their changes from year-1 to year 2 (see Table 15.5): the

TABLE 15.3 Impacts of self-expectation and math and science efficacy on STEM major intentions

	Year-1 (9th Grade)[a]			Year-2 (10th Grade)		
	B	Sig.	Exp(B)	B	Sig.	Exp(B)
Gender	0.251*	.041	1.285	.440**	.003	1.552
Lunch Status	0.117	.258	1.124	-.007	.917	.993
College Degree in the US	0.121	.360	1.129	-.069	.575	.933
Asian	.651**	.001	1.918	.742***	.000	2.320
African American	-0.347	.118	0.707	.290	.164	1.337
White	-0.229	.313	0.795	.293	.098	1.341
Self-Expectation	-0.032	.407	.968	.173**	.001	1.189
Math Efficacy	0.285***	.000	1.330	.491***	.000	1.570
Science Efficacy	0.456***	.000	1.578	.081	.213	1.085
Constant	-2.042***	.000	0.130	-1.617***	.000	.198

***$p < .001$, **$p < .01$, *$p < .05$

a Source: Sahin et al. (2017a)

number of science fair projects and level of self-expectation. Although, the number of science fair projects completed dropped from 9th to 10th grade for all students, the drop was significantly less for *positive* changers than it was for the negative changers. In addition, *positive* changers had increased levels of self-expectation whereas the *negative* changers had decreased levels of self-expectation.

Regarding the qualitative component for exploring the changes in students' decisions, we identified two major themes in the interview data: (1) *motivation/behavior* and (2) *contextual/environment*. These themes are parallel with what Lent et al. (1994) proposed in their SCCT framework. All the themes, codes, definitions, and related quotes are provided in Appendix B. The first overarching theme was about students' *motivation* or *behavior-related* factors. Under this theme, we could only develop one code: student's self-interest/expectation.

Our analyses revealed that one of the influential factors in students' development of STEM major interest was related to their *self-interest* and *self-expectation* about their future educational attainment. One of the critical points among STEM major choosers was how confident students were about what they want to do in college and their future lives. Students continually expressed

TABLE 15.4 Independent samples t-test comparing positive and negative changers' average changes in school, Pygmalion, and motivational factors

	Df	Sig.[a]	Mean difference in changes from 9th to 10th Grade[b]	S.E.	95% C.I. Lower	95% C.I. Upper
STEM SOS Project	101	.59	−0.24	0.45	−1.14	0.66
STEM SOS Website Contest	223	.27	−0.11	0.10	−0.32	0.09
DISTCO Contest	227	.36	0.07	0.07	−0.08	0.21
STEM Summer Camp	227	.08	−0.19	0.11	−0.40	0.02
STEM Internship	227	.51	−0.06	0.09	−0.24	0.12
STEM AP Courses	86	.16	−0.33	0.23	−0.79	0.13
Science Fair	222	.04*	0.92	0.45	0.02	1.81
GPA	228	.42	−0.13	0.16	−0.46	0.19
Parent Expectation	227	.59	0.07	0.12	−0.17	0.31
Teacher Expectation	227	.17	0.21	0.15	−0.09	0.52
Self-Expectation	161	.02*	0.62	0.26	0.10	1.13
Math Efficacy	185	.96	0.01	0.15	−0.28	0.30
Science Efficacy	202	.85	0.04	0.19	−0.34	0.42

a   2-tailed
b   Mean Difference = Mean change (positive changers) − Mean change (negative changers)

how important their own decisions and interests in what they would like to pursue in college. It seems that students already had high expectations for themselves in life and this could only be achieved by choosing a STEM major in college as they implied.

The second overarching theme was *contextual and environment* where students expressed (and coded by us as) the roles of their *parents' expectation*, *courses* they took, and *in-and-out of school STEM activities* they participated in their contemplation of choosing a STEM major in college.

The first identified code was *parents' expectations* about their children's future. Students emphasized either their parents' high expectations about their coursework or what their parents were doing for living as a profession affected students' decision to what area to study in college. The second identified code was related to *courses* students have taken either during 9th or 10th grade. This was seen to affect their decision of choosing or dropping their intentions for selecting STEM major in college. They switched to STEM because either they liked how the particular STEM course they took was taught or they found that they were good at handling that course. On the other hand, there were students who switched from STEM to non-STEM because either they did not like how the course was taught or mostly they found that they were not strong enough to handle that STEM course. In either case, teachers' way of teaching or the content and curriculum of the course might have played critical role in students' developing of availing or non-availing intentions towards STEM major selection. The final code identified in the interview data was about students' participation to *in-and-out of school STEM activities* where students underlined the importance of extracurricular STEM activities in developing their career interest and finding out what they wanted to pursue in college. They appreciated the STEM opportunities offered by their schools including project lead the way (PLTW) courses, organization of science fair events, STEM festivals, and summer camps where they discovered their interests towards STEM disciplines.

6   Implications for K-12 Education

The results of this study lead to several implications for policymakers, administrators, and educators. First, it suggests that school districts and schools can have a positive impact of students' intentions to major in STEM areas. For example, HPS's STEM clubs and science fair competitions help its students develop positive interest towards STEM subjects. Overall, our findings are especially encouraging because they suggest that schools can implement programs and practices that may impact students' future careers in STEM.

Second, we have found that specific district and school-based programs such as science fairs, STEM internships, summer camp participation, and STEM clubs influence students' STEM career aspirations. Systematic evaluations of these programs may be needed to inform us how and why they may positively or negatively influence students' career aspirations. For example, the negative effects we found for STEM summer camp participation on students' STEM career aspirations suggests that more investigation of that program is needed. Other school districts may want to explore implementing some of the beneficial HPS STEM-related activities/programs that HPS develops for all of its students. It may also be important for school districts to examine some of the STEM practices that were *not* found or that were *not consistently* (between 9th and 10th grade) found to increase the likelihood of students' choosing a STEM major. Participating in DISTCO contest or taking AP courses, for example, were not found to significantly influence students' interest of STEM career. On the other hand, participating in STEM internships or science fairs, for instance, had inconsistent results between year 1 (9th grade) and year 2 (10th grade). As understood from the student interviews, some of these school and out-school STEM activities have a negative impact on students – i.e., students discover that the STEM is not for them. This may account for some of the inconsistent results across the years. On the other hand, as an implication of these findings, school districts may want to further explore why these activities do not appear to increase *all* students' aspirational STEM interests.

Finally, the results highlight the importance of parent, teacher, and student self-expectations for promoting career aspirations in STEM fields. It is important to discuss this specific finding with teachers and parents so that they can continue to increase their encouragement of their children. These findings suggest that districts and schools need to continually focus on the affective dimensions of STEM so that all students feel confident that they can be successful in math and science.

7   Limitations and Future Steps

One of the limitations of this study is that we do not know how students' aspirations in 10th grade actually predict the college major they will eventually choose in 2019 when they graduate high school. Another limitation of this study is that it was only conducted in one large school district that was implementing an integrated STEM curriculum for several years. Although this large charter school system includes a majority of highly-diverse and economically-disadvantaged students and schools across the state of Texas, it would be interesting to include other large school districts in future studies so that we could possibly compare across districts and even across different school types on how the factors of (a) demographics, (b) school and out-of-school factors, and (c) Pygmalion effect variables differentially affect students' choosing STEM majors in college. Another limitation of the study relates to the measurement of variables. Although student self-report measures have been in many STEM studies (Gottfried & Williams, 2013; Lee et al., 2015), there are some concerns about the use of such measures due to the measurement properties of the instrument (i.e., construct validity and reliability) and/or social-desirability bias (Fisher & Katz, 2008). Future studies could address this issue by either (a) including more similar items so that construct validity and internal consistency reliabilities could be calculated, or (b) developing a strategy to resurvey a small sample of the same students after a few weeks in order to determine the test-retest reliability of the instrument. There are some concerns about using either approach, however, since they both may result in lowering the response rate of the survey.

A final limitation of the study relates to the semi-structured interviews of 10th grade students. We only interviewed seven negative and three positive changers in this study. A larger and more in-depth interview with more students might provide us with "richer" data and provide more insight on how and why students changed their aspirations between the 9th and 10th grades.

Future studies should also extend the SCCT framework to include the STEM teachers' background and qualifications as well as their motivational beliefs for teaching as part of the contextual/environmental factors. This may produce important discoveries, since teacher quality is considered among the most important contextual factors in student academic outcomes (Hattie, Masters, & Birch, 2016).

8 Conclusion

This study makes several important contributions to the research on students' persistence in the STEM pipeline. First, the design of the study included a large, representative sample of students from one of largest multi-ethnic charter school districts in the U.S. The sample represents over 50 schools located in both urban and rural cities and includes a majority of students who are underrepresented minorities and from low socio-economic households. The large sample size and the diversity of this sample is one of strengths of this study.

Second, this is one of the few studies that focused on high school students' interest in STEM. Most prior research on STEM persistence has focused on college students (Sass, 2015), yet the findings of this study suggest that student intentions for entering the STEM pipeline begins as early as 9th grade. The implications of our findings highlight the need for more STEM research in high school and perhaps even extending to middle and elementary schools.

Third, this is one of the few longitudinal studies in this field. This is an important strength of this study because this panel design allows us to make two-year comparisons for the same individuals and determine *why* some students changed their STEM intentions from one year to the next. Interestingly, we found that about 77% of the students' intentions for STEM remained stable for the two-year period. However, about 17% already changed their mind and decided that they were no longer interested in pursuing a STEM career in college. Furthermore, about 6% of the students changed their mind and decided that they were now interested in pursuing a college major in STEM.

Students' self-expectations and science fair participation were the only two significant factors accounting for the difference between positive (non-STEM to STEM) and negative (STEM to non-STEM) changers. This significant finding has important implications for schools who may want to develop specific programs or provide more professional development to teachers to promote students' self-expectations to succeed and promote science fairs across schools.

A final significant aspect of this study is that it lends support to the social cognitive theory (Bandura, 1986) and Lent et al.'s (1994) social cognitive career theory (SCCT). The SCCT is an elaborate theory and most prior research has only tested part of the theory and its assumptions (van Tuijl & van der Molen, 2015). The present study, however, tests all of the relationships in the theory and found that at least one of individual, environmental, Pygmalion, and motivational factors significantly influenced students' choice of STEM major. In other words, it suggests that research on students' interest in STEM should try to address all four of these factors in order to comprehensively examine the theory.

References

Andersen, L., & Ward, T. J. (2014). Expectancy-value models for the STEM persistence plans of ninth-grade, high-ability Students: A comparison between black, Hispanic, and white students. *Science Education, 98*(2), 216–242.

Archer, L., DeWitt, J., Osborne, J., Dillon, J., Willis, B., & Wong, B. (2012). Science aspirations, capital, and family habitus: How families shape children's engagement and identification with science. *American Educational Research Journal, 49*(5), 881–908.

Bandura, A. (1986). *Social foundations of thought and action: A social cognitive theory.* Englewood Cliffs, NJ: Prentice Hall.

BusinessEurope. (2011). *Plugging the skills gap – The clock is ticking (science, technology and maths).* Brussels: Author. Retrieved from https://www.businesseurope.eu/sites/buseur/files/media/imported/2011-00855-E.pdf

Crosnoe, R., & Muller, C. (2014). Family socioeconomic status, peers, and the path to college. *Social Problems, 61*(4), 602–624.

DeWitt, J., Archer, L., Osborne, J., Dillon, J., Willis, B., & Wong, B. (2011). High aspirations but low progression: The science aspirations–careers paradox amongst minority ethnic students. *International Journal of Science and Mathematics Education, 9*(2), 243–271.

Eccles, J. S., & Wang, M. T. (2016). What motivates females and males to pursue careers in mathematics and science? *International Journal of Behavioral Development, 40*(2), 100–106.

Eccles, J. S., & Wigfield, A. (2002). Motivational beliefs, values, and goals. *Annual Review of Psychology, 53*(1), 109–132.

Ehrenberg, R. G. (2010). Analyzing the factors that influence persistence rates in STEM field, majors: Introduction to the symposium. *Economics of Education Review, 29*(6), 888–891.

Faber, M., Unfried, A., Wiebe, E. N., Corn, J., & Townsend, L. W. (2013). *Student attitudes toward STEM: The development of upper elementary school and middle/high school student surveys.* Paper presented at the 120th American Society for Engineering Education Annual Conference and Exposition. Retrieved from http://www.asee.org/public/conferences/20/papers/6955/view

Fisher, R., & Katz, J. E. (2008). Social-desirability bias and the validity of self-reported values. *Psychology & Marketing, 17*, 105–120.

Frank, K. A., Muller, C., Schiller, K. S., Riegle-Crumb, C., Mueller, A. S., Crosnoe, R., & Pearson, J. (2008). The social dynamics of mathematics coursetaking in high school. *American Journal of Sociology, 113*(6), 1645–1696.

Garriott, P. O., Navarro, R. L., & Flores, L. Y. (2017). First-generation college students' persistence intentions in engineering majors. *Journal of Career Assessment, 25*(1), 93–106.

Gottfried, M. A., & Williams, D. (2013). STEM club participation and STEM schooling outcomes. *Education Policy Analysis Archives, 21*(79), 1–27.

Gunderson, E. A., Ramirez, G., Levine, S. C., & Beilock, S. L. (2012). The role of parents and teachers in the development of gender-related math attitudes. *Sex Roles, 66*(3–4), 153–166.

Hardin, E. E., & Longhurst, M. O. (2016). Understanding the gender gap: Social cognitive changes during an introductory stem course. *Journal of Counseling Psychology, 63*(2), 233–239.

Hattie, J., Masters, D., & Birch, K. (2016). *Visible learning into action: International case studies of impact.* New York, NY: Routledge.

Hui, K., & Lent, R. W. (2018). The roles of family, culture, and social cognitive variables in the career interests and goals of Asian American college students. *Journal of Counseling Psychology, 65*(1), 98–109.

Ing, M. (2014). Gender differences in the influence of early perceived parental support on student mathematics and science achievement and STEM career attainment. *International Journal of Science and Mathematics Education, 12*(5), 1221–1239.

Jackson, K. M., Suizzo, M. A., & Harvey, K. E. (2017). Connecting teacher perceptions to stem occupational goals in low income adolescents of color. *Journal of Women and Minorities in Science and Engineering, 23*(1), 73–86.

Le, H., & Robbins, S. B. (2016). Building the STEM pipeline: Findings of a 9-year longitudinal research project. *Journal of Vocational Behavior, 95*, 21–30.

Lee, S. W., Min, S., & Mamerow, G. P. (2015). Pygmalion in the classroom and the home: Expectation's role in the pipeline to STEMM. *Teachers College Record, 117*(9), 1–36.

Lent, R. W., & Brown, S. D. (1996). Social cognitive approach to career development: An overview. *The Career Development Quarterly, 44*(4), 310–321.

Lent, R. W., & Brown, S. D. (2006). On conceptualizing and assessing social cognitive constructs in career research: A measurement guide. *Journal of Career Assessment, 14*(1), 12–35.

Lent, R. W., & Brown, S. D. (2013). Social cognitive model of career self-management: Toward a unifying view of adaptive career behavior across the life span. *Journal of Counseling Psychology, 60*(4), 557–568.

Lent, R. W., & Brown, S. D. (2017). Social cognitive career theory in a diverse world: Guest editors' introduction. *Journal of Career Assessment, 25*(1), 3–5.

Lent, R. W., Brown, S. D., & Hackett, G. (1994). Toward a unifying social cognitive theory of career and academic interest, choice, and performance. *Journal of Vocational Behavior, 45*(1), 79–122.

Lohr, S. (2017, November 1). Where the STEM jobs are (and where they aren't). *The New York Times*. Retrieved from https://www.nytimes.com/2017/11/01/education/edlife/stem-jobs-industry-careers.html

Maltese, A. V., & Tai, R. H. (2011). Pipeline persistence: Examining the association of educational experiences with earned degrees in STEM among US students. *Science Education, 95*(5), 877–907.

Means, B., Wang, H., Young, V., Peters, V. L., & Lynch, S. J. (2016). STEM-focused high schools as a strategy for enhancing readiness for postsecondary STEM programs. *Journal of Research in Science Teaching, 53*(5), 709–736.

Meyers, L. S., Gamst, G., & Guarino, A. J. (2006). *Applied multivariate research: Design and interpretation* (3rd ed.). Thousand Oaks, CA: Sage Publications.

Morgan, P. L., Farkas, G., Hillemeier, M. M., & Maczuga, S. (2016). Science achievement gaps begin very early, persist, and are largely explained by modifiable factors. *Educational Researcher, 45*(1), 18–35.

Mujtaba, T., & Reiss, M. J. (2014). A survey of psychological, motivational, family and perceptions of physics education factors that explain 15-year-old students' aspirations to study physics in post-compulsory English schools. *International Journal of Science and Mathematics Education, 12*(2), 371–393.

National Science Board. (2018). *Science and engineering indicators 2018* (NSB-2018-1). Alexandria, VA: National Science Foundation. Retrieved from https://www.nsf.gov/statistics/indicators/

National Science Foundation. (2017). *Women, minorities, and persons with disabilities in science and engineering: 2017 digest* (NSF 17-310). Arlington, VA: Author. Retrieved from www.nsf.gov/statistics/wmpd/

Naviance. (2016). *Connecting learning & life*. Retrieved from http://www.naviance.com/

Noonan, R. (2017). *STEM jobs: 2017 update* (ESA Issue Brief # 02-17). Washington, DC: Office of the Chief Economist, U.S. Department of Commerce. Retrieved from http://www.esa.gov/reports/stem-jobs-2017-update

President's Council of Advisors on Science and Technology. (2012). *Engage to excel: Producing one million additional college graduates with degrees in science, technology, engineering, and mathematics*. Retrieved from https://obamawhitehouse.archives.gov/sites/default/files/microsites/ostp/pcast-engage-to-excel-final_2-25-12.pdf

Riegle-Crumb, C., Farkas, G., & Muller, C. (2006). The role of gender and friendship in advanced course taking. *Sociology of Education, 79*(3), 206–228.

Rincon, R. (2017, December). Is there a shortage of STEM jobs to STEM graduates? It's complicated. *Society of Women Engineers Blog*. Retrieved from http://alltogether.swe.org/2017/12/is-there-a-shortage-of-stem-jobs-to-stem-graduates-its-complicated/

Rozek, C. S., Svoboda, R. C., Harackiewicz, J. M., Hulleman, C. S., & Hyde, J. S. (2017). Utility-value intervention with parents increases students' STEM preparation and career pursuit. *Proceedings of the National Academy of Sciences, 114*(5), 909–914.

Sahin, A. (2013). STEM clubs and science fair competitions: Effects on post-secondary matriculation. *Journal of STEM Education: Innovations and Research, 14*(1), 7–13.

Sahin, A., Ekmekci, A., & Waxman, H. C. (2017a). Collective effects of individual, behavioral, and contextual factors on high school students' future STEM career plans. *International Journal of Science and Mathematics Education*, 1–21. doi:10.1007/s10763-017-9847-x.

Sahin, A., Ekmekci, A., & Waxman, H. C. (2017b). The relationships among high school STEM learning experiences, expectations, and mathematics and science efficacy and the likelihood of majoring in STEM in college. *International Journal of Science Education, 39*(11), 1549–1572.

Sahin, A., Erdogan, N., Morgan, J., Capraro, M. M., & Capraro, R. M. (2012). The effects of high school course taking and SAT scores on college major selection. *Sakarya University Journal of Education, 2*(3), 96–109.

Sahin, A., Gulacar, O., & Stuessy, C. (2014). High school students' perceptions of the effects of science Olympiad on their STEM career aspirations and 21st century skill development. *Research in Science Education*. doi:10.1007/s11165-014-9439-5

Sass, T. R. (2015). *Understanding the STEM pipeline* (CALDER #125). Washington, DC: American Institutes for Research.

Schunk, D. H., Meece, J. R., & Pintrich, P. R. (2014). *Motivation in education: Theory, research, and applications* (4th ed.). New York, NY: Pearson.

Sevdalis, C., & Skoumios, M. (2014). Non-formal and informal science learning: Teachers' conceptions. *International Journal of Science in Society, 5*(4), 13–25.

Shumow, L., & Schmidt, J. A. (2013). Academic grades and motivation in high school science classrooms among male and female students: Associations with teachers' characteristics, beliefs and practices. *Journal of Education Research, 7*(1), 53–72.

Steenbergen-Hu, S., & Olszewski-Kubilius, P. (2017). Factors that contributed to gifted students' success on STEM pathways: The role of race, personal interests, and aspects of high school experience. *Journal for the Education of the Gifted, 40*(2), 99–134.

Tai, R. H., Liu, C. Q., Maltese, A. V., & Fan, X. (2006). Planning early for careers in science. *Science, 312,* 1143–1144.

Taskinen, P. H., Dietrich, J., & Kracke, B. (2016). The role of parental values and child-specific expectations in the science motivation and achievement of adolescent girls and boys. *International Journal of Gender, Science and Technology, 8*(1), 103–123.

Uitto, A. (2014). Interest, attitudes and self-efficacy beliefs explaining upper secondary school students orientation towards biology related careers. *International Journal of Science and Mathematics Education, 12,* 1425–1444.

van Tuijl, C., & van der Molen, J. H. W. (2016). Study choice and career development in STEM fields: An overview and integration of the research. *International Journal of Technology and Design Education, 26*(2), 159–183.

Wang, X. (2013). Why students choose STEM majors: Motivation, high school learning, and postsecondary context of support. *American Educational Research Journal, 50*(5), 1081–1121.

Wigfield, A., & Eccles, J. S. (2002). The development of competence beliefs, expectancies for success, and achievement values from childhood through adolescence. *Development of Achievement Motivation, 91*(120), 91–120.

Xue, Y., & Larson, R. C. (2015, May). STEM crisis or STEM surplus? Yes and yes. *Monthly Labor Review* (U.S. Bureau of Labor Statistics). Retrieved from https://www.bls.gov/opub/mlr/2015/article/stem-crisis-or-stem-surplus-yes-and-yes.htm

Yu, S. L., Corkin, D. M., & Martin, J. P. (2017). STEM motivation and persistence among underrepresented minority students: A social cognitive perspective. In J. T. DeCuir-Gunby & P. A. Schutz (Eds.), *Race and ethnicity in the study of motivation in education* (pp. 67–81). New York, NY: Taylor & Francis.

Appendix A: Interview Protocol

First, let me say thank you very much for agreeing to be here for this study today. We value your time and promise not to go over the allotted time. We expect this interview to last about 30 minutes.

The purpose of this interview is to understand your experiences as a student who has been taking a survey for last two years as part of a longitudinal study to investigate factors that affect high school students' selection a STEM major in college. We randomly chose you among those who changed their majors from STEM to non-STEM or non-STEM to STEM majors between 9th and 10th grades. Your input will be very valuable in not only knowing the factors that affect students' STEM major selection but also in developing more effective and sustainable STEM programs for the larger education community. You are free to decline any interview questions or the entire interview if you'd like. Do you understand? Would you like to proceed?

You have been taking a survey entitled "Tracking Class of 2019" since 2015–2016 school year (9th grade) through Naviance. In that survey, you answer questions like the ones below:

*8. State the number of LEVEL I essential investigation STEM SOS projects you have complet

*9. State the number of LEVEL II (year-long, as an individual or group interdisciplinary project)

*10. State the number of LEVEL III (advanced, year-long, self-driven projects) science, technolo school (including 9th and 10th grade):

*11. In which subject(s) have you completed a STEM SOS project(s) in high school (including 9t
- Algebra I
- Geometry
- Algebra II
- Biology
- Chemistry
- Physics
- AP Biology
- AP Chemistry
- AP Environmental Science
- AP Physics 1
- AP Physics 2
- AP Physics C
- AP Calculus AB
- AP Calculus BC
- AP Computer Science A
- AP Computer Science Principles
- AP Statistics
- Other (please specify)

*12. Did you participate in Digital Storytelling Competition (DISTCO) with your STEM SOS proje
- 2015-2016
- 2016-2017
- 2017-2018

One of the questions you've answered so far was asking you about whether you want to major in STEM-related area in college:

> Do you have intention to declare a science, technology, engineering, and mathematics (STEM)-related major in college?

Do you remember this question and survey you've been taking last two years?

(usually in March/April through Naviance, college counselors helps track survey-takers)

How would you response to this question right now? (i.e., do you want to major in STEM in college?)

More specifically, do you have a particular major in mind? Do you happen to know what STEM stands for? Do you remember how you responded to STEM major selection questions in 9th grade? What about your response for the 10th grade?

OK, then, let's talk about why we wanted to meet with you:

When we analyzed two years' data, we realized that some students changed their statements or college major intensions. Some of the students changed from STEM majors to non-STEM majors in year 2 (10th grade) – 2016–2017 school year – while others stated they want to major in STEM-related areas although they did not choose STEM in year 1 (9th grade). Some of your friends did not change their decisions.

Now, we are going to ask you some open-ended questions to try to understand why you changed your decisions. THERE ARE NO RIGHT OR WRONG ANSWERS!

1. You changed your major selection decision from [STEM/non-STEM] to [non-STEM/STEM]. We are not here to judge, we just want to know what made you change your mind. What made you change your intensions about college majoring from 9th grade to 10th grade? Do you remember why you responded differently across the years? Or were you just not sure which one to select and just selected one for the sake of answering? If that's the case, it is OK. You are not going to be judged or penalized for that, we just need to explore some of the reasons behind changes in student responses.

[Repeat the Question if Necessary]: *What was the reason you switched your major selection?*

2. Research shows and we believe that the reasons affecting career-related choices fall into three main categories: individual, environmental, and motivational. Related to these factors which one of the followings might have affected your decision change and why?

    [encourage student to articulate for each of the prompts]
    (a) Parent expectation;
    (b) Teacher expectation;
    (c) Self-expectation;
    (d) School related STEM activities (please be as specific as possible)

3. What science, technology, engineering, and/or math classes are you in?

(a) What STEM classes did you take during 10th grade?
(b) What STEM classes did you take during 9th grade?
4. What would you tell us about your grades in STEM courses and non-STEM courses? In which group do you think you have higher GPA?
5. What extracurricular STEM activities (e.g., science fairs, STEM clubs, summer camps or internships) are you participating this year?
(a) What extracurricular STEM activities did you participate in 10th grade?
(b) What extracurricular STEM activities did you participate in 9th grade?
6. Tell us about STEM SOS project you have been working on this school year (Level 1, 2, &3). How long do they take?
7. Can you give an example of Level 1 project you have completed this year?
8. Can you describe your Level 2/3 project?
9. Across years (i.e., 9th grade, 10th grade, and now), how would you say your involvement in STEM projects (STEM SOS in particular) changed?
10. What, if anything, do you wish you had at your school regarding learning opportunities in science, technology, engineering, or math?
11. What kind of jobs do you want to have after college, if you have any idea?
12. Is there anything else you would like to tell us about science, technology, engineering, and math education, anything we may have neglected to ask you about that you would like to share?

Appendix B: 11th Grade Students' Reasons of Switching Majors Between STEM and Non-STEM From 9th to 10th Grade: Interview Themes

Code	Definition	Quote
		Motivational/Behavior
Self-interest/ Self-expectation	The degree to which students expressed that they have an ongoing interest to majoring in STEM.	I have a lot of expectation for myself, I want to go to college, and get a good job. I have high expectations for myself. (N-10) Well I switched it because, I'm always interested in new things, I like to study new things and I started getting more into technology and engineering, and I was into computers a lot that's what got me into the technology and engineering. (N-10) [Although] all teachers are in favor of STEM majors, I don't think they affected my decision. I felt like if I liked something I wanted to do it.

Code	Definition	Quote
	Contextual/Environment	
Parents	The extent to which students indicated how their mothers and/or fathers affected their reasons of showing interest to STEM-related fields.	*They definitely have more expectations, as I get farther in high school, they want me to get better grade. They want to make sure I have straight A's in the core areas. They just want me to have straight A's in everything.* (N-9) *I feel like my parents always wanted me to be STEM related. ... I feel like they've always wanted me to do STEM.* (N-8) *Yes, uhm my dad is always talking about growing up strong, so when I started to look into the Navy I started looking at it.* (N-3) *I considered studying computer science because my mom was an accountant and she was good with numbers.* (N-6) *I knew I wanted to be a doctor because my mom went to nursing school but she didn't get to finish so I thought I can do that.* (N-4)
Courses/ Teachers	The extent to which students indicated how their teachers/courses affected their reasons of showing interest to STEM-related fields.	*I was taking PLTW engineering so I was focused on that class, and in 10th grade I dropped because I was not comfortable, with like I felt like I didn't get out a lot of it.* (N-8) *Probably my art teacher. My art teacher greatly improved my art skills, I am doing really good in my art class, I really like my teacher and I am very good.* *I took a concepts of engineering course, and I didn't like it all. I didn't feel like any good at it, or familiar with any of the concepts I really felt overwhelmed about it and I was doing really good in my art class.* (N-7)

Code	Definition	Quote
In-and-out of School STEM Activities	The degree to which students expressed how school-related and other extracurricular STEM activities affect their decisions of majoring in STEM.	*I think they [STEM activities that happen within school and out of school] affect it (STEM interest decision) pretty great, I think that is where I found out what I wanted to do engineering, and find better solutions and brainstorm better ways. Well this school has a lot of STEM science fairs, and STEM expos, and taking the PLTW, I think those classes and expo help find that out.* (N-8) *During 9th grade I wasn't sure but then I took a STEM program during the summer, and we went over there and it helped me realize I wanted to do engineering or computer science.* (P-3)

CHAPTER 16

Reaching Youth with Science: A Look at Some Data on When Science Interest Develops and How it Might Be Sustained

Robert H. Tai

Abstract

This chapter examines data from three of studies on the theme of science interest development and how science might be developed or sustained. The first analysis examines when science interest among professionals in science and engineering first develops. Using these results, the subsequent analysis focuses on youth in grades 3–8 by comparing parallel analyses on data sets collected in 2012 and 2017 that examine youth preferences for learning activities commonly applied in both formal and informal science curriculum. The results offer insights into how youth who report aspiring to science and engineering careers have learning preferences that differ from their peers who do not aspire to science and engineering career. The discussion offers some insight into how these different groups might be engaged so that science and engineering aspirations may be sustained in one group and engendered in another.

1   Introduction

In 2010, the President's Committee of Advisors on Science and Technology for President Barack Obama issued a report entitled "Prepare and Inspire" that described the importance of engaging youth in science and engineering learning (PCAST, 2010). While much work prior to this report focused on science interest engagement among youth in terms of scientific workforce and career development, since 2012 more attention has been paid to generating interest and excitement in science and engineering in general. Given the ever-increasing role that science and engineering play in our everyday lives, especially in the form of technology, public understanding has grown in importance. While interest in science and engineering may begin at any age, beginning with young people offers the opportunity to both engage future generations of scientists and engineers and engender science and engineering interest and curiosity among youth with nonscience-related aspirations.

This chapter presents results from data analysis from three different research studies carried out in 2008, 2012, and 2017. The connecting strand among these three studies is an aspiration to engage in science and engineering at an early age, i.e., prior to high school. The first of these studies explored when scientists and graduate students first recalled developing an interest in science and when they recalled developing an interest in their career disciplines among other things (*Project Crosover*). In this chapter, I examined the results from the responses to these two questions from this survey study and their implications. Next, I compare results from two projects carried out five years apart with some overlapping research survey questions on young people's preferences for different types of learning activities. Using a validated survey instrument originally developed in 2010 (Ryoo, Tai, & Skeeles-Worley, 2018), I discussed the findings from the survey data on youths' preferences across seven different types of learning activities collected in 2012 (*Spark 2 Flame Project*) and in 2017 (*Science Everywhere Project*) across grades 3–8 for both. The survey instrument queries youth on their preferences for engaging in seven different types of learning activities. The results from these youth-reported learning activity preferences are compared for youth with science and engineering aspirations to youth who do not have these aspirations.

Project Crossover was sponsored by the National Science Foundation and carried out in between 2004–2009. This study proposed to examine the transition of individuals as they progress from student to scientist and the educational experiences they report having. The study included interviews of graduate students, scientists, engineers, and those who chose to leave a career in science/engineering as well as a large-scale national survey that collected data from 3000 scientists and engineers, as well as 1000 graduate students in the disciplines of chemistry and physics. Initially, the interview segment of the project informed the development of 2 survey questionnaires, one for professional scientists and engineers and another for graduate students in the disciplines of chemistry and physics. The survey participants were randomly selected from among the membership of professional organizations in these disciplines. The respondents were contacted through letters and postcards over the course of 8 months and offered the option of completing the survey in a physical form or online. The response rate for both surveys were slightly above 30%, which is comparable to similar surveys carried out by professional science/engineering societies (e.g., 2008 Sigma Xi survey). Overall, the surveys collected responses from 3083 professional scientists and engineers and 1021 graduate students in the disciplines of chemistry and physics. The purpose of collecting data on the educational experiences of both professional scientists and engineers as well as graduate students is to provide very similar, but clearly non-overlapping data sets that offer the opportunity for confirmatory analyses.

Among the questions asked on this survey were Questions 18 and 19 shown in Figure 16.1. The results from the responses comparing scientists/engineers to graduate students are shown in Figures 16.2 and 16.3.

The match-up between the responses across the two data sets is clear and offers strong confirmatory evidence of these results. These results were originally reported to the Chemical Sciences Roundtable of the National Academies of Sciences (National Research Council, 2009, pp. 10–11). It appears that among those who choose careers in science and engineering, general interest in science begins before high school while disciplinary interest develops later. These results, when coupled with a longitudinal analysis spanning 12 years that found among individuals who professed an interest in a science-related career in 8th grade had 2–3 times greater odds of graduating with a degree in the sciences and engineering than individuals who did not (Tai, Liu, Maltese, & Fan,

18. When did you first become interested in science, in general?
        ○ K–5th Grade   ○ 9th – 10th Grade   ○ During first 2 years of undergraduate study
        ○ 6th–8th Grade   ○ 11th – 12th Grade   ○ After second year of undergraduate study

19. When did you first become interested in chemistry/physics?
        ○ K–5th Grade   ○ 9th–10th Grade   ○ During first 2 years of undergraduate study
        ○ 6th–8th Grade   ○ 11th–12th Grade   ○ After second year of undergraduate study

FIGURE 16.1   Project crossover questions 18 and 19

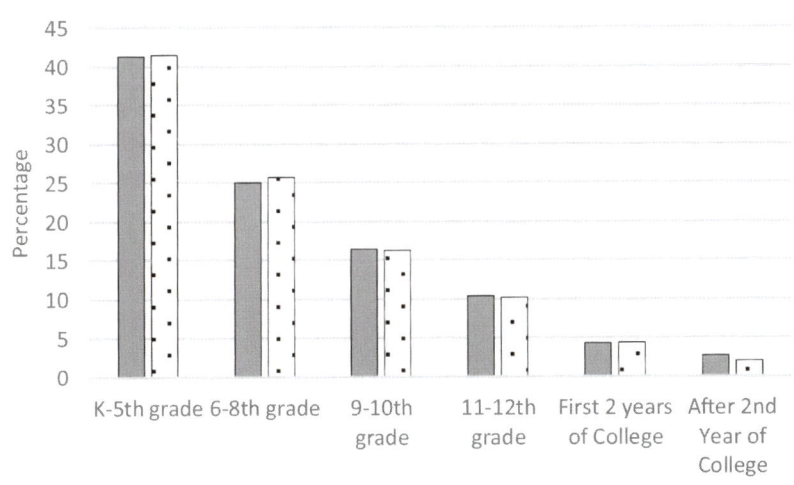

FIGURE 16.2   Question 18, When did you first become interested in science, in general?

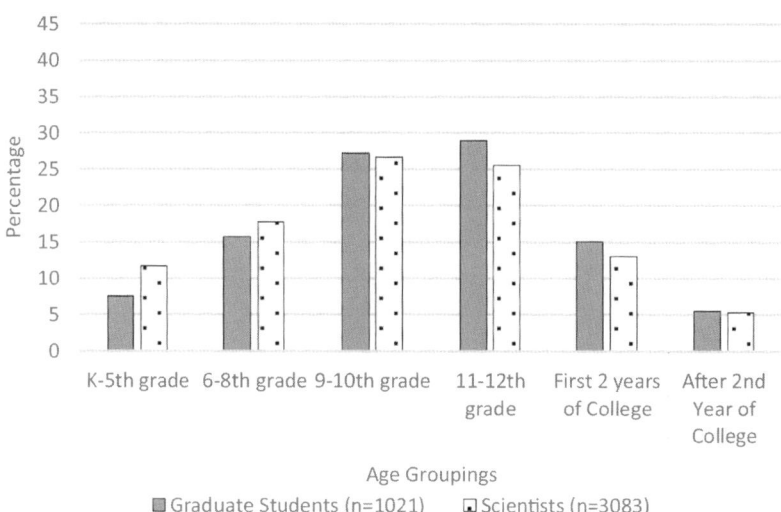

FIGURE 16.3   Question 19, When did you first become interested in chemistry/physics [your career disipline]?

2006), offer strong evidence that developing at least a general interest in science among youth prior to high school is important for potentially impacting adult career choice.

While a number of different questions may follow from these results, the question I chose to pursue in this discussion is related to the topic of science learning engagement among young people. Specifically,

1. What types of learning activities might youth who have science/engineering aspirations prefer?
2. Do their preferences differ from youth who do not have science/engineering aspirations?
3. How might this information be used by science educators?

2   Methods

For the purposes of this research I chose to focus on seven different types of learning activities that I identify as: (1) collaborating, (2) competing, (3) discovering, (4) creating/making, (5) performing, (6) caretaking, and (7) teaching/tutoring. These seven learning activity types are commonly practiced and often applied in various combinations by science curriculums. For examples, see Project WILD,[1] *Curriculum and Resources* offered by the Smithsonian Science Education Center,[2] and *Instruction Materials* from BSCS Science Learning.[3]

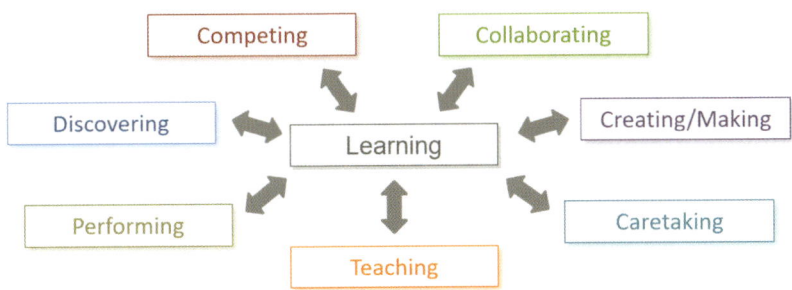

FIGURE 16.4   FOCIS learning activities

My research group and I developed a survey question to collect information on youth preferences for engaging in these seven types of learning activities that we have come to refer to as a framework for the observation and categorization of instruction strategies (FOCIS, pronounced fō-kiss). Our work examining the validation, reliability, and longitudinal invariance of this instrument as well as the entire survey maybe found in Ryoo, Tai, and Skeeles-Worley (2018). The FOCIS instrument has 28 items with 3–5 items used to form a composite preference score for each learning activity applying a Likert-type scale ranging from 1–5. A score of 5 indicates a strong positive preference while a score of 3 is neutral/no preference and 1 is a strong negative preference/aversion. The FOCIS instrument was included in two different research surveys carried out in 2012 (Spark 2 Flame Project) and 2017 (Science Everywhere Project).

The Spark 2 Flame Project carried out a survey study of school children in four different school districts in the United States. These four districts represented four different types of locales: urban, suburban, small town, and rural. All students from Grades 3–12 were included in the study which began by surveying Grades 3–11 in the Fall and Spring of Year 1 and followed up by surveying students in Grades 4–12 in the Fall and Spring in Year 2. The survey involved surveying all students within each school in these grade levels. This research design is referred to as an accelerated longitudinal study (Miyazaki & Raudenbush, 2000). This study ultimately collected data from 8933 individual students across Grades 3–12. Each student was given an ink pen for their participation. Given the study limitations, we were not able to follow students who left these school districts. However, students who transferred between schools within each district were tracked. For the purposes of this particular analysis, we will only be examining one wave of the survey, Fall Year 1.

The Science Everywhere Project was carried out in collaboration with the DonorsChoose.org organization. DonorsChoose.org is similar to GoFundMe.

We want to know how you feel about different activities. (Please check only 1 box for each activity listed below.)	I feel ... ☹ ☺ ☺
When I find out that an activity involves ...	
Being in a group	1 2 3 4 5
Being in a competition	1 2 3 4 5
Making or building things	1 2 3 4 5
Discovering and learning new things	1 2 3 4 5
Presenting in front of lots of people	1 2 3 4 5
Taking care of animals	1 2 3 4 5
Helping people learn things	1 2 3 4 5

We want to know what you think about each of the statements below. If you strongly agree, then choose 5. If you strongly disagree, then choose 1. (Please circle only 1 number for each statement below)	☹ ☺
Working with others is more fun than working alone	1 2 3 4 5
I like being part of a team	1 2 3 4 5
I learn better when I am working with others	1 2 3 4 5
I get excited when I hear there will be a competition	1 2 3 4 5
I enjoy competing against other people	1 2 3 4 5
I like to focus on my own goals, rather than competing with others	1 2 3 4 5
I like figuring out how things work	1 2 3 4 5
I like taking things apart to see what is inside	1 2 3 4 5
I like trying different ways to figure things out	1 2 3 4 5

	☹ ☺
I like solving problems	1 2 3 4 5
Helping others to learn things is fun	1 2 3 4 5
I like teaching things to others	1 2 3 4 5
Having a pet is big responsibility, but something I like to do	1 2 3 4 5
I like to take care of things like plants and aquariums	1 2 3 4 5
I feel good when people depend on me	1 2 3 4 5
Performing in front of people is fun	1 2 3 4 5
I like telling people about my work	1 2 3 4 5
I like presenting my work to my class	1 2 3 4 5
I like doing projects where I make things	1 2 3 4 5
Whenever I can, I make the things I need	1 2 3 4 5
I like building things	1 2 3 4 5

FIGURE 16.5 FOCIS instrument items

Com in concept with a focus on school teachers. Teachers who wish to purchase school supplies for class activities or projects may set-up a donors page. While parents and local supports contribute funding for activities, DonorsChoose.org also seeks out national level corporate sponsorship for these class activities. Our project was supported by the Overdeck Family Foundation and the Simons Foundation, which gave over $500,000 to fund over 600 teachers across the country. All participating teachers were asked to voluntarily participate in surveying their students with the FOCIS instrument before and again after their class projects. We were able to collect data from over 60% of the classes participating in this project. We provided two reports to each participating teacher on their students learning activity preferences, a pre-activity report and a post-activity report comparing pre-post student preferences. A total of 5899 students participated in the FOCIS research survey across Grades PK-12. For this analysis, we will only examine data from the pre-activity survey.

Though the *Spark 2 Flame* Project and the *Science Everywhere* Project were carried out five years apart, the survey instruments used in each study contained the same 28 FOCIS items and the question asking students to report on their future career aspirations. The career aspirations questions offered 12 different career categories and an option for students to write-in a career option category that was not listed. One of the career option categories, was "science and engineering" which offered job options such as "scientists, engineers, biomedical researchers, computer programmers." The youth were encouraged to select as many categories that applied to them.

For this analysis, we created an identifier variable where youth selecting the "science and engineering" category was assigned the value of "1." Youth not

What kind of job do you expect to be doing when you grow up? *(Please check the ONE job category that comes closest to what you EXPECT to be.)*
☐ Agriculture & Natural Resources (like park rangers, farmers, gardeners)
☐ Architecture & Construction (like builders, planners, architects)
☐ Arts, Communications, Journalism, & Tourism (like artists, designers, news reporters, travel agents)
☐ Business and Finance (like accountants, bankers, tellers, managers, insurance agents)
☐ Education & Counseling (like teachers, librarians, psychologists, social workers)
☐ Entertainment and Sports (like musicians, actors/actresses, athletes, coaches, announcers)
☐ Medicine (like nurses, doctors, physical therapists, dentists)
☐ Government, Law, Security, & Military (like lawyers, police, mail carriers, soldiers, sailors)
☐ Science & Engineering (like scientists, engineers, biomedical researchers, computer programmer)
☐ Transportation (like pilots, truckers, mechanics)
☐ Veterinary Care (like veterinarians, animal care)
☐ I Don't Know Right Now
☐ If the job you have in mind is not on this list, please write your answer on the line below:

FIGURE 16.6   Future career aspiration question

selecting this category were assigned the value of "0." The "written" option was individually reviewed and assigned (0, 1) values based on their direct connection to science and engineering.

Given the difference in grade/age ranges in the data collected between the two research projects and in light of the importance of science and engineering engagement prior to high school, we focus our analysis on youth in Grades 3–8.

3   Findings

We begin our analysis with some descriptive statistics of the two narrowed data sets. Table 16.1 shows that the distribution of males versus females is roughly even for both data sets. There appears to be a significantly greater number of youth with science and engineering aspirations in the 2017 *Science Everywhere* data compared with the 2012 *Spark 2 Flame* data. Proportionately, the percentage of youth with science and engineering aspirations is more than double. Given that the schools in which these youth were attending were not selected based on any science and engineering curricular focus with respect to either study, this simple comparison suggests that science and engineering future career aspirations among youth is larger in 2017 compared with 2012. With the significant investment in science, technology, engineering, and mathematics (STEM) programs in both formal and informal settings within the past decade, it appears these programs may be having the intended positive impact.

Next, the mean composite scores are compared across the seven different types of learning activities for the two data sets in Grades 3 – 8 between science and engineering aspiring (SEA) youth versus non-science and engineering aspiring (non-SEA) youth. Table 16.2 displays this information including the t-test results significant at the $\alpha = 0.05$ level or better.

TABLE 16.1   Comparison of descriptive statistics among youth participants in Grades 3–8

	Spark 2 Flame 2012		Science Everywhere 2017	
Males	2641	51.8%	1895	49.2%
Females	2462	48.2%	1958	50.8%
Science & Eng. Aspirations (SEA)	332	6.5%	527	13.7%
Non-SEA	4771	93.5%	3326	86.3%
Total	5103		3853	

TABLE 16.2 Learning activity preference comparisons between Science & Engineering Aspiring Youth (SEA) versus non-Science & Engineering Aspiring (non-SEA) Youth

Learning Activity	Comparison Group	Spark 2 flame 2012				Science everywhere project 2017			
		N	Mean (s.d.)	t-test SEA vs. non-SEA F	Sig.	N	Mean (s.d.)	t-test SEA vs. non-SEA F	Sig.
Collaborating	SEA	332	3.72 (1.07)	5.28	0.022	530	3.83 (0.93)	9.25	0.002
	non-SEA	4745	3.98 (0.99)			3335	4.10 (0.84)		
Competing	SEA	332	3.31 (1.11)	6.32	0.012	530	3.35 (0.97)		
	non-SEA	4738	3.19 (1.18)			3335	3.35 (0.96)		
Creating/Making	SEA	332	4.36 (0.76)	37.36	0.000	530	4.33 (0.75)	9.06	0.003
	non-SEA	4751	4.04 (0.96)			3334	4.20 (0.83)		
Discovering	SEA	332	4.23 (0.70)	33.23	0.000	530	4.19 (0.67)	8.87	0.003
	non-SEA	4734	3.84 (0.87)			3335	4.06 (0.75)		
Performing	SEA	332	2.97 (1.25)			530	3.01 (1.13)		
	non-SEA	4743	2.91 (1.24)			3335	3.06 (1.13)		
Caretaking	SEA	332	4.02 (0.95)			530	4.03 (0.83)	7.32	0.007
	non-SEA	4745	4.17 (0.93)			3335	4.19 (0.77)		
Teaching	SEA	332	3.53 (1.00)			530	3.67 (1.05)		
	non-SEA	4739	3.72 (1.01)			3335	3.89 (1.02)		

These results show that statistically significant differences exist across both data sets for three types of learning activities when comparing SEA youth and non-SEA youth: Collaborating, Discovering, and Creating/Making. The differences also appear to be consistent in this comparison. With respect to Collaborating, non-SEA youth report higher preference scores compared with SEA youth. This result suggests that SEA youth are somewhat less positively predisposed to work with others. For both Discovering and Creating/Making, SEA youth show significantly higher positive preferences. Discovering activities commonly involve problem-solving and learning new things, while Creating/Making activities commonly include making or building things or deconstructing objects. Both of these activities rely heavily on imagination and creativity. One important non-consistently significant result is youth preferences for Competing. It is important to note that for youth in general, Competing preference scores appear to hover near 3. This result suggests that while half of youth may have a positive predisposition to competition-based activities, half are negatively predisposed. Given that science and engineering-based competitions appear to be widely proliferated and highly publicized, these types of activities do not appear to be well-suited to for engaging significant numbers of both SEA and non-SEA youth. The least preferred among the seven types of learning activities appears to be Performing, an activity that commonly involves making presentations or speaking in front of groups.

For the two remaining learning activity types, Caretaking and Teaching both appear to have more positive preferences among non-SEA youth and SEA youth. Caretaking activities involve stewardship and require youth to take responsibility for things that may include animals, plants, and places. Teaching activities typically involve helping others learn or sharing ones knowledge about things. Many science and engineering-related careers directly draw in these types of activities. As a result, science educators hoping to reach out and engage non-SEA youth in science and engineering may wish to develop curricular activities that put these types of learning activities to greater use.

4 Limitations and Future Research Directions

The current analysis offers some insight into the youth career aspirations and their learning preferences, lacking in this study is a longitudinal examination of how the aspirations and preferences of youth may be impacted over time. What shifts might occur in their thinking about their future? What activities my impact their educational and career trajectories as they mature into adulthood? While this analysis does offer some insight into potential approaches to impacting non-SEA youth engagement, the degree of impact is unexamined. Clearly further study is necessary.

5 Conclusions

A general approach common among many science educators is to focus on science content and target the imagination and fascination of youth. Explosions, fires, fossils, cognitively dissonant demonstrations. This approach has been very effective for gaining the initial engagement of youth, but long-term engage in science learning requires effort on the part of each youth. After a while, the explosions and fires of riveting demonstrations fade, and more thoughtful and individualized efforts take over. As a result, it is important for science educators to develop a deep understanding of youth in their care. This chapter offers some potentially useful insight into the various types of learning activity preferences that different groups of children may possess. From the findings discussed here, we can surmise that Discovering and Creating/Making activities might be popular with SEA youth. As a result, if an educator chooses to engage non-SEA youth with science and engineering activities, it is important to understand that some degree of scaffolding must be included. While youth who have a strong positive predisposition toward a particular activity may require little to no supervision, youth who do not have a preference for an activity would likely need some guidance and encouragement. The same is true for Collaborating activities. In this case, SEA youth would likely require guidance and encouragement over their non-SEA peers. One should bear in mind that all of these categories of learning activities offer real and important benefits to youth who come to understand and engage in them. As a result, among youth reporting that they do not prefer to engage in a learning activity implies that it is incumbent for the educator to find means the engaging youth in these activities, shifting attitudes and preferences to more positive dispositions as a matter of practice is both important and necessary.

Notes

1   Association of Fish and Wildlife Agencies, https://www.fishwildlife.org/projectwild
2   https://ssec.si.edu/explore-our-curriculum-resources
3   https://bscs.org/site-categories/products/instructional-materials

References

Miyazaki, Y., & Raudenbush, S. W. (2000). Tests for linkage of multiple cohorts in an accelerated longitudinal design. *Psychological Methods, 5,* 44–63. doi:10.1037/1082-989X.5.1.44

National Research Council. (2009). *Strengthening high school chemistry education through teacher outreach programs.* Washington, DC: National Academies of Sciences. doi:10.17226/12533

President's Council of Advisors on Science and Technology. (2010). *Prepare and inspire: K-12 education in Science, Technology, Engineering, and Mathematics (STEM) for America's Future* (Document No. 117803). Retrieved from https://www.nsf.gov/attachments/117803/public/2a--Prepare_and_Inspire--PCAST.pdf

Ryoo, J. H., Tai, R. H., Skeeles-Worley, A. D. (2018). Examination of longitudinal invariance on a framework for observing and categorizing instructional strategies. *Research in Science Education* (Advanced online publication). doi:10.1007/s11165-018-9698-7

Tai, R. H., Liu, C. Q., Maltese, A. V., & Fan, X. (2006). Planning early for careers in science. *Science, 312,* 1143–1144. doi:10.1126/science.1128690

PART 5

Community Partnerships and Innovation to Improve K-12 Students' 21st Century Skills

CHAPTER 17

Crossing Borders and Stretching Boundaries: A Look at Community-Education Partnerships and Their Impact on K-12 STEM Education

Brett Criswell, Theodore Hodgson, Carol Hanley and Kimberly Yates

Abstract

The complexity of designing and implementing STEM educational experiences that meet the needs of a diverse population of students while also preparing them for careers in the 21st-century cannot be understated. Partnerships between the various stakeholders invested in providing such experiences may be the only way to respond to that complexity. In this chapter, we review a sample of the spectrum of business-education/community-education (B-E/C-E) partnerships to the nature, structure, outcomes, and impact of those partnerships. We also describe two such partnerships that were enacted in Kentucky and were centered around teacher externships. In order to draw the most powerful insights from the literature and the discussion of the two Kentucky-based projects, we employ a conceptual framework that includes the concept of a STEM ecosystem and the construct of communities of practice (COP). Related to the construct of COPs, we consider the nature of the border crossings that were embedded in the partnerships, the way brokers were involved in building bridges between stakeholders, and the manner in which boundary objects were utilized to connect practices across different stakeholder groups. This allows us to describe what seems to have been effective in allowing these B-E/C-E partnerships to accomplish their goals, as well as to understand the challenges that these partner- ships must overcome. The conceptual framework also allows us to point clearly to future directions with regards to both the practice and research that needs to be done around B-E/C-E partnerships.

1 Previous Literature

The following statement appears in the federal Science, Technology, Engineering and Mathematics (STEM) Education 5-Year Strategic Plan (Holdren, Marrett, & Suresh, 2013): "Maintaining America's historical preeminence in the STEM fields will require a concerted and inclusive effort to ensure that the STEM workforce is equipped with the skills and training needed to excel in these fields." Given that this report was released over five years ago, it is critical to consider the state of STEM education and whether progress has been made in meeting its call. In particular, it seems imperative that those invested in STEM education consider the extent to which "a concerted and inclusive effort" has occurred in this country to improve the quality of our students' STEM education experiences. Related to this concern, we will focus this chapter on examining the nature of business-education and community-education (B-E/C-E) partnerships, and what has and could be happening with those partnerships to strengthen STEM education in the U.S. This will include looking at two B-E/C-E partnership projects implemented in Kentucky, both organized around teacher externships. The questions we will ask are (1) What were the nature of those partnerships?, (2) What were the goals/expected outcomes?, (3) How were they implemented?, (4) What have they actually accomplished?, and (5) What can we learn from these partnerships that could inform related endeavors in the future?

1.1   Conceptual Framework

To provide a common language and a clear lens to examine a diverse set of B-E/C-E partnerships, we will present a two-component conceptual framework. The first component is the notion of a *STEM ecosystem*,[1] which describes the entities, interactions and relationships that need to be considered when analyzing such partnerships. In their paper on cross-sector collaborations, Traphagen and Traill (2014) explain that STEM ecosystems involve (1) individual entities (they identify schools and communities) that (2) become interconnected in order to (3) create synergy and provide opportunities that would not otherwise be available ("harnesses the unique contributions of all these different settings in symbiosis") so as to (4) provide the best STEM learning experiences possible for all children (p. 2). The focus of these authors was on STEM ecosystems that linked the formal and informal STEM education worlds, and so their description noticeably omits businesses. However, the four core features identified above can be applied to broader ecosystems than they describe. In our review of partnerships examined in the literature and in our discussion of our own partnership projects, we will consider these core features and how the partnership has (or has not) functioned to maximize them.

Traphagen and Traill provide recommendations for different areas of work around STEM ecosystems. Among their recommendations for *Practice*, they indicate the need to "create a *community of practice* for STEM learning ecosystems"

(p. 6, italics added). The second component of our conceptual framework is the construct of a *community of practice* (Wenger, 1998; Wenger, McDermott & Snyder, 2002). Wenger (1998) conceptualized learning in four ways: *learning as* (1) *becoming* (through identity), (2) *experience* (through meaning), (3) *doing* (through practice), and (4) *belonging* (through community). A community of practice, then, provides pathways through those different processes of learning by (1) developing certain identities, (2) establishing certain meanings, (3) supporting certain practices, and (4) creating a certain kind of community. Since the focus of STEM education is on learning that enables one to be ready for careers, colleges, and citizenship, we can understand how B-E/C-E partnerships are promoting STEM education in terms of developing identities, establishing meanings, supporting practices, and creating communities.

Crucial to the content of this chapter is the fact that B-E/C-E partnerships represent, in theory, *an intersection* of *different communities of practice*. Aikenhead (1996) realized that part of the challenge of successful learning in school science is a function of negotiating *border crossings*: "I shall argue that science educators ... need to recognize the inherent border crossings between students' lifeworld subcultures and the subculture of science, and that we need to develop curriculum and instruction with these border crossings explicitly in mind ..." (p. 2). In a similar vein, collaborations among business, community, and education partners will involve border crossings between different communities of practice represented by those stakeholders. Success of those collaborations, then, will depend on the extent to which those border crossings are made visible and effectively navigated.

Wenger offered his notion of *boundary objects* as one means of navigating the boundaries/borders between communities of practice. Citing sociologist Leigh Star, Wenger defines *boundary objects* as things that "serve to coordinate the perspective of various constituencies for some purpose" (p. 106). He also defined *brokering* as the process of providing connections within a community of practice (between different practices) or between different communities of practices (to link their practices). Wenger proceeded to note how challenging this work may be: "It involves the process of translation, coordination, and alignment between perspectives. It requires enough legitimacy to influence the development of a practice, mobilize attention, and address conflicting interests" (p. 109). Thus, as we review the nature, goals, implementation, and outcomes of various B-E/C-E partnerships, we will consider the way in which the challenges created by the intersection of different communities of practice were acknowledged, addressed, and overcome (or not) through boundary objects and brokers.

1.2 Overview of B-E/C-E Partnerships: Nature, Purposes and Outcomes

Within this review, we will sample from the wide-ranging set of B-E/C-E partnerships that have been developed to show a spectrum of types of partnerships. We will be focusing extensively on those that involve teacher in-/externships, since those are the centerpiece of our own projects. For each representative in this sample, we will discuss the nature, structure and outcomes of the partnership, as well as describe the partnership through the lenses of STEM ecosystems and communities of practice.

1.2.1 Kantrov's Teacher Externship Typology

Kantrov (2014) stated that a common form of B-E/C-E partnerships is employer externships, in which teachers directly experience industry trends and skill requirements related to their area of expertise to bring relevance to student learning. Kantrov contended that teachers who participate in such externships develop strong relationships with employers that help them provide career opportunities for students and aid them in designing high-quality classroom experiences. Furthermore, increasing teachers' externship experiences could ensure that teachers have "up-to-date knowledge of industry needs and the ability to translate that understanding into their instruction to best prepare their students for college and career success" (Kantrov, 2014, p. 4). According to Kantrov, teacher externships can take many forms; however, they share four key characteristics: (1) engage teachers in learning regarding the nature of the workplace; (2) familiarize teachers with academic, technical, and 21st century knowledge, skills, and dispositions required for career success; (3) inform teachers about current and emerging career opportunities; and (4) expand teachers' knowledge of the education and training requirements required for the various positions available.

Kantrov (2014) proposed a typology of externships: (1) Extended individual externships with limited professional development; (2) Extended individual externships with intensive professional development; (3) Short-term team externship with limited professional development; and (4) Short-term team externship with intensive professional development. Because of the limited data regarding the impact of teacher externship experiences, questions remain regarding what configurations have the greatest impact on STEM education. Data from one program showed that even with substantial support during individual externships, in the absence of ongoing support, "teachers' intent to change their practice immediately following their externship experience tended to exceed their capacity to implement those changes" (Kantrov, 2014, p. 18). Data from another program, in which teachers received intensive professional development support over an entire school year, indicated that teachers struggled to sustain new practices when faced with resistance from colleagues. Consequently, Kantrov (2014) suggests that a team approach, combined with sustained district and business partner support, may be preferable to an individual externship approach.

1.2.2   The Math English Science Technology Education Project (MESTEP)
The first project to be reviewed began 30 years ago. The Math English Science Technology Education Project (MESTEP) was chosen because the partnership involved a teacher preparation program (at the University of Massachusetts at Amherst) and businesses in the Boston area. Clark (1990) noted that MESTEP involved an associated teacher preparation program, of which he was the director, that won the Distinguished Achievement Award in Teacher Education in 1986 (p. 71).

The driving force behind MESTEP was the concern that corporations in the Boston area had for filling STEM teaching positions in area schools with talented candidates – a concern prompted by the need for schools to prepare workers for the "technology-dominated economy" that existed in this area (p. 71). Thus, MESTEP was a *recruiting tool* for the teacher preparation program, as well as a *training tool* for the corporations – in a sense, serving as a boundary object between the university and business communities of practice. Candidates in the MESTEP program worked "for one semester as a full-time, paid teaching intern at a secondary school and for another semester as a full-time, paid corporate intern in a local industry" (p. 72). The MESTEP program was able to function as a boundary object because corporate representatives were involved in the admission process (p. 73) and because "[W]hen the internship ended, an advocacy relationship continued between one or more people in the organization and the intern" (p. 74). Further, the teacher preparation program director (Clark) provided on-site briefings at the corporate placement and involved corporate supervisors in selection and placement of the teacher interns, as well as the design of the internship experience. Clark suggested that MESTEP had an impact on the retention rate of teachers coming through the program, as they often continued their work at the corporations following program completion (p. 75). He also indicated that the program helped teachers better understand how to engage their students in the kind of "tool-based thinking" that businesses desire in their employees (p. 75).

A more recent iteration of MESTEP is the Educators in Industry: K-12 Externship Program, a project that placed in-service teachers at various grade levels in a four-week summer externship with a company that focuses on engineering, continuous improvement, or manufacturing (Bowen & Shume, 2018). Unlike MESTEP, Educators in Industry was not a teacher preparation program, but worked with practicing teachers at a variety of grade levels and in a variety of subjects. The boundaries crossed during this externship, therefore, were the K-12 classroom (as implemented by the teacher-participant) and the problem-based industrial and work environments. Educators in Industry is particularly noteworthy because of the nature of the externship work (teachers were engaged in authentic, problem-driven situations), the funding model (funded primarily by the host businesses), and efforts to assess the impact of the four-week experience on teachers' practice. According to the authors, teachers that completed the externship reported a much greater appreciation for process objectives – e.g. softs skills such as communication – and commitment for classroom contexts that allowed students to develop these processes (Bowen & Shume, 2018). The authors did not assess the impact of project activities on classroom practices, but noted the need for such a focus in future research.

1.2.3   Office of STEM Education Partnerships (OSEP)

Jona (2014) described the Office of STEM Education Partnerships at Northwestern University[2] as "a bridge between these diverse stakeholders to help leverage, maximize, and effectively utilize resources and expertise, build capacity, and catalyze cooperation" (p. 2). Thus, OSEP serves a brokering role in a broad STEM ecosystem that involves the university, university research facilities, schools throughout Illinois, and industry collaborators.

Included in the goals that Jona identified for OSEP were (1) increase student and teacher access to STEM research, (2) illustrate the diverse STEM careers available, and (3) increase equity in STEM education. These goals draw attention to the needs and potential contributions of the variety of stakeholders existing in this ecosystem. The breadth and complexity of the activities undertaken by OSEP were demonstrated by the three different models described in Jona's report: Northwestern University Biology Investigations in Oncofertility (NUBIO[3]); the Research and Development Learning Exchange (RDLE) and the Mentor Matching Engine[4]; and FUSE.[5] A critical aspect of the approach employed by OSEP is the identification of a boundary object that can be a focal point for the differing perspectives of the stakeholders that it brings together. RDLE, which provides research opportunities for teachers, demonstrates this: "Through these research programs [supported by RDLE], they [in-service teachers] improve their understanding of the nature of science and develop important skills, such as critical thinking, problem solving, communication, collaboration, and creativity" (Jona, 2014, p. 9). Like Educators in Industry, the intersection of science and engineering practices with 21st-century skills becomes the boundary object that various stakeholders can act upon and improve. As Jona later notes, parallel research programs for students "directly align with and support the new Next Generation Science Standards (NGSS) and its focus on practices" (p. 10).

OSEP draws attention to the value of a dedicated entity serving a brokering role between the varied communities of practice that constitute more complex STEM ecosystems. Of course, establishing and operating such a dedicated entity requires a tremendous commitment of resources, one that will require continuous and dependable contributions from various stakeholders. Jona, though, explains why this is worth the investment: "As a bridge builder, convener, translator, and catalyst for and between diverse stakeholders, we are able to effectively align efforts and interests, maximize resources and expertise, leverage existing assets and mobilize additional support, and utilize and further grow existing networks and partnerships" (p. 13).

2   Examination of Two B-E/C-E Partnership Projects

In this section, we will provide an overview of two STEM partnership projects, both involving teacher externships and both involving business-school-university collaborations. For each project, we provide a description of the goals/objectives and the structure, followed by a discussion of the outcomes and insights for successfully implementing such collaborations.

2.1   *STEM Pride: Partnering with Research & Industry to Develop (STEM) Educators*

2.1.1   Goals/Objectives and Structure

STEM Pride was a two-year project funded by an Improving Educator Quality grant from the Kentucky Council of Post-secondary Education (CPE). The project involved partnerships between the University of Kentucky (UK) and schools and businesses in the Bluegrass region of Kentucky. STEM Pride had the following objectives: (1) Develop pre- and in-service teachers' understanding and application of STEM practices,[6] (2) Increase pre-service and in-service teachers' integration of STEM practices into the development of learning experiences, (3) Develop and implement a graduate course for pre-service teachers that will help them understand the needs of the 21st-century STEM workforce and count toward their degree program, (4) Increase students' understanding and application of STEM practices, and (5) Increase students' awareness, understanding and interest in STEM careers.

Several project activities were enacted to achieve these objectives. Participating pre- and in-service teachers were brought together for a one-day orientation to the project in spring of the first year. The teachers were going to be engaging in a 40-hour summer externship at a UK research facility during the first summer, so the expectations for and structure of that externship were overviewed. The teachers then engaged in the externship based on a schedule negotiated by the teachers and the university research mentors. Most teachers completed the experience over a single week, although some spread their time over several weeks. The teachers were brought back together for two days at the end of the summer to debrief about the experiences and work on translating what they had learned into new pedagogy and curriculum. The teachers were convened again twice during the academic year for follow-up professional development (PD). Additionally, student field trips to the University of Kentucky – both to visit the research lab in which their teacher was placed and to visit the campus – occurred throughout the school year. In the second year of the project, the schedule just described was repeated, except this time the pre- and in-service teachers were placed in industry externships.

One challenge of STEM Pride was developing a shared understanding of the project with all the stakeholders. Teachers were recruited through communication with school administrators, who nominated candidates. Further, teachers from both the more traditional public high schools and the career technical education (CTE) schools in the same county were recruited. While these participants were all teachers, the different contexts of their teaching – traditional and CTE schools – meant that they really represented different communities of practice. For instance, while the traditional public school teachers were generally familiar with the science and engineering practices (SEPs), the CTE teachers had largely not been exposed to these prior to involvement in the program. A third group of teachers – pre-service teachers (PSTs) enrolled in a summer graduate course – also participated. Part of the goal of involving the PSTs was to begin to inculcate them into the science teaching community of practice; thus, they were partnered with experienced teachers in their research experiences.

There were two other significant groups of stakeholders: the university scientists and industry researchers who mentored the teachers in their summer experiences. Prior to summer one, two members of the project leadership team (PLT) visited all of the university scientists to discuss the project with them; the same discussions took place the following spring with the industry researchers. Part of the conversation focused on the responsibilities of the university/industry mentors which included hosting the participating teachers in their facilities and also hosting students during campus visits. Additionally, the university/industry mentors were encouraged to attend at least one of the PD days to see the work being done around translating experiences into classroom content.

The conversation with the university/industry mentors also focused on the science and engineering practices (SEPs), which were positioned as the boundary object of this project. Four practices were emphasized: (1) asking questions/defining problems, (2) developing and using models, (3) using mathematics and computational thinking, and (4) arguing from evidence (National Research Council, 2013). For the university scientists, the PLT discussed how those practices connected with the research process in which they engage and how they could make that process transparent to the teachers. For the industry researchers, the PLT explored similar connections, but also talked about the intersections between the SEPs and 21st-century skills (National Research Council, 2010). University and industry mentors were expected to explicitly talk about how the SEPs were embedded in their work; teachers were expected to look for the use of the SEPs in the work that they were observing/doing.[7]

At the opening day of STEM Pride, the PLT and participating teachers were joined by administrators from participating schools, as well as university and industry mentors. With all of the stakeholders in place, the PLT tried to frame the importance of the different border crossings that would be occurring. A quote from Steven Johnson's *Where Good Ideas Come From* (2011) was presented: "All of us live inside our own private versions of the adjacent possible ... we are surrounded by potential new configurations, new ways of breaking out of our standard routines ... Unlocking a new door can lead to a world-changing scientific breakthrough, but it can also lead to a more effective strategy for teaching second-graders" (p. 40).[8] The idea of STEM Pride was to create myriad adjacent possibles for the different stakeholders represented at this opening day and, through these, transform the way science was being experienced in schools and the way that businesses and communities interacted with these schools. In the next sub-section, we will consider the extent to which this outcome was achieved.

2.1.2   Outcomes and Insights

Two of the objectives of STEM Pride related to having the pre- and in-service teachers better understand and be able to use the SEPs in their teaching. As noted above, the SEPs served as a boundary object to connect the different communities of practice represented in this project. The SEP that figured most prominently in the experiences of the teachers was *developing and using models*. Teachers who worked with Dr. Nancy Webb at the UK Medical Center were made aware of the central role that *model organisms* play in cancer research (Sharpless & DePinho, 2006). Dr. Webb noted that being able to determine the extent to which research findings on model organisms[9] extrapolate to humans is central to her work; thus she exposed teachers to a critical aspect of *translational science* (National Center for Advancing Translational Sciences, 2017). Related to that, Dr. Nancy Bosserman, who works at a research facility for Evolva,[10] discussed the use of *model molecules* in her research. More powerfully, Dr. Bosserman highlighted one of the main challenges of her research: "The scientists here struggle with perfecting a model or process at one scale and not being able to have it work at a different scale. Every time we increase the scale, more variables are introduced into the system." Through this statement, Dr. Bosserman was connecting the SEP of developing and using models to the crosscutting concept (CCC) of scale, proportion and quantity. One of the participating teachers was motivated to incorporate some element of developing and using models in every unit in her introductory biology course.

The learning that occurred around the boundary object of the SEPs was multi-directional. The PLT visited each of the research/externship sites both summers of the project. During a visit to the Center for Applied Energy Research (CAER[11]) at UK in summer 1, Dr. Jack Groppo, lead mentor for the teachers, commented to the visiting PLT member that, "I'm kind of stunned because the group meeting seemed like the highlight of the week [for the teachers]." Dr. Groppo attended one of the PD days that followed the research experience, and the STEM Pride group was focused on how supporting students in arguing from evidence: the features of scientific argumentation (Erduran, Simon, & Osborne, 2004), what those features might look within a science classroom discussion, and how teachers might scaffold those features. Following the session, Dr. Groppo approached the lead presenter and stated, "I didn't fully realize why our group meetings worked so well with the structure that we use, but I understand now." The session helped him to see how the structure of CAER's group meetings facilitated productive argumentation; he also understood why our teachers found the weekly group meeting so valuable.

Objective 4 of the STEM Pride program was to increase students' awareness, understanding and interest in STEM careers. Students of the participating teachers were given a pre-survey containing items related to this objective at the beginning of the year and a post-survey at the end of the year; the survey was a modified version of the STEM Career Interest Survey (Kier, Blanchard, Osborne, & Albert, 2014). A paired t-test was conducted in SPSS and a statistically significant difference at $\alpha = .05$ was found pre/post on a set of items including (1) I would feel comfortable talking to people who work in mathematics careers, (2) I am able to do well in activities that involve technology, and (3) I have a role model in an engineering career. Critically, there was no statistical significance to items related to students changing their college major to one related to science or math. It was beyond the scope of the project to explore why students' *attitudes* towards STEM showed statistical gains, while their *interest* in STEM careers did not exhibit such changes; this would be important for future research to consider.

2.2 NKY-FAME: Experiences in Career & Technical Preparation for Teachers & Students

2.2.1 Goals/Objectives and Structure

As with STEM Pride, NKY-FAME was funded by the Kentucky Council of Postsecondary Education as an Improving Educator Quality project. This was a one-year project that began in the summer of 2017 and continued through the 2017–2018 school year. The project emerged from an existing externship program piloted by the Kentucky Federation for Advanced Manufacturing Education (KY FAME). KY FAME seeks to enhance the availability of a skilled manufacturing workforce through apprentice-style educational programs for secondary students. Concerned about teachers' and students' awareness of apprentice programs and wanting to enhance teachers' perceptions of contemporary manufacturing settings, KY FAME piloted one-week teacher externships in the summer of 2016 in four northern Kentucky businesses. The externship was four days with the manufacturer and a one day workshop to explore ways that the externship could change classroom practice.

Post-externship feedback on the pilot program indicated that, although the externship acquainted teachers with the KY FAME program and provided teachers with a new perception of advanced manufacturing, it had little impact on classroom practice. As one participant observed, "I ended the externship with wonderful intentions to incorporate the ideas into my classroom, and then the school year started." NKY-FAME recognized that more debriefing time was needed to better assist teachers in the development of materials and incorporation of manufacturing principles into the classroom. The NKY-FAME project was founded on the following objectives: (1) Acquaint NKY-FAME teachers with the mathematics and science content, and critical thinking processes that are embedded in modern manufacturing; (2) Assist NKY-FAME teachers with the development and implementation of classroom lessons that are aligned with the content, problem solving, and modeling recommendations of standards (CCMS/NGSS); (3) Evaluate the effectiveness of the standards-based application lessons in the NKY sites, with a focus on students' mastery of the content targets, use of critical-thinking processes, and overall awareness of the applications of mathematics and science; (4) Intentionally equip all participants to provide professional development in their own school and district.

To meet these objectives NKY-FAME provided a revised version of the KY FAME externship program in the summer of 2017. The NKY-FAME project consisted of the same format for the externship experience, four days with the manufacturer followed by a one-day workshop to debrief. The most notable difference in the NKY-FAME project was the addition of supports for teachers to develop and implement lessons related to their externship experience, as well as follow up from the project leadership team (PLT) during the academic year. Externships began in early June and ran from Monday to Friday. Teachers were asked to return to NKU for a 2-day workshop in late June. During this time teacher participants were presented with sample lesson ideas from the PLT that incorporated skills related to modern manufacturing, relevant content areas, and aligned to math and/or science standards. Teachers were given time to brainstorm ideas for their own lessons and were instructed to develop ideas for 3 manufacturing related lessons that could be implemented in their classrooms in the upcoming school year. A standardized lesson plan template was provided for teachers with the intention of creating an online repository of manufacturing lessons that would be made available on a teacher resource website. Teachers were instructed to continue to develop and finalize lessons for feedback and presentations during a second 2-day planning workshop in late July. This provided teachers with structured time to brainstorm ideas and receive feedback as well as time on their own to reflect, modify and formalize their lessons. Upon returning to the July workshop teachers received feedback on their lesson plans, finalized and submitted the lesson plans to the PLT for review. Each teacher was also given an opportunity to present one lesson to the group. By presenting their lesson, the teachers were able to pilot a lesson prior to presenting the lesson to students. For further accountability and to ensure follow through, all participating teachers were observed by a member of the PLT during the academic year. A standardized observation rubric was created to evaluate each lesson. The observation form included items such as, included CCMS standards, included NGSS standards, included soft skills, etc. The project would conclude with a mini conference for teacher participants to share their experience with their colleagues, area teachers and administrators, and the manufacturer partners.

2.2.2 Outcomes and Insights

NKY-FAME began in summer 2017, with lesson implementation during the 2017–2018 school year. Data on the project, therefore, has yet to be fully collected or analyzed. Initial observations of teachers' lessons, however, confirms the patterns that were observed by Bowen and Shume (2018). In particular, the externship experience did not seem to alter the content of teachers' lessons. Alternatively, many of the observed lessons included an explicit focus on process objectives. A consistent theme among manufacturers, for instance, is the importance of soft skills: communication, teamwork, adaptability, problem solving, dependability, conflict resolution, and leadership. Several of these skills are also prominent in state and national teaching standards. With a newfound awareness for the importance of soft skills in the workforces, therefore, it seemed that the objectives of the classroom lessons had expanded – from a primary focus on content to a more balanced focus on content and process.

NKY-FAME illustrated that reflection and time are needed to promote change in classroom practices. In the KY-FAME pilot project, teachers devoted one day to reflection on their externship experience and developing curriculum. With limited time to reflect and devote to lesson development, the KY FAME externship pilot project had limited impact on teachers' practice. NKY-FAME enabled teachers to collaborate with one another and devote more time to lesson planning and development. In turn, lessons were implemented, students' awareness of advanced manufacturing seems to have been enhanced, and process skills have assumed a more prominent place in the classroom. One wonders, therefore, if even greater impact could be realized with greater intensity and duration (NKY-FAME was a one-year project). Bowen and Shume (2018) report that an extended, immersive externship enhanced teachers' knowledge of the problem-solving process. The authors did not report on the impact on teachers' problem-solving skills or content understanding, but one would expect that significant time-on-task would be required to promote the growth needed to actually change classroom outcomes.

3   Conclusions: Successes and Challenges, and Future Directions for Research

In her list of the challenges faced by the diverse stakeholders who partner to try to improve STEM Education, Jona (2014) notes a couple of factors relevant to the discussion in this chapter: (1) Broad cultural divide between K-12, industry, and higher education, (2) Schools and school districts are very large, complex organizations – difficult to navigate as an outsider and build effective partnerships, and (3) Partner relationship building is extremely time intensive (p. 13). Using our conceptual framework, we will discuss the ways in which STEM Pride and NKY-FAME were able to meet some of those factors successfully, while unsuccessfully responding to others. This will allow us to close the chapter by discussing future work and research that needs to be done to fully address such challenges.

Designing B-E/C-E partnerships using a STEM ecosystem perspective clearly has value. In the STEM Pride grant, the project leadership team (PLT) forged relationships and designed activities with 'teachers' as a collective. The team did not consider that the nature of relationships and the structure of activities might need to be different for the career technical education (CTE) teachers than they were for the traditional public school teachers. The team also failed to recognize that a significant percentage of CTE teachers come from business/industry backgrounds, and understood and had relationships with that stakeholder group that could have been leveraged in engineering a sustainable STEM ecosystem. In the NKY-FAME project, the PLT was surprised when a number of non-STEM (special education and foreign language teachers) applied for the externships. In the end, these teachers brought a significant benefit to the project as they found it easier to build learning experiences around soft skills (e.g. communication). This provided a model for the STEM teachers, for whom this seemed more difficult, and it also was very much appreciated by the businesses, for whom a focus on soft skills was a critical outcome. In addition to the unanticipated focus on soft skills, some teacher participants also worked to continue a relationship with the manufacturers for future collaboration and student field trips. This will no doubt result in a long term benefit for the manufacturer, teacher and students.

The community-of-practice (COP) view also has benefits related to designing and studying B-E/C-E partnerships. For example, there had been some debate as the STEM Pride grant was being developed regarding which of the experiences – the university research or the industry externship – should come first in the sequence. The university research experience was put first, largely for logistical reasons. In retrospect, the benefit of this sequence can be understood from a COP perspective: Universities and research labs are more closely related to the COPs of the traditional public school teachers than the industries and their externships. The PLT found that this gave teachers the opportunity to develop a 'researcher identity' (Muhammad et al., 2015), which gave them more confidence moving into the industry externship in year 2. For the NKY-FAME team, taking a COP perspective has helped them consider what the ideal amount of time might be related to teacher participation in externships and in the accompanying PD. Monitoring the extent to which teachers seem to grasp what is similar and different between business/industry COPs versus school COPs is one way to make this determination. Future research could actually track such changes in teachers' thinking as a way to make proposals about both the length and structure of teacher externships.

A COP perspective also causes STEM partnership designers and researchers to give adequate attention to how boundary objects can help strengthen relationships between COPs. A university scientist in the STEM Pride project noted that, "although research scientists work on bigger or more significant problems, we draw on the same principles and techniques that students learn while in high school. The problems we solve are big, and when we go about solving them we go back to high school science principles." Thus, using the science and engineering practices (SEPs) as a boundary object between K-12 schools and university research COPs is effective, but only if the teachers and researchers involved in these partnerships have a common view of them. One of the role of the PLT in both STEM Pride and NKY-FAME was to serve as brokers and communicate between the participants to find such common languages and understandings. An example of this was helping industry mentors and teachers see the intersection of 21st-century skills and SEPs: the different stakeholders were shown how *empathy* (part of the 21st-century skill of communication) is linked to *arguing from evidence* (an SEP) by the fact that both require individuals to *take the perspective of others* (Bohlin, 2009).

The authors noted that studies reviewed for this chapter focused on the *products* of B-E/C-E partnerships to determine the effectiveness of the partnership. The boundary objects and the brokers represent the scaffolds that have a significant impact on how those products are created and, therefore, what the long-term impact of the partnership is. Research, as a result, needs to focus on the boundary objects and the brokers *themselves*, recognizing that these function as a means to the end of the products. This would allow researchers to more meaningfully identify how the challenges of the various border crossings endemic to the partnership were met, and why some were successfully negotiated and others were not. For instance, in the STEM Pride and NKY-FAME projects, it was learned that having the SEPs as a boundary object was not sufficient; having teachers specifically reflect on what they learned about the use of those SEPs in university/industry settings and having lesson/unit plan templates that highlighted the use of SEPs was critical to translating from university/industry to schools.

Finally, it is important to highlight the need to find a way to disseminate and archive the work being done around B-E/C-E partnerships. In preparing to write this chapter, the authors found no central repository of projects or studies. This work seems to be fragmented across different levels of government and institutions. In her 2014 report, Jona brings this issue to the fore: "... my key message to the committee is that what has been missing from the recent discussions about the proper role of federal STEM policy and funding is a recognition of the importance of creating robust dissemination mechanisms that support the scalability and sustainability of those high quality STEM education programs" (p. 1). The authors hope the brief review and discussion of a few critical points concerning B-E/C-E partnerships begins a conversation that eventually leads to the creation of a central repository of work in this area to aid in the dissemination effort. We also strongly recommend that a national body, such as the National Council for Advanced Manufacturing or the National Science Foundation, initiate a convening that could bring together representatives from stakeholder groups to better determine where we are and where we want to go relative to using these partnerships to improve STEM education. Given that the period of the Federal Science, Technology, Engineering and Mathematics (STEM) Education 5-Year Strategic Plan has passed, it seems time to revisit the recommendations in that report and consider future directions of efforts in the area of business/community and education partnerships.

Notes

1. The STEM Ecosystem Initiative is a national effort to support the development and maintenance of STEM ecosystems that was launched in Denver at the Clinton Global Initiative. The program's web site can be accessed here: http://stemecosystems.org
2. Home page for the Office of STEM Education Partnerships at Northwestern: https://osep.northwestern.edu
3. Home page for Northwestern University Biology Investigations in Oncofertility NUBIO: http://nubio.northwestern.edu
4. Home page for the Mentor Matching Engine: https://www.istcoalition.org/education-programs/mentor-matching-engine/
5. Home page for FUSE: https://www.fusestudio.net
6. STEM practices in this project referred to the mathematical practices in the Common Core Math Standards and the science & engineering practices in the Next Generation Science Standards.
7. Most of the summer experiences allowed teachers to actively participate in research work, but some were limited to passive observation because of safety concerns, etc. Regardless, all teachers maintained a STEM Pride journal recording the work that they did/observed, uses of the SEPs they made/encountered, and possible connections to their classrooms.
8. The notion of the *adjacent possible* was actually first suggested by theoretical biologist Stuart Kauffman. A discussion of this concept in Kauffman's own words can be found here: https://www.edge.org/conversation/stuart_a_kauffman-the-adjacent-possible
9. A wonderful web resource for science teachers/teacher educators related to model organisms is https://www.pbslearningmedia.org/resource/hew06.sci.life.gen.modelorg/model-organisms/#.Ws6AfC-lmqA
10. Evolva's home page is https://www.evolva.com
11. CAER's home page is http://www.caer.uky.edu

References

Aikenhead, G. S. (1996). Science education: Border crossing into the subculture of science. *Studies in Science Education, 27*(1), 1–52.

Bohlin, H. (2009). Perspective-dependence and critical thinking. *Argumentation, 23*(2), 189–203.

Bowen, B., & Shume, T. (2018). Educators in industry: An exploratory study to determine how teacher externships influence K-12 classroom practices. *Journal of STEM Education, 19*(1), 5–10.

Camasso, M. J., & Jagannathan, R. (2018). Nurture thru nature: Creating natural science identities in populations of disadvantaged children through community education partnership. *The Journal of Environmental Education, 49*(1), 30–42.

Clark, R. J. (1990). Extending the boundaries of teacher education through corporate internships. *Journal of Teacher Education, 41*(1), 71–76.

Erduran, S., Simon, S., & Osborne, J. (2004). TAPping into argumentation: Developments in the application of Toulmin's argument pattern for studying science discourse. *Science Education, 88*(6), 915–933.

Holdren, J. P., Marrett, C., & Suresh, S. (2013). Federal Science, Technology, Engineering, and Mathematics (STEM) education 5-year strategic plan. *National Science and Technology Council: Committee on STEM Education.*

IBM Corporation (2015). *IBM SPSS statistics for windows, Version 25.0.* Armonk, NY: IBM Corp.

Johnson, S. (2011). *Where good ideas come from: The seven patterns of innovation.* London: Penguin.

Jona, K. (2014). *Promising models for private sector engagement in STEM.* Testimony to the U.S. House of Representatives, Committee on Science, Space, and Technology, Subcommittee on Research and Technology Hearing: "Private Sector Initiatives that Engage Students in STEM".

Kantrov, I. (2014). *Externships and beyond: Work-based learning for teachers as a promising strategy for increasing the relevance of secondary education.* Waltham, MA: Education Development Center, Inc.

Kier, M. W., Blanchard, M. R., Osborne, J. W., & Albert, J. L. (2014). The development of the STEM Career Interest Survey (STEM-CIS). *Research in Science Education, 44*(3), 461–481.

Muhammad, M., Wallerstein, N., Sussman, A. L., Avila, M., Belone, L., & Duran, B. (2015). Reflections on researcher identity and power: The impact of positionality on Community Based Participatory Research (CBPR) processes and outcomes. *Critical Sociology, 41*(7–8), 1045–1063.

National Center for Advancing Translational Sciences. (2017). *Translational science spectrum.* Retrieved December 4, 18, from https://ncats.nih.gov/translation/spectrum

National Research Council. (2010). *Exploring the intersection of science education and 21st century skills: A workshop summary.* Washington, DC: National Academies Press.

National Research Council. (2013). *Next generation science standards: For states, by states.* Washington, DC: National Academies Press.

Sharpless, N. E., & DePinho, R. A. (2006). Model organisms: The mighty mouse: genetically engineered mouse models in cancer drug development. *Nature reviews Drug discovery, 5*(9), 741.

Star, S. L. (1998). The structure of ill-structured solutions: Boundary objects and heterogeneous distributed problem solving. In L. Gasser & M. N. Huhns (Eds.), *Distributed artificial intelligence* (pp. 37–54). Amsterdam: Elsevier.

Traphagen, K., & Traill, S. (2014). *How cross-sector collaborations are advancing STEM learning.* Los Altos, CA: Noyce Foundation.

Wenger, E. (1998). *Communities of practice: Learning, meaning, and identity.* New York, NY: Cambridge University Press.

Wenger, E., McDermott, R., & Snyder, W. A. (2002). *Cultivating communities of practice: A guide to managing knowledge.* Boston, MA: Harvard Business School Press.

CHAPTER 18

What Skills Do 21st Century High School Graduates Need to Have to Be Successful in College and Life?

Kristina Kaufman

Abstract

Expectations for students and employees have evolved and increased over the past few decades as digital technologies and global connectedness have raised awareness of the types of skills individuals should possess to effectively navigate the 21st century landscape in school and beyond. In some quarters education has been at the center of criticism from the business community and others for not adequately preparing graduates for college, career and life. Solutions to improving education have often considered the need for students' development of a set of skills that include, most notably, communication, collaboration, creativity and technical skills, also referred to as 21st century skills. The concept of 21st century skills is broad, historically vague to implement in the classroom, and evolving. To unpack such a vast topic, this chapter incorporates a literature review on 21st century skills combined with data from a research study on perceptions of high school principals and business executives. Feedback from 92 public high school principals in Illinois is examined to discuss the role of high school and student outcomes for the 21st century.

1   Previous Literature

In the book *The 21st Century Principal* (2003), Milli Pierce presents a question that public schools and their principals ultimately must ask themselves: "Are we doing a good enough job getting all children prepared for the world we live in?" (p. 91). This is an important question to reflect on as Pierce suggests that part of what is not preparing students for the world we live in, as suggested by a persistent achievement gaps and international benchmarking (OECD), is education's inability to come to a consensus around three more questions: What should students know? What should they be able to do? And, finally, what should students be like as "we send them into the world at the end of the twelfth grade?" (pp. 91–92).

There is an argument to be made that embracing what will help society meet challenges of tomorrow, will redefine what education looks like. Vockley, for the Partnership for 21st Century Skills, (2006) boasts that high schools should "be designed, organized and managed with a relentless focus on the results that matter in the 21st century" (p. 2) and businesses are looking to be collaborators in such efforts.

The realities and demands of the workplace have changed since years past and the business community is fueled by this development to bring education to the forefront of the discussion as the nexus from which success should stem (Barton, 2006; Hall, 2013; Murray, 2010; Vockley, 2006). Criticism of education and high school graduates' preparedness for college and career is well documented (Crew, Garcia, & Castro, 2016; O'Sullivan & Dallas, 2010; Ravitch, 2010 Rumberger, 2011). Despite this, as data from this study uncovers, some high school principals acknowledge an ignorance that those from the business community or "outsiders" have pertaining to the challenges, pace and difficulties schools face in order to "keep up." The reality is that it is very challenging to prepare adolescents for college and life in a relentless fast-paced digital world. But, as headlines continue to present themselves that other countries are performing better than American students (OECD, 2015; Sahlberg, 2014) and the imperative for the next generation to develop technologies and solutions to 21st century problems (environment, energy, technology, etc.) it is hard not to think we can do better in the United States, a country that spends more money on average per pupil than any other nation (OECD, 20017).

The study discussed in this chapter aimed to ask principals the heart of this discussion: *What is the purpose of high school education?* What is so significant about this is that by finally explicitly stating what the role of high school is, backwards planning could be applied to align high school learning experiences to best prepare students for college, career and personal development and fulfillment.

Viewing education through a lens in which "The business community is the number one consumer of the public education system" (U.S. Chamber of Commerce, 2012, para. 1), is one perspective that stands to promote and affirm that high school graduates should possess sound skill for postsecondary education and employability. Indeed, there is a desire for high schools to teach and assess competencies that would be valued in "the real world" (Committee for Economic Development, 1998) thus preparing and testing for learning that could be applied in the workplace or postsecondary education (Van Der Lind, 2000).

The reason these questions become so important is because high school is a critical time and the last stop for compulsory education. High school is the time when students make important decisions about which paths they will pursue pertaining to their futures, and while a career is not the only purpose of life, it is, however, as Burstyn (1986) acknowledges, important to one's contributions to society, self worth, livelihood, and quality of life. Right or wrong, a job is what people spend most of their adult life doing. Because of this, finding a path that is of interest to students and exposing them to what careers and entrepreneurial opportunities exist is worthwhile time spent especially as post-secondary tuition continues to climb and the cost for students to figure it out as they go along will ultimately price them out. Most recent data published by the U.S. Department of Education for the 2014–2015 academic year, cites that the average annual current dollar prices for undergraduate tuition, fees, room, and board were estimated to be $16,188 at public institutions, $41,970 at private nonprofit institutions, and $23,372 at private for-profit institutions (National Center for Education Statistics, 2016). Cost is something not to forget as one principal in this study articulated this fact by stating that in the context of this principal's school-business partnership, students are "getting valuable experience without having to endure financial hardships that would occur without [their internship provided] training."

Historically, the purposes of secondary schooling in general have been "less able to be resolved than at the elementary level" (Campbell & Sherington, 2006, p. 3) thus leaving room to negotiate the goal of high school and what types of experiences students should have by the time they graduate. The term 21st century skills was developed to promote aspirational skills students would exercise and develop to be successful beyond graduation whatever their path may be. These skills are defined in various ways, but a consensus seems to describe them as a set of skills that shift the focus of learning from memorization of specific content knowledge to applied, use learning experiences that challenge students to showcase problem solving and reasoning. These sills are explored in the subsequent sections.

1.1   *21st Century Skills – The Literature*
In the book *The Second Machine Age: Work, Progress, and Prosperity in a Time of Brilliant Technologies* (Brynjolfsson & McAfee, 2016), the authors highlight that given developments in technology and telecommunications globally, people have more access to information than ever before. This evolution has leveled the playing field in terms of information acquisition, therefore skills to critically evaluate, examine, apply and solve problems – 21st century skills – are deemed more valuable.

A review of peer reviewed literature, 21st century skills focused organizations, and current books in the field highlights several core "skills" that are essential to 21st century success. These texts range from current embraced standards within the education system (ISTE/NETS performance standards, Common Core State Standards focusing on higher order thinking skills) to academic research articles and the organizations The Partnership for 21st Century Skills, National Education Association, National Governors Association, Education Commission of the States, and others. Overridingly, essential skills include: creating and producing (Nichols, 2016; P21, 2018), curiosity & imagination (Claxton, Costa, & Kallick, 2016; Crew et al., 2016; Wagner, 2018), analyzing & evaluating financial information (financial literacy) (Erner, Goedde-Menke, & Oberste, 2016; Greenfield, 2016; Lusardi, 2015), collaboration skills and engagement in participatory culture (P21, 2018; Tombleson & Wolf, 2017; Urbani, Roshandel, Michaels, & Truesdell, 2017), communication skills (presenting, publishing, writing, speaking) (Jacobson-Lundeberg, 2016; P21, 2018; Robb, 2017) entrepreneurial and innovation skills (ISTE/NETS, year; P21, 2018), technology literacy and digital citizenship (ISTE/NETS; Lapek, 2017). These skills provide a well-rounded ability set for an individual to navigate many different social environments and task-related challenges in modern society.

2 Methods

A research study was conducted to assess principals' perceptions of what the role of high school education is in their own words, if they feel schools are meeting that standard, and what skills they believe students develop in high school in the context of a school-business partnership.

Inspiration for the study first came from a newly developed connection with a principal at a high school in central Illinois. Through a series of conversations, I learned that this principal was partnering with a locally based yet nationwide insurance company, that, according to this principal, "approached" his school to improve curriculum and the education these students were receiving to be better prepared for college and the workforce. Quite frankly, the insurance company was dissatisfied with the level of preparedness that students possessed when leaving high school. This bluntness both surprised and intrigued me to delve further.

The extant literature on school-business partnerships especially from the viewpoints of school leaders is a relatively new area of concentration with limited qualitative and even less quantitative data to pull from (Bennett & Thompson, 2011; Muhlenberg, 2011; Murray, 2010; Rivkin, 2013).

Moreover, there were no studies found that specifically address the foci of this study including examining high school principals' perceptions of business partnerships, the role of high school education in the context of 21st century learning, and students' workplace readiness and preparation for post-secondary education. This work represents the perspectives of 105 public high school principals in Illinois.

A mixed methods approach to data collection was chosen as the appropriate methodology for this research. Past research on school-business partnerships has also employed a mixed methods strategy. In addition to a review of the literature, there are four sources of data collection in this study chosen to provide a rich data set and multiple perspectives with depth pertaining to partnerships between schools and businesses. Interviews were first conducted with 10 high school principals in Illinois and 10 business professionals who work at Illinois-based companies. These interviews provided a baseline of understanding from both the educational realm and the business world regarding partnerships and expectations of the two entities. While this study primarily focuses on the perceptions of high school principals, it was important to include some perspective from the business community in Illinois to more fully understand their motivations, and involvement with schools. These combined 20 initial interviews provided the first set of qualitative data that was coded for themes and helped to revise a working survey.

Upon the completion of the interviews, five principals evaluated a 35 question draft survey that in revised form would ultimately be sent to 622 high school principals throughout Illinois. The pre-test of the survey provided further understanding of current partnerships between high schools and businesses in Illinois and refined survey questions and wording. An electronic survey was made available to potential participants for several weeks with three rounds of data collection harnessing the Dillman Method (Dillman, 1978). Ninety-two usable surveys were collected. Statistical analyses were applied to the survey data using SPSS. Upon analysis of the survey data, post-survey interviews were conducted with five high school principals in Illinois to discuss select data findings and ask follow-up questions that arose.

3   Research Findings

3.1   *Study of High School Principals' Perceptions*
The data presented here is from a research study of perceptions of high school principals and business professionals regarding the purpose of high school and outcomes students may develop as a result of school-business partnerships.

First, high school principals were surveyed and asked to describe in their own words what they believed to be the primary role of high school education. Themes arose pertaining to preparation for events after high school – college, the workforce, a trade, the military, and being a social, contributing, active adult in a democratic society. These survey comments were consistent with comments from initial interviews from principals and business professionals with high school's major role seen as preparing students for college and career.

3.2 Twenty-First Century Learning as Seen by High School Principals

Principals were also specifically asked to define, in their own words, what they viewed 21st century learning to be. Again, participants provided their own written comments to this question. Data was analyzed and coded for themes of which many emerged. By far, the most frequently mentioned theme was that of technology. Principals mentioned technology training, using technology, applying technology, and knowing technology 46 different ways. One principal stated, "Our nation and national economy are developing into a technology integrated, networked and global set of interdependent societies. Our public education system must likewise evolve to teach students to understand and navigate in this world."

A second theme was that of collaboration being important to 21st century learning, cited 18 times. "Twenty-first century learning causes students to work collaboratively with others to solve problems using technology and thinking skills," noted one principal. A second principal noted collaboration more broadly stating that regarding 21st century learning, "Its collaboration with many stakeholders to educate students. It is allowing student to collaborate, learn independently, and have high level conversation with their peers and teachers. It's the district giving the students the educational and technological skills to succeed in life."

Also mentioned 18 times was the importance of students to exercise problem solving skills by developing research abilities and then apply knowledge to solve problems. The next most highly cited theme, mentioned 15 times, was critical thinking, closely followed by students' preparation for a job or career, cited 14 times. The theme of preparation for career overlapped with technology as one principal stated, "Preparation for what is next in the world of work and to be successful in many ways including being versed with technology" is 21st century learning. The development of skills needed for life after college was noted 13 times. Overall, the theme of "preparation" for something whether it be college, career, application of technology, students' futures, ability to work with potential others, and solve future problems, is evident.

Minor data themes include the importance for students to become independent thinkers, engage in hands-on activities for learning, develop strong written and oral communication skills, become globalized citizens, develop self-direction, become lifelong learners and learn how to learn, the development of

TABLE 18.1 Principals' perceptions on partnership outcomes

Statement	Mean	By the numbers
This partnership has helped students to foster 21st century skills more than if there was no partnership in place.	1.79	85.9% agree/strongly agree; only 1 person disagreed; 12% neutral
This partnership has helped to foster students' workplace readiness more than if there was no partnership in place.	1.60	50.9% strongly agree; 89.5% agree/strongly agree; 0% disagreed/strongly disagreed
This partnership has helped students to be prepared for post-secondary education more so than if there was no partnership in place.	1.86	82.5% agree/strongly agree; 14% neutral, 3.5% disagree; 0% strongly disagreed
Student learning is greater due to your school-business partnership.	2.16	66.7% agree/strongly agree; 28.1% neutral; 2 principals disagree; 1 principal strongly disagreed

Note: Statements were ranked from 1–5 across five levels of agreement from strongly agree (1), neutral, disagree to strongly disagree (5).

content knowledge or skills for particular jobs, and adaptability. One principal summed up 21st century learning as "Providing the most relevant and realistic learning experiences for students so they can apply skills and knowledge to new settings."

The top three themes of 21st century learning that emerged from survey participant comments include: (1) technology – working with technology/developing technology skills, (2) collaboration/working with others, and, (3) problem solving techniques. Given this is how 92 principals from across one state described 21st century learning, interviewees were asked to indicate whether they believed school-business partnerships could address these three areas. Essentially, if we know principals think these three areas are important for 21st century learning, then can business partnerships help to foster or enhance this kind of learning? All five interviewed principals believed that a partnership could address students gaining technology aptitude, practice working with others, and engage in problem solving. One principal noted that while businesses "are very good at addressing it, they are not so good at addressing it with what kids do in school," highlighting a lack of understanding by businesses of the classroom context in which students are situated. Another principal noted that these three themes are what businesses will need to stay employed and what will make students employable. Another noted that some businesses are better than others at addressing these areas and each business has its focus or strength depending on who the students, teacher and business professionals are interacting in or outside of the classroom.

Those schools with partnerships were asked to indicate their level of agreement with five statements pertaining to their partnership's outcomes. Feedback from this series of questions aimed to gauge any partnership effects on 21st century skills, workplace readiness, preparation for post-secondary education, and student learning. Table 18.1 displays the results from these respondents.

Principals with a partnership were asked to identify two specific 21st century skills that they believed students acquired from their school's business partnership. Out of all skills provided five major themes emerged. Workplace skills including those specifically of communication and workplace etiquette were the most frequently cited skills developed, noted 23 times. This theme included comments such as a conceptualization by students of "how to present themselves," "an understanding of what is valued in the workforce," and, "skills to do a job." The second most mentioned skills developed were people skills and the ability to work with others. The third most cited skill was that of work ethic. Following work ethic, principals noted students' development of financial literacy including skills of financial planning and fiscal responsibility as skills developed by students as a result of their partnerships. Time management and knowledge about different career fields were the fourth and fifth most mentioned skills acquired by students. Other skills acquired included the following: personal responsibility, technical/computer skills, networking, self-worth/confidence, task management, perseverance, problem solving, business knowledge, and the development of vocabulary/terminology in a specific field.

3.3   *Skills Desired by Employers*

Interviews with business professionals are a part of this study to complement the discussion. This is important to understand especially as one of the strongest themes from both high school principals and business professionals regarding the role or purpose of high school education was preparation for a career. This aligns with the first interviewee's comment employed in the healthcare industry who stated: "there are certain areas where we have a shortage of talent." Each of the 10 interviewees was asked, "What skills are vital in today's marketplace?" The most frequently cited theme was communication skills with skills of writing and speaking lacking from most students.

A sub-theme of communication skills from the data is the *way* in which one communicates. These interviewees were clear in that, "people don't know how to talk face to face," and that, "many young people I interact with do not know how to just have a conversation with someone older than them." The interviewees from the drug retail and sunglasses industries indicated their disdain with texting and would like younger people to "pick up the phone!" as the business professional from a sunglasses company quoted. Two other interviewees from the grocery manufacturing and food processing industry, and construction industry indicated a need in strengthened phone communication skills by students.

A second theme of skills deemed important to employers was self-directedness. Self-directedness was directly stated by two of the 10 interviewees but addressed by four more interviewees in one's ability to "follow instructions," "take direction," exhibit "drive" and "initiative," and demonstrate "leadership." A third theme emergent from the interviews was that of social skills. These interviewees placed high value on one's ability to "work with others," "interact with other individuals," show "respect," "listen to others," and be able to "talk to people we don't know." Probably one of the most meaningful comments shared was from a business professional in the fast food industry. She expressed that, "as an employer, you can teach some functional things, but you don't have a lot of time to teach some of these things like getting along with others or dependability. Even basic reading and math. That's one thing that we struggle with. We see many high school graduates where their reading level is low."

3.4   Summary

The data presented here from a research study surveying 92 public high school principals, a subset of business professionals and a review of the literature on 21st century skills provides a consistent outline for the types of skills desired for high school graduates to develop for life and career. Study findings and literature are in agreement that high school education should help students to foster an ability to be flexible and work with different types of people, communicate ideas in written and oral format, nurture, exercise and exhibit creativity, be adaptable to the relentlessly changing landscape of technology and the workplace (Friedman, 2017), develop the mentality of a lifelong learner to continually embrace and utilize digital technologies as they evolve for personal and professional purposes, and, as a core component to 21st century learning is critical thinking which goes beyond recalling content knowledge (traditional schooling model) but applying learnings, making connections and solving problems utilizing higher order thinking skills. Therefore, the roadmap for curriculum is illuminated by the development of these 21st century skills.

4   Limitations

While this study contributes new understanding about the insights of public high school principals in Illinois pertaining to school-business partnerships, this study could have been conducted in different ways which may affect the perception of data findings. First, interviews were conducted over the phone with principals and business professionals as the researcher took field notes. An audio recording device was not used to capture every word from interview conversations while data findings relied upon the ability of the researcher to capture as much specific language as possible. While a recording device would have captured every comment shared by an interviewee, participants may not have felt as comfortable with their candid feedback being recorded (Glesne, 2011) or provided more general statements. As the researcher, I felt that positioning the interview as more of a conversation and not a recorded study helped to gather as much accurate insights as possible.

Principals were also asked to describe in their own words what 21st century learning is. I would have liked to have asked them what 21st century learning is not. This comparison would be extremely helpful to identify outdated ways of thinking and provide comparison to what schools are currently doing, have done and could do.

Future research would be valuable to seek concrete ways in which teachers are or could be creating opportunities for students to develop these skills of creativity, collaboration and technology to name a few while also meeting state or other "standards." By creating high experience and engaging learning experiences that also meet curriculum requirements, perhaps high schools won't have to make a choice between students learning facts and also developing collaboration, problem-solving, and technology skills of students demanded by 21st century employers. Flipping the classroom may be one vehicle in which to attain this. Prior to class students review basic concepts and in class time is focused on experiential learning. Engaged students may also have more motivation to complete school and suffer fewer social problems such as anxiety which is steadily growing.

To operationalize this, can we not only identify what students need, but also how to teach and assess that students have acquired such abilities? How do we best prepare students for post-secondary education and the workforce? What does this look like? While we can say we want students to foster creativity, that will likely look vastly different across schools and content areas. A beneficial roadmap for educators preparing students with 21st century skills would contain detailed rubrics and use of examples and case studies that document in detail instructional sequences to assess soft skills. To develop both efficiency and effectiveness in the education space, these materials ideally should be available online at no cost to educators. Integrity of materials could be assured by a minimum number of teacher reviews with reviewer's profile available to all.

5   Conclusion

Twenty-first century skills are defined as a vast skill set that can translate to an individual's success in any facet of life – from the workplace to personal finances to schoolwork and in the ability to effectively communicate with others – driven by global competition and universal benchmarks for success. Twenty-first century skills most notably include skills of creativity and innovation, problem-solving skills, leadership skills, global awareness and social responsibility.

School-business partnerships are becoming a reality for more and more public high schools across the nation and in Illinois. There is an important connection between education and the workforce that the literature has echoed from the voices of high school principals and businesses.

Essentially, high school should afford the time for students to explore future options and be able to leverage that time to find out what they want to do with their lives after high school. Finding ways in which high school can tangibly, explicitly and implicitly teach the outlined 21st century skills will exercise students' abilities to be deft, confident and employable.

References

Barton, P. (2006). *High school reform and work: Facing labor market realities*. Princeton, NJ: Educational Testing Service Policy Information Center.

Bennett, J., & Thompson, H. (2011). Changing district priorities for school-business collaboration: Superintendent agency and capacity for institutionalization. *Education Administration Quarterly, 47*(5), 826–868.

Brynjolfsson, E., & McAfee, A. (2016). *The second machine age: Work, progress, and prosperity in a time of brilliant technologies*. New York, NY: W. W. Norton & Company.

Burnett, J. (2017). Employers having a tough time finding qualified candidates, survey says. *Ladders*. Retrieved from https://www.theladders.com/career-advice/employers-trouble-finding-qualified-candidates-glassdoor-survey

Burstyn, J. (Ed.). (1986). *Preparation for life? The paradox of education in the late twentieth century*. Philadelphia, PA: The Falmer Press.

Campbell, C., & Sherington, G. (2006). *The comprehensive public high school.* New York, NY: Palgrave Macmillan.

Clark, A., DeLuca, B., Ernst, J., Kelly, D., & Ridgeway, J. (2018). Designing standards-based STEM. *Technology & Engineering Teacher, 77*(4), 30–35.

Claxton, G. G., Costa, A., & Kallick, B. K. (2016). Hard thinking about soft skills. *Educational Leadership, 73*(6), 60–64.

Committee for Economic Development. (1998). *Employer roles in linking school and work: Lessons from four urban communities.* New York, NY: CED.

Crew, R., Garcia, P., Escalante, M., & Castro, F. (2016). *Restoring the value of a high school diploma in the United States using 21st century skills as pedagogy: a case study of 21st century skills development and preparation for the global economy.* Los Angeles, CA.

Dillman, D. (1978). *Mail and telephone surveys: The total design method.* New York, NY: John Wiley and Sons.

Erner, C., Goedde-Menke, M., & Oberste, M. (2016). Financial literacy of high school students: Evidence from Germany. *Journal of Economic Education, 47*(2), 95–105.

Friedman, T. (2017). *Thank you for being late: An optimist's guide to thriving in the age of accelerations.* New York, NY: Picador.

Glesne, C. (2011). *Becoming qualitative researchers: An introduction* (4th ed.). Boston, MA: Pearson Education, Inc.

Greenfield, J. S. (2015). Challenges and opportunities in the pursuit of college finance literacy. *High School Journal, 98*(4), 316.

Hall, S. (2013). Putting college and career readiness at the forefront of district priorities in Dallas. *Voices in Urban Education, 38,* 6–9.

Jacobson-Lundeberg, V. (2016). Pedagogical implementation of 21st century skills. *Educational Leadership and Administration: Teaching and Program Development, 27,* 82–100.

Kendall, J. S. (2007). *High school standards and expectations for college and the workplace.* National Center for Educational Evaluation and Regional Assistance, Institute of Education Sciences, U.S. Department of Education.

Lapek, J. (2017). 21st century skills: The tools students need. *Children's Technology & Engineering, 21*(3), 24–26.

Lusardi, A. (2015). Financial literacy skills for the 21st century: Evidence from PISA. *Journal of Consumer Affairs, 49*(3), 639–659.

Muhlenberg, E. (2011). *Who benefits? A comparison of school-firm partnerships in Chicago and Berlin* (Doctoral dissertation). Retrieved from http://library.illinoisstate.edu/

Murray, M. (2010). *The nature of the liaison in developing and sustaining successful business partnerships with high schools* (Doctoral dissertation). Retrieved from http://library.illinoisstate.edu/

National Center for Education Statistics. (2012). *Public high school graduation rates.* U.S. Department of Education. Retrieved from http://nces.ed.gov/programs/coe/indicator_scr.asp

National Center for Education Statistics. (2016). Tuition costs of colleges and universities. *Digest of Education Statistics, 2015*, U.S. Department of Education.

Nichols, S. (2016). Open-ended projects: 21st century learning in engineering education. *Technology & Engineering Teacher, 76*(3), 20–25.

OECD. (2015). *PISA 2015 technical report*. Retrieved from http://www.oecd.org/

OECD. (2017). *Education expenditures by country*. Retrieved from https://nces.ed.gov/programs/coe/indicator_cmd.asp

O'Sullivan, M., & Dallas, K. (2010). A collaborative approach to implementing 21st century skills in a high school senior research class. *Education Libraries, 33*(1), 3–9.

P21. (2018). *Framework for 21st century learning*. Retrieved from http://www.p21.org/about-us/p21-framework

Ravitch, D. (2010). *The death and life of the great American school system: How testing and choice are undermining education*. New York, NY: Basic Books.

Robb, L. (2017). Read talk write: Developing 21st-century skills. *Voices from the Middle, 24*(4), 19–23.

Rivkin, J. (2013). *Partial credit: How America's school superintendents see business as a partner*. Boston, MA: Harvard Business School U.S. Competitiveness Project.

Rumberger, R. (2011). *Dropping out: Why students drop out of high school and what can be done about it*. Boston, MA: Harvard University Press.

Sahlberg, P. (2014). *Finnish lessons 2.0: What can the world learn from educational change in Finland?* New York, NY: Teachers College Press.

Stone, J., & Lewis, M. (2012). *College and career ready in the 21st century: Making high school matter*. New York, NY: Teachers College Press.

Tombleson, B., & Wolf, K. (2017). Rethinking the circuit of culture: How participatory culture has transformed cross-cultural communication. *Public Relations Review, 43*(1), 14–25.

Urbani, J., Roshandel, S., Michaels, R., & Truesdell, E. (2017). Developing and modeling 21st century skills with preservice teachers. *Teacher Education Quarterly, 44*(4), 27–50.

U.S. Chamber of Commerce. (2012). *Business education network*. Retrieved from http://icw.uschamber.com/sites/default/files/Business%20Education%20Network.pdf

Vockley, M. (2006). *Results that matter: 21st century skills and high school reform*. Tucson, AZ: Partnership for 21st Century Skills.

Wagner, T. (2018). *Tony Wagner's seven survival skills*. Retrieved from http://www.tonywagner.com/7-survival-skills/

CHAPTER 19

An International View of STEM Education

Brigid Freeman, Simon Marginson and Russell Tytler

Abstract

Science, technology, engineering and mathematics (STEM) education and research are increasingly recognized globally as fundamental to national development and productivity, economic competitiveness and societal wellbeing. There has been a global turn to STEM that is clearly evident in government efforts worldwide to elaborate STEM policy governing school science and mathematics, and tertiary level education and research in the STEM disciplines. This shift is also reflected in emerging research priorities that are most frequently conceived in STEM terms, underpinned by commitments to internationalization and multidisciplinarity. This chapter explores STEM policies and programs from an international perspective extending from the Anglosphere, East Asia, Western Europe and Latin America to the Middle East. We identify discernible trends and parallels regarding government STEM policy and structural responses, school and tertiary level STEM education participation, comparative performance measured by international assessments such as PISA and TIMMS, STEM research and innovation, and issues concerning gender and under-represented groups. The chapter examines various programs and solutions including school-level curriculum and pedagogy reform to enhance science and mathematics participation and performance, teaching-related initiatives, and strategies at the tertiary-level to redress current systemic disparities.

1   Previous Literature

Science, technology, engineering and mathematics (STEM) education and research are increasingly recognized globally as fundamental to national development and productivity, economic competitiveness and societal wellbeing (Marginson et al., 2013). There has been a global turn to STEM (Freeman, Marginson & Tytler, 2015) that is clearly evident in government efforts worldwide to elaborate STEM policy governing school science and mathematics, and tertiary level education and research in the STEM disciplines. In recent years awareness of the ubiquity and impact of technology has grown as the influence of artificial intelligence, automation and big data on the world of work is imagined, and increasingly realized.

2 Context of the Study

This chapter discusses the findings of the STEM: Country Comparisons project initiated by Australia's Chief Scientist, and funded by the Australian Council of Learned Academies (ACOLA). The project commissioned 23 reports that investigated attitudes towards STEM, the perceived relevance of STEM to economic growth and wellbeing, patterns of STEM provision in school and tertiary education, student uptake of STEM programs, factors affecting student performance and motivation, and strategies, policies and programs to enhance STEM. Country and regional reports spanned the Anglosphere (United States, Canada, New Zealand, United Kingdom, Australia), Europe (Western Europe, Finland, France, Portugal, Russia), Asia (China, Taiwan, Japan, Singapore, South Korea), Latin America (Argentina, Brazil), the Middle East (Israel), and South Africa. The project also commissioned a small number of special interest reports focused on Indigenous peoples and STEM, the Australian labour market, gender and 'identity' and international agencies involved in international assessments and reporting. The project was overseen by an expert working group comprising fellows of Australia's learned academies.

3 Research Findings

3.1 *Secondary School Science and Mathematics Education Participation*

Participation in school level biology, chemistry, physics and mathematics reflects students' attitudes, especially interest and self-efficacy, and ability, with gender playing an important role (Palmer, Burke, & Aubusson, 2017). Other factors, such as socioeconomic status and geographic location, also impact on school student participation and learning. Exploring secondary school participation in science and mathematics programs is problematic as there are no international, standardized datasets for school level fields of study. In addition, there are fundamental systemic differences regarding curriculum and schooling structures, including streaming practices. The number of years of schooling, and level of compulsion also add complexity. For example, in a small number of instances, such as China, Taiwan, Finland, Israel and Brazil participation in mathematics is compulsory in one or both of the final years of schooling (see Marginson et al., 2013). Despite these constraints, some examples from high performing school systems and Anglosphere systems are illustrative.

School students in both Singapore and Japan perform very well in international science and mathematics assessments. While performance is not necessarily simply a function of participation, in Singapore it is notable that participation in school level mathematics and sciences is high. Singapore's secondary school students enrolled in the academic express course participate in mathematics and sciences, while those enrolled in the normal (academic) and normal (technical) course undertake mathematics (Ministry of Education Singapore, 2017). In Japan, government curriculum reforms dating from the 1990s impacted total school hours dedicated to mathematics and science, and course content. The government's 'relaxed education' policy, which aimed to alleviate unnecessary student pressure while simultaneously enhancing student motivation, was reversed after Japan's 'PISA-shock' in 2003. More recent reforms have emphasized reading literacy and increased school hours for mathematics and science (Ishikawa, Moehle, & Fujii, 2015).

In the United States, consistently over the period 1990–2009 more high school graduates completed early mathematics courses (algebra I, algebra II/trigonometry and geometry I) than advanced mathematics courses (algebra/pre-calculus, statistics/probability and calculus). However, the proportion of high school graduates completing all of these courses increased over this period. In relation to school science, more high school graduates completed biology than chemistry, while fewer again completed physics. During the same period, the proportion of high school graduates completing biology, chemistry and physics courses progressively increased (National Center for Education Statistics, n.d.; Maltese et al., 2015). By contrast, in Australia participation in both senior secondary school science and mathematics experienced long term decline (Ainley, Kos, & Nicholas, 2008; Kennedy, Lyons & Quinn, 2014; Lyons & Quinn, 2015); however, the preference of Year 12 students for elementary mathematics rather than either intermediate or advanced level mathematics (Barrington & Evans, 2017) was consistent with the trend observed in the United States.

Concerns regarding participation in and motivation for school science and advanced mathematics have impacted school curriculum, pedagogy and resourcing, and post-school options. With curriculum, systems have sought to balance a focus on lifting all students' STEM capabilities, and a focus on nurturing an elite STEM workforce. This is also related to the tension between tightly scripted curricula and testing and a desire to nurture critical and creative STEM capabilities. The impact at the post-school level is most evident in relation to tertiary education program pre-requisites and admissions systems, and the necessity for foundational or preparatory programs, particularly for STEM disciplines.

3.2 Tertiary Education STEM Participation

At the tertiary education level, countries have focused attention on their population's preparedness for a rapidly changing, globally interconnected world requiring increased scientific literacy, and high-level STEM research skills. Accordingly, participation in tertiary education STEM disciplines has been closely monitored, as have efforts to increase graduates' transferable skills and to develop curricula responsive to industry needs. Science and technology advances have encouraged a reimagination of the future world of work, and the place of disciplinary knowledge in preparing for this.

Participation in broadly defined STEM disciplines, including engineering, sciences, information technology, health and agriculture, varies by country/territory and region, over time. For the period 2011 to 2015, participation was highest in some Western European (Finland, Germany, Sweden, United Kingdom) and East Asian (South Korea, China) economies, as well as Singapore. Comparatively, the United States and Australia lagged behind (UNESCO Institute of Statistics [UNESCO], 2018; Ministry of Education of the People's Republic of China, 2015). Large numbers of tertiary education students enrolled in these STEM programs are located in the three largest higher education systems, that is, China, India and the United States (UNESCO, 2018; Ministry of Education of the People's Republic of China, 2015).

Over the period 2011 to 2015, participation remained relatively static in most countries across the STEM disciplines (Table 19.1); however, there were some exceptions. Enrolments in natural sciences, mathematics and statistics tertiary education programs increased in the United Kingdom, India and France. At the same time, enrolments in information communication technologies increased in Brazil and Israel, and enrolments in agriculture, forestry, fisheries and veterinary increased in Brazil. Greater volatility was recorded in engineering, manufacturing and construction, where enrolments dropped marginally in Brazil and Finland, and considerably in India, while increasing in Norway (UNESCO, 2018).

In these selected countries, typically a larger proportion of women than men enrolled in health and welfare tertiary education programs. For many countries, a marginally higher proportion of women than men enrolled in agriculture, forestry, fisheries and veterinary programs, with notable exceptions including India, the United States and Israel. In other STEM disciplines, a larger proportion of men enrolled. Gender disparity is most evident in engineering,

TABLE 19.1   Percentage of students in tertiary education enrolled in STEM tertiary education programs (both sexes) (2011–2015)

Country	2011 (%)	2012 (%)	2013 (%)	2014 (%)	2015 (%)
Finland	53	53	52	53	54
South Korea	47	47	48	48	47
Singapore	49	47	–	–	–
Germany	–	–	46	46	46
Sweden	44	45	46	46	44
United Kingdom	41	41	45	46	44
Denmark	41	41	42	43	43
Switzerland	39	40	40	40	42
France		42	38	38	42
Norway	37	38	35	35	40
New Zealand	37	38	39	39	39
Israel	35	36	37	38	39
Brazil	–	34	–	38	39
India	–	43	37	38	38
Australia	–	–	–	–	37
Japan[a]	–	–	–	–	37
United States	32	32	36	36	36

a   The OECD data for Japan excludes agriculture, forestry, fisheries and veterinary.
SOURCE: UNESCO (2018), OECD (2017)

manufacturing and construction programs, and information and communication technologies programs (UNESCO, 2018). Typically a larger proportion of students enrolled in engineering, manufacturing and construction, and health and welfare programs than other STEM disciplines (Table 19.2).

Germany, Finland and Switzerland, which each had a high percentage of students enrolled in STEM tertiary education programs, also had the highest percentage of students enrolled in doctoral level programs (UNESCO, 2018).

3.3   *Issues Concerning Gender and Under-Represented Groups*
There is also considerable variability within countries and territories in terms of demographics (gender, ethnicity, socio-economic status, religion) and distribution (metropolitan/non-metropolitan). Gender disparities persist in STEM education (Marginson et al., 2013), and women typically remain

TABLE 19.2 Percentage of students in STEM tertiary education programs (both sexes), by discipline (2015)

Country	Engineering, manufacturing and construction (%)	Health and welfare (%)	Natural sciences, mathematics and statistics (%)	Information and communication technologies (%)	Agriculture, forestry, fisheries and veterinary (%)
South Korea	25	12	6	3	1
Germany	21	7	10	6	2
Finland	19	18	6	9	2
Israel	19	8	6	5	0
Sweden	18	18	5	2	1
Japan	16	16	3	2	–
Switzerland	15	15	8	3	1
India	14	3	15	5	1
France	13	16	10	3	1
Norway	12	18	5	4	1
Brazil	12	5	2	8	12
Denmark	10	22	5	4	1
United Kingdom	9	14	15	4	1
Australia	9	18	5	4	1
New Zealand	8	14	9	6	2
United States	7	18	6	4	1

Note: Any discrepancies in totals between Tables 19.1 and 19.2 result from rounding.
SOURCE: UNESCO (2018), OECD (2017)

under-represented in the STEM workforce, including the United States (Beede et al., 2011) and Australia (Office of the Chief Scientist, 2016). Similarly, while there is a growing body of literature regarding cross-cultural science and technology (Aikenhead & Jegede, 1999) and indigenous ways of knowing (see Borden & Wiseman, 2016), disparities persist in indigenous participation in STEM education and the STEM workforce (Marginson et al., 2013). Redressing systemic inequities would go some considerable way to meeting STEM-focused human capital targets, and broader social agendas. Accounts of the challenges, and encouraging policy and practice approaches to indigenous participation in STEM, can be found in the commissioned reports by

Aikenhead, Nelson-Barber, and in the New Zealand report (https://acola.org.au/wp/stem-consultants-reports/).

3.4 Comparative Performance in School Level Science and Mathematics

International assessment and reporting regimes, while contested in terms of the impact on student learning and government and institutional intervention, provide some indication of the variation between countries and territories with respect to student performance in school level science and mathematics. In some instances, the results from the international assessments generated renewed government policy and societal attention on students' learning generally, and school-based science and mathematics specifically. The most notable example of this was the experience of 'PISA-shock' in Germany (Bank, 2012) and Japan (Ishikawa, Moehle, & Fujii, 2015) in response to decreasing comparative and absolute performance.

The 2015 Programme for International Student Assessment (PISA) exercise undertaken by the Organisation for Economic Co-operation and Development (OECD) illustrates the dominance of East Asia, Singapore, Western Europe and Canada in 15 year old school student proficiency in science, reading, mathematics and collaborative problem solving. In relation to science, Singapore, Japan, Estonia and Finland are the highest-performing OECD countries, while school students from Taiwan and Macao (China) perform very well. In relation to mathematics, Singapore, Hong Kong (China), Macao (China), Taiwan, Japan and Beijing-Shanghai-Jiangsu-Guandgdong (China) dominated. Equity in education outcomes is most readily observed in relation to the science and mathematics performance of school students in Canada, Denmark, Estonia, Hong Kong (China) and Macao (China) (OECD, 2018).

Similarly, the 2015 International Association for the Evaluation of Educational Achievement Trends in International Mathematics and Science Study (TIMSS) also reveals the strength of East Asia in fourth and eighth grade school student science and mathematics performance. Particularly strong performance was reported for Singapore, South Korea, Japan, Hong Kong (China) and Chinese Taipei. Russia ranked well in relation to performance by grade four science students (Martin et al., 2016; Mullis et al., 2016).

3.5 Government STEM Policy

Recognition of the growing imperative for foundational scientific literacy, STEM skills and research and development (R&D) capacity has seen the establishment of legislation and policy governing STEM, science and technology, industry innovation and commercialization. These policies typically focus on areas most receptive to government intervention, including education and R&D. Variations in terms of policy responses have been observed between high performing Anglosphere countries and a number of Western European countries, dynamic East and South East Asian countries with a Post-Confucian heritage, and emerging economies and education systems (Freeman, Marginson, & Tytler, 2015).

High performing Anglosphere countries such as the United Kingdom, Canada, Australia and New Zealand, and a number of Western European countries have national STEM policies aimed at addressing unmet labour market demand for STEM skills, and securing international competitiveness within an increasingly globalized economy. Several of these policies have emerged within a narrative of 'STEM crisis' and declining relative performance in international science and mathematics assessments, along with growing emphasis on industry human resource requirements and innovation. Notable examples include the United Kingdom's *Science & Innovation Investment Framework 2004-2014* and 2017 paper, *Industrial Strategy: Building a Britain Fit for the Future*. In Australia, STEM education, R&D and industry innovation-related policy includes the *National STEM School Education Strategy* released in 2015, and 2018 plan, *Australia 2030: Prosperity Through Innovation*. New Zealand's *National Statement on Science Investment 2015-2025* aims to establish a coherent vision for the country's science system. In the United States, leading reports such as *Rising Above the Gathering Storm* and *Revisiting the STEM Workforce* have generated interest in the development of a national STEM workforce strategy (see National Academies of Sciences, Engineering, and Medicine, 2016). In Western Europe, STEM or science policies have been adopted in Germany, France, Ireland, the Netherlands and Spain. They typically address public perceptions and knowledge of science, school-based mathematics and science teaching, participation and performance, and tertiary-level participation in STEM disciplines.

In contrast, East and South East Asian countries with very high performing education systems, including Japan, South Korea, China and Taiwan, typically have national science and technology policies and plans. Examples include the Japanese *Science and Technology Basic Law* (S&T Law) and attendant five year *Science and Technology Basic Plan*, and China's *Science and Technology Development Goal* (2006-2020) and *National Mid and Long-term Education Reform and Development Framework* (2010-2020). These policies emphasize university science and technology, industry-driven R&D and innovation.

Finally, emerging economies and education systems, including Brazil, Argentina, and arguably South Africa, have established national policies focused on quality education and emerging industry development. For example, Brazil's

Education Development Plan 2011–2020 emphasizes school education, teaching quality and teacher career pathways; and Argentina's *National Plan of Science, Technology and Innovation: Argentina Innovadora 2020* prioritizes research and innovation, general scientific capacity, and development of biotechnology and health. South Africa *National Development Plan 2030* of the National Planning Commission aims to redress injustices of the past, facilitate economic growth, and improve education, health and social protection.

As these examples illustrate, there is a great deal of variability with respect to STEM policy objectives, and these variations in part reflect different economic, cultural and social contexts. Some policies seek to promote a positive image of science while others aim to increase public engagement with and knowledge of science, through increasing scientific literacy and understanding of the scientific method. Policy aimed at the education sectors frequently focus on strategies to enhance student engagement and ultimately, consolidate the STEM pipeline. Policy may focus on some or all aspects of increasing participation and performance in school-based mathematics and science, tertiary level STEM-disciplines and high-end STEM R&D, including doctoral training and post-doctoral research. Policy may be aimed specifically at encouraging transition into the STEM labor-market. In many instances STEM policy seeks to redress disparities based on gender, ethnicity or race for minority groups including Indigenous peoples, and geographical location. STEM policy may establish mechanisms for co-ordination across STEM-related ministries, agencies and organizations (including scientific agencies, and R&D funding agencies). Policies may articulate annual and long-term objectives, common metrics or performance indicators to monitor progress, and establish an evaluation strategy.

Government STEM policy increasingly complements national research priorities that aim to focus public investment in global science. These priorities are typically framed in STEM terms as wicked problems or grand challenges, such as energy, water, resources, agriculture, climate change, and security. Concurrently, governments have acknowledged the emerging importance of internationalization and interdisciplinarity, including collaborations involving scholars from the humanities, arts and social sciences disciplines. For developing economies, governments have also prioritized the shift from the adoption of foreign technologies to indigenous technological innovation.

3.6   *Structural Responses*

The STEM agenda is diverse, spanning different education and training sectors, global science and the labour market. In the school education sector, there are a range of structures that have responsibility for supporting science and mathematics teaching, student learning and engagement, and scientific literacy more broadly. Structures may be at regional level, for example connecting education ministries across Europe through the European Schoolnet. Numerous other regional structures, such as the STEM Alliance (which superseded inGenious) co-ordinate STEM education initiatives. At the national and local level, there are structures that disseminate science and mathematics curriculum and teaching resources, and promote innovative science and mathematics pedagogy such as the United Kingdom's Science Learning Centres. In many countries, including Russia, Japan and Singapore, specialist or selective schools focus on science learning. There are numerous examples or organizations that conduct complementary or after-class STEM initiatives including competitions (Science, Mathematics and Engineering Olympiads), enrichment activities (museums, festivals, planetaria) and science centres, such as San Francisco's Exploratorium. Structures have also been established to communicate science and creativity more broadly to the population such as the Korea Institute for the Advancement of Science and Creativity (KOFAC). These examples illustrate the concerted and multilayered efforts at regional, national and local levels to engage young people in science and mathematics education and experiences. These structures are shaped by histories, traditions, geopolitics, priorities and public perceptions.

3.7 *Programs and Solutions*

The STEM: Country Comparisons project highlighted the need to broaden and deepen school students' STEM learning. This involves providing some STEM education for all school students, improving the engagement and performance of students from under-represented groups, and increasing participation in, and improving achievement through, intensive STEM education. The project report suggests consideration be given to mandating mathematics and/or at least one science subject for the final one or two years of senior secondary education. In relation to school curriculum and pedagogy, the international comparative analysis illustrated the importance of inquiry, reasoning, and creativity and design in school science and mathematics curriculum. The project highlighted the importance of career pathways for STEM teachers, discipline-specific professional development, and the negative impact of 'out of field' teaching, particularly in science and mathematics. In relation to early childhood and primary education, the project recommended increasing the confidence and competence of primary teachers in science and mathematics. Finally, the project discussed strategies to increase the engagement and participation of girls and indigenous students (Marginson et al., 2013).

4 Limitations

The research explored sectors and solutions where national government intervention was most prevalent, including school and higher education. Less emphasis was given to the important vocational education and training (VET) sectors, and the STEM labour market.

5 Implications for K-12 Education

The STEM: Country Comparisons project suggests that engaging K-12 school students in science and mathematics will involve reforms to curriculum, pedagogy and the learning environment to ensure a strong focus on disciplinary knowledge, as well as creativity, reasoning and innovation. It will involve resourcing schools with teachers who have science and mathematics disciplinary training, and progressive professional development opportunities.

6 Future Research

The STEM: Country Comparisons study arose from a desire to identify successful strategies that could be implemented to enhance participation in STEM at a range of levels. One needs to be mindful of the dangers of any simplistic 'policy borrowing' that may be tempting, looking at the STEM policies and practices of the more dynamic economies. The STEM focus in these different countries is shaped by cultural, historical and economic factors. Nevertheless there are lessons to be learned from these policy/practice drivers, and patterns have emerged, as described in this chapter, that help clarify the choices that are available in relation to STEM. Much of the focus on international comparative tests such as PISA is driven by the correlation between results on these tests and the degree of dynamism of a country's economy, but the direction of the causal connection is not clear. Further clarification is needed on the nature of relations between different aspects of education policy (age of focus, relative focus on elite students compared to disadvantaged populations, subject focus), R&D policy, and industry drivers. As workplaces change in response to technological advances, and internationalization, and economic drivers shift, there is an increasing focus in education on core competences such as critical and creative thinking or collaborative reasoning, transferable across professions, that are associated with both disciplinary and interdisciplinary thinking. There is a need for research that examines the nature of these competences

AN INTERNATIONAL VIEW OF STEM EDUCATION 361

and how to address them in pursuing an engaging and rigorous education in STEM.

The countries represented in the STEM: Country Comparisons study are in the main successful in STEM education, research, and industry. They are mostly developed economies. There is a need for a similar study that represents a range of developing economies, to identify the policies and practices and their attendant conditions that lead to a STEM educated citizenry and to an expanding economy.

Last, the report focused mainly on the education pathways leading from school to tertiary STEM studies. There is a need to more closely examine the relative roles of mainstream and vocational education in supporting STEM and the relationship between the variety of work in STEM in different economies, and the training needs that this implies.

7   Future Research Questions

1. What is the nature of the relationship between particular types of focus in STEM education, on traditional disciplinary knowledge compared to higher order reasoning competences, or of the relative emphasis on inclusivity in STEM education policy, and growth factors in a STEM based economy?
2. What is the nature of the causal links between the quality of STEM education and the dynamism of economies at different stages of growth?
3. What policy and practice settings in STEM lead to successful economic outcomes for emerging economies?
4. What is the role of the training sector in supporting STEM education, research and industry development?

8   Conclusion

STEM is an increasing focus for governments around the world, with concerns driven mainly by the links made between STEM education and research, and wealth creation. The STEM country comparison study identified strong commonalities across countries in their focus on STEM participation and quality, but differences in policy and practice that could be broadly grouped according to economic regions. Major groupings in this study were the emerging East Asian economies, Anglophone countries, and Western Europe. Concerns about STEM participation differed in intensity, but the focus followed broadly similar patterns, including quality of education participation and outcomes in STEM, public perceptions of and engagement with STEM, recruitment into targeted STEM professions, supporting disadvantaged as well as elite groupings, and developing coherent policy that coordinated STEM effort. The particular focus and strategies depended on historical, cultural and economic factors, with developing economies having distinctive foci for policy framing.

There is a need to extend the comparative analytic work of this study to pursue research that identifies the particular links between STEM education foci, education more generally, and the nature and needs of emerging work futures. We also need to understand better the links between STEM education, research, and the economic wealth of a country and wellbeing of its citizens.

References

Aikenhead, G. S., & Jegede, O. J. (1999). Cross-cultural science education: A cognitive explanation of a cultural phenomenon. *Journal of Research in Science Teaching, 36*(3), 269–287.

Ainley, J., Kos, J., & Nicholas, M. (2008). *Participation in science, mathematics and technology in Australian education* (Research Monograph No. 63). Melbourne: Australian Council for Educational Research.

Bank, V. (2012). On OECD policies and the pitfalls in economy-driven education: The case of Germany. *Journal of Curriculum Studies, 44*(2), 193–210. doi:10.1080/00220272.2011.639903

Barrington, F., & Evans, M. (2017). *Year 12 mathematics participation in Australia 2007–2016*. Australian Mathematical Sciences Institute.

Beede, D. N., Julian, T. A., Langdon, D., McKittrick, G., Khan, B., & Doms, M. E. (2011). *Women in STEM: A gender gap to innovation* (Economics and Statistics Administration Issue Brief 04-11). Washington, DC: US Department of Commerce.

Borden, L. L., & Wiseman, D. (2016). Considerations from places where Indigenous and western ways of knowing, being, and doing circulate together: STEM as artifact of teaching and learning. *Canadian Journal of Science, Mathematics and Technology Education, 16*(2), 140–152.

Freeman, B., Marginson, S., & Tytler, R. (2015). Widening and deepening the STEM effect. In B. Freeman, S. Marginson, & R. Tytler (Eds.), *The age of STEM* (pp. 23–43). Oxon: Routledge.

Ishikawa, M., Moehle, A., & Fujii, S. (2015). Japan: Restoring faith in science through competitive STEM strategy. In B. Freeman, S. Marginson, & R. Tytler (Eds.), *The age of STEM* (pp. 81–101). Oxon: Routledge.

Kennedy, J. P., Lyons, T., & Quinn, F. (2014). The continuing decline of science and mathematics enrolments in Australian high schools. *Teaching Science, 60*(2), 34–46.

Lyons, T., & Quinn, F. (2015). Understanding declining science participation in Australia: A systemic perspective. In E. K. Henriksen, J. Dillon, & J. Ryder (Eds.), *Understanding student participation and choice in science and technology education* (pp. 153–168). Dordrecht: Springer.

Maltese, A. V., Potvin, G., Lung, F. D., & Hochbein, C. D. (2015). STEM and STEM education in the United States. In B. Freeman, S. Marginson, & R. Tytler (Eds.), *The age of STEM* (pp. 102–133). Oxon: Routledge.

Marginson, S., Tytler, R., Freeman, B., & Roberts, K. (2013). *STEM: Country comparisons: International comparisons of Science, Technology, Engineering and Mathematics (STEM) education. Final report.* Melbourne: Australian Council of Learned Academies.

Martin, M. O., Mullis, I. V. S., Foy, P., & Hooper, M. (2016). *TIMSS 2015 international results in science.* Retrieved from http://timssandpirls.bc.edu/timss2015/international-results/

Mullis, I. V. S., Martin, M. O., Foy, P., & Hooper, M. (2016). *TIMSS 2015 international results in mathematics.* Retrieved from http://timssandpirls.bc.edu/timss2015/international-results/

Ministry of Education Singapore. (2017). *Education statistics digest 2017.* Ministry of Education Singapore. Retrieved from https://www.moe.gov.sg/docs/default-source/document/publications/education-statistics-digest/esd_2017.pdf

Ministry of Education of the People's Republic of China. (2015). *Educational statistics for 2015.* Retrieved from http://en.moe.gov.cn/Resources/Statistics/edu_stat_2015/2015_en01/index.html

National Academies of Sciences, Engineering, and Medicine. (2016). *Developing a national STEM workforce strategy: A workshop summary.* Washington, DC: The National Academies Press. https:doi.org/10.17226/21900

National Center for Education Statistics. (n.d.). *The condition of education 2017. Student effort, persistence and progress. High school coursetaking. Table 225.40 (Digest 2013)* [Dataset]. Retrieved from https://nces.ed.gov/programs/coe/current_tables.asp

Office of the Chief Scientist. (2016). *Australia's STEM workforce: Science, technology, engineering and mathematics.* ACT: Australian Government.

Organisation for Economic Co-operation and Development. (2017). *Education at a glance 2017. OECD indicators* [Dataset]. Retrieved from http://www.oecd.org/education/education-at-a-glance-19991487.htm

Organisation for Economic Co-operation and Development. (2018). *PISA 2015 results in focus.* Retrieved from https://www.oecd.org/pisa/pisa-2015-results-in-focus.pdf

Palmer, T. A., Burke, P. F., & Aubusson, P. (2017). Why school students choose and reject science: A study of the factors that students consider when selecting subjects. *International Journal of Science Education.* doi:10.1080/09500693.2017.1299949

UNESCO Institute for Statistics. (2018). *Education* [Dataset]. Retrieved from http://data.uis.unesco.org/Index.aspx

CHAPTER 20

Conclusions and Next Directions

Margaret J. Mohr-Schroeder and Alpaslan Sahin

While the field STEM Education is still relatively new, its tenets have been around for many years. STEM Education officially burst into reality in the early 1990s (see Mohr-Schroeder, Calvalcanti, & Blyman, 2015). Since then, it has grown exponentially. In this book we aimed to take stock of how far we have come as a field of STEM education and the implications it has had in K12 education. While core elements of STEM education have been previously studied (e.g., Bybee, 2013; Johnson, Peters-Burton, & Moore, 2015; Sahin, 2015), there are still varied conceptions of what STEM education should and does look like. The chapters presented in this book were not meant to be an exhaustive list of the vast field of STEM education, but rather touch on key elements and ideas.

In the first part, we set the stage by examining the current status of the field of STEM education. The importance of integrating STEM is well recognized now. It is essential that all students have a strong understanding of STEM to compete in today's global economy. Some proponents of STEM (e.g., U.S. Department of Education, 2015) focus on future job opportunities, noting that integrated STEM jobs (e.g., biomedical engineer) are projected to increase at more than double the rate of jobs in non-integrated STEM fields (e.g., mathematician) (Chapter 1). Others, such as those presented in Chapter 4, focus on how the integration component of STEM and improves the quality of learning, especially as it relates to workforce readiness. But while workforce readiness and creating a STEM literate society are key ideas presented in reports and media, it still remains highly-debated as to whether or not the United States is truly experiencing a shortage, or even predicted shortage, in the STEM fields. While STEM is currently a small portion of the workforce in the U.S., it is expected to greatly increase with new jobs that are not even known right now. What's even more important than understanding the current and predicted workforce, is how we are preparing our K-12 students for the unpredictable workforce and even more importantly, ensuring that all of our students, especially those who are underrepresented, have opportunity and access to high quality, integrated STEM experiences (Chapters 3 and 4). Personal interest and motivation are key components to inspiring K-12 students to pursue careers and pathways in STEM learning. Moreover, students' interest and motivation contributes to their success in learning and retaining STEM content

(Bell, Lewenstein, Shouse, & Feder, 2009). Therefore, it is essential we create and maintain high-quality integrated STEM experiences for all our students. Thanks to a robust research field in education in the U.S., we have a multitude of reports that can help guide us in what those experiences should look like and what the potential impact of those experiences are (Chapter 5).

The synthesis of those reports and their implications for K-12 education set the stage for Part 2 where we look more deeply at research-based strategies that yield positive outcomes in STEM education and what those look like in the innovative classrooms and schools of today. In today's climate of increased accountability along with the pressure to use "proven" research based methods, the teaching of STEM topics to students who are typically developing can be a challenging task. Moreover, meeting the educational needs of an extremely diverse group of learners is no simple task. Even further, research has shown the importance of motivating students to learn STEM content in the middle grades. "Students who express interest in STEM in eighth grade are up to three times more likely to ultimately pursue STEM degrees later in life than students who do not express such an interest" (PCAST, 2010, p. 19). This has placed pressure on K-12 education to be nimble, responsive, and innovative in providing opportunity and access to high-quality STEM education experiences for all students, all the while tending to the pressures that come along with accountability and high-stakes testing. Chapters 7 and 8 take a deeper dive into these innovative school models and the what the future holds for our students. Industry and businesses have more recently begun a stronger investment in these innovative education models in an effort to help push innovation forward at a more rapid pace than education has traditionally moved.

One major theme threaded through innovative school models is the use of project-based instruction. The use of project-based instruction has allowed for a more comprehensive, integrated approach to education, especially in the STEM areas. Further it has allowed material to be better contextualized and allowed K-12 students a more realistic look at collaboration and teamwork as they will likely play out in their workplace (Chapter 6). Even further, technology has played a major role in our rapidly changing society. K-12 students today have infinitely more opportunity and access to technology than even their immediate prior generation. The role technology plays in our increasingly technology-dependent society has lots of implications for K-12 education, especially as they relate to intentional integration and opportunity and access (Chapter 10). While the U.S. is moving away from a traditional school model to a more student-centric, integrated model, we're also realizing the importance and key role out-of-school learning plays in K-12 education. Informal learning environments have been around for many years, but many focused just on science and history as they were mostly situated in libraries and museums. As the field of STEM education has burst out into the open, so has informal learning opportunities. Many organizations, industries, businesses, and even institutes of higher education have embraced the critical role they play in creating high quality STEM experiences for all students. While informal learning environments are not intended to replace the formal classroom, they do provide a safe, open-ended, low-stakes environment for students to freely engage with STEM content and other disciplines (Chapter 9). It is important to provide K-12 students with the opportunity to engage in out-of-class activities, which in turn can increase their motivation and interest in STEM, while having a positive impact on content and achievement in the formal classroom environment.

While STEM content and even the idea of interdisciplinary education has been around for many years, the primary focus has been on science and mathematics content. The field of engineering has long been known and purported as a truly interdisciplinary field for those interested in both mathematics and science. However, the field has struggled with recruiting a diverse workforce for a variety of reasons, including opportunity and access K-12 students have to even basic tenets of engineering (Chapter 11). Much of this struggle has been around the how to meaningful integrate engineering into K-12 schools, especially since they tend to be discipline-specific courses. The *Next Generation Science Standards* took a bold leap in 2013 and integrated engineering and engineering design throughout its K-12 standards. But even then, some states decided it was too much would not adopt the engineering components. With the field of engineering slated to continue its rapid increase in workforce needs, it remains vital to the conversation how to continue the intentional integration of engineering into K-12 schools and what that might look like. The rapid rise of technology and the role it plays in society today has provided one avenue for engineering, specifically in the field of robotics. The rising popularity of robotics is due in part to its flexibility in adapting to the environment (formal and informal settings) and due to its interdisciplinary, project-based nature (Chapter 15). And through the rise of popularity of robotics and integration of engineering into K-12 education, we have begun to think of more ways to cultivate creativity and innovation in our younger generations. While play-based education remains largely popular in early childhood education, the rise of accountability and high-stakes testing has largely pushed these ideas out of K-12 classrooms. However, the maker movement seeks to bring back environments that tap into students' curiosity and cultivate the engineering design process (Chapter 14).

While the innovation occurring inside and outside of the classrooms are essential to developing a STEM literate society, it is important to fully understand the factors that influence and motivate K-12 students to pursue a career in STEM. For example, social cognitive career theory provides a comprehensive picture of the multiple factors that influence career decisions. But what does that look like when applied to a longitudinal context of K-12 schools? Chapter 16 takes a deeper dive into this very idea by providing quantitative and qualitative research backing why students make the career choices they do and who or what their strongest influencers were. Even further still, engagement in high quality STEM experiences remains one of the top variables that influence students' career choices (Chapter 17). While it is not feasible or necessary for every K-12 student to choose a STEM career, it is important that they become STEM literate, especially as technology becomes an even stronger driving force in society.

Much of the success we have seen in the K-12 STEM education sector, especially over the past 15 years, has been driven much in part by partnerships. Partnerships with the community, industry, businesses, and other stakeholders who are driven to help provide K-12 students with opportunity and access to high quality STEM experiences. These partnerships are important as they provide a bridge between the often non-contextualized formal school STEM content and opportunities to engage in curricular development, experiences, and mentorship that directly relate to workforce development and readiness (Chapter 18). The results of these partnerships between business and community has led to more robust research and clarification on what exactly K-12 students need to be prepared to do when they graduate. College and career readiness topics and research has been thrust into the spotlight as the drive for innovation continues to be imperative in the U. S. Literature around college and career readiness and 21st century skills have been around for the past 20 years, but they have largely remained a conversational piece. The more recent community, business and industry partnership has help to contextualize the readiness skills necessary (Chapter 19). And while the U.S. still remains one of the world's top leaders in innovation and education, there is synergy internationally to remain competitive with the U.S. and for countries to create their own niches. Research and trends outside of the U.S. are wide and varied (Chapter 20). One theme remains true throughout though, the emphasis on STEM and its impact on education is of high importance and many resources are being put towards its success.

Throughout each of the twenty chapters in this book, it is clear we, as a field, have moved past STEM as a buzzword to STEM as an established field. The implications for this in K-12 education are wide and varied as highlighted throughout the book. Overall, we know that exposure to a variety, high quality STEM opportunities has been shown to have a positive, long-term effect on individuals and the overall STEM education community (e.g., Wai, Lubinski, Benbow, & Steiger, 2010). Research has also shown that K-12 students who have an increased interest in STEM early, especially before eighth-grade, are more likely to pursue a STEM-related career (Afterschool Alliance Report, 2011). Regardless of the pursuit of a STEM career or not, the research presented in this book shows the importance of including student-centered STEM activities to promote positive dispositions toward STEM.

As the U.S. strives to become and remain a leader in STEM and innovation in a global market, we must continue to identify and create innovative models and opportunities for our K-12 students. We must continue to produce research and evaluation on such experiences, especially the longitudinal impacts. As former President Obama declared, "We must educate our children to compete in an age where knowledge is capital, and the marketplace is global."

References

Bybee, R. W. (2013). *The case for STEM education: Challenges and opportunities.* Arlington, VA: National Science Teachers Association Press.

Johnson, C. C., Peters-Burton, E. E., & Moore, T. J. (2015). *STEM road map: A framework for integrated STEM education.* New York, NY: Routledge.

Mohr-Schroeder, M. J., Cavalcanti, M., & Blyman, K. (2015). STEM education: Understanding the changing landscape. In A. Sahin (Ed.), *A practice-based model of effective Science, Technology, Engineering and Mathematics (STEM) education teaching: STEM Students on the State (S.O.S) model* (pp. 3–14). Rotterdam, The Netherlands: Sense Publishers.

Sahin, A. (2015). *A practice-based model of effective Science, Technology, Engineering And Mathematics (STEM) education teaching: STEM Students on the State (S.O.S) model.* Rotterdam, The Netherlands: Sense Publishers.